The Radiance of France

Inside Technology
edited by Wiebe E. Bijker, W. Bernard Carlson, and Trevor Pinch

The Radiance of France
Nuclear Power and National Identity after World War II

Gabrielle Hecht

The MIT Press
Cambridge, Massachusetts
London, England

Set in New Baskerville.
Printed and bound in the United States of America.

Library of Congress Cataloging-in-Publication Data

Hecht, Gabrielle.
    The radiance of France : nuclear power and national identity after World War II / Gabrielle Hecht.
        p.   cm.
    Includes bibliographical references and index.
    ISBN 0-262-08266-7 (hardcover : alk. paper)
    1. Nuclear engineering—France—History.   2. Nuclear engineering—Social aspects—France.   3. Nationalism and technology—France.   4. National characteristics, French.   I. Title.
TK9071.H43   1998   94-13590
621.48′0944—dc21   CIP

For Paul

Contents

Acknowledgments

There is a sense in which this book began in 1975. That was the year my parents and I moved to Paris for the first time, and the year that France became a real home in my continually itinerant life. My mother, Maria Teresa Lamarche Hecht, died before I could begin my research, but this project owes much to her. Two years before we moved to France, she gave me my first French lesson. By unconscious example, she taught me to love Paris. By conscious act, she gave me the courage and determination I would later need to see this book through to the end. In these and many other ways, she will be with me always.

I conducted the research for this book in France in 1989, 1990, 1994, and 1996. When I began, in 1989, the official archives of the nuclear program were closed. Without the goodwill of dozens of nuclear engineers and workers, as well as that of residents of the Touraine and the Gard, I would have never been able to do this research. A complete list of the people I interviewed appears in the bibliography. I am especially grateful to those who went out of their way to be hospitable, find documents, give me tours of nuclear facilities, and in some instances even read my work: Claude Bienvenu, Yves Chelet, Pierre Constantin, André Crégut, Paul Delpeyroux, Phililppe Filhol, Claude Heurteau, Mireille Justamond, Bernard Laponche, Henri Loriers, M. Mièvre, Marie-Anne Sabatier, Joël Sorin, Jean Thomas, Marie-Lise Grémy, Bernard Tourillon, Jacques and Raymonde Trélin, Jean-Claude Zerbib, and the workers and managers of EDF's Saint-Laurent-des-Eaux site.

The Centre de Sociologie de l'Innovation provided a professional home in Paris at all stages of my research. I wrote the first chapter of this book in a tiny office there with a magnificent view of the Eiffel Tower, the quintessential symbol of French technological prowess in the nineteenth century. The researchers and staff of the CSI provided me with unfailing intellectual stimulation and professional assistance of all kinds. Special thanks to Madeleine Akrich, Michel Callon, Florian Charvolin, Jean-Pierre Courtial, Penelope Dunning, Antoine Hennion, Juliette Hubert, Cécile Méadel, and Vololona Rabeharisoa. And of course to Bruno Latour, who was particularly generous with his time, his advice, and his car.

Academics at other institutions also helped me find my way through the maze of French scholarly resources. At various stages of this project I received encouragement and counsel from Alain Beltran, Patrick Fridenson, Pascal Griset, Dominique Pestre, Jean-François Picard, Antoine Prost, Brigitte Schroeder-Güdehus, and Françoise Zonabend. My thanks also to the many archivists and *documentalistes* I encountered: Louisette Battais, Annie Eskenazi, Odile Frossard, Helga Grunenwald, Marie-Hélène Joly, Mireille Justamond, Jean-Marie Palayrat, José Parreira, Santhi Pascal, Catherine Renon, Henri Sinnot, and Rose-Marie Wendling. These trips to France were possible thanks to financial support from the National Endowment for the Humanities, the National Science Foundation, the Institute for Electrical and Electronics Engineers, the Mellon Foundation, the University of Pennsylvania, and the Hewlett and MacNamara funds at Stanford University.

Previous versions of some of the chapters appeared in *Technology and Culture, Social Studies of Science, French Historical Studies,* and the *Journal of Contemporary History;* full citations are in the bibliography.

My intellectual and professional debts outside France extend far and wide. Sharon Traweek was my first mentor in this field: without her example and encouragement, I might never have become an academic. The History and Sociology of Science department at the University of Pennsylvania was a wonderful place to get a graduate education. Faculty, students, and staff built an exhilirating and supportive environment there. Special thanks to Julie Denenberg, Marta Hanson, Julie Johnson, Pat Johnson, Henrika Kuklick, Rob Kohler, Judy McGaw, Joyce Roselle, Eric Schatzberg, Dave Shearer, Lynne Snyder, Susan Speaker, Raman Srinivasan, and Keith Wailoo. Nina Lerman has been my close intellectual companion in the sometimes terrifying adventures of the past decade. Her support and our laughter have been precious in equal measure.

Arne Kaijser, Svante Lindqvist, and Hans Weinberger of the Department of History of Science and Technology at the Royal Institute of Technology in Stockholm hosted a memorable visit there at a crucial stage of my dissertation. Stanford University has provided a stimulating environment in which to write this book. Many thanks to the colleagues who commented on various portions of the manuscript and who have encouraged me in my professional development: Keith Baker, Joel

Beinin, Bart Bernstein, Joe Corn, Stephen Hastings-King, Paula Findlen, Estelle Freedman, Norman Naimark, Lou Roberts, Richard Roberts, Karen Sawislak, Laura Smoller, Jim Sheehan, Bob Tatum, and Walter Vincenti. I am grateful to Tim Lenoir, who directed my first readings course on nuclear weapons at Penn, and who has since become a generous and sympathetic colleague at Stanford. His counsel and support have been invaluable.

I greatly appreciate the wonderful graduate students who have helped me with my research, commented on my drafts, and inspired my thinking: David Kirsch, Jennifer Lee, Angus Lockyer, Carlos Martín (who also drew many of the illustrations in this book), Sara Pritchard, Heather Schell, and Phillip Thurtle. My thanks also to Carolyn Commiskey, Jennifer Milligan, and Mary Beth Nikitin for their research assistance. The members of the Bay Area technology and culture discussion group cheerfully endured several drafts of chapter 8. Many other scholars also provided thoughtful comments on pieces of the manuscript: Ken Alder, Michael Bess, Herrick Chapman, Laura Downs, Bob Frost, Cathy Kudlick, Arwen Mohun, Ted Porter, Erik Rau, Gene Rochlin, Helen Rozwadowski, Michael Smith, John Staudenmaier, and Tyler Stovall. The MIT Press has been a pleasure to work with, thanks to Larry Cohen, Trevor Pinch, and Paul Bethge.

Thomas Parke Hughes and Agatha Chipley Hughes sheperded me through this project from its inception, with endless patience, humor, and wisdom. Agatha's kindness and wit continues to sustain me after her tragic and untimely death. Tom heroically read even the penultimate draft of my manuscript; his astute sense of technology and history helped me through all the stages of revision. I could not have asked for better mentors.

I have also been sustained by warm friendships outside academia with Ivan Deutsch, Jamie Henderson, Florence Le Gal, Todd Jones, Karin Nilsson, Marie-Laure Paris, Dori Schack, Elaine and Don Singer, and Rick and Eli Zinman. Jay Slagle has been more involved with this project than he ever imagined he would be, to the tremendous benefit of my book and my sanity.

My family, new and old, have been a vital source of nourishment. The Edwardses have shown me the delights of a large clan. Lauren Singer has given me steadfast support in the best and worst of times. Her daughter Maya Emlin Delaney brought me joy before she even drew breath.

The unwavering faith and love of my father, Otto Hecht, has sustained me through all the phases of my education. He taught me how to make a great deal from very little, and has always had the grace and courage to let me learn to fly on my own.

Finally, Paul Edwards has been at my side for the last five years of this project. He has endured much and given even more, always with infinite gentleness and understanding. His own work has been an inspiration to me, and his keen editorial eye has improved every page of this book. He has kept me whole, and for that I dedicate my book to him.

The Radiance of France

The EDFI reactor at Chinon. Source: EDF Photothèque.

Introduction

France cannot be France without grandeur.

—*Charles de Gaulle*[1]

It would be good if it were French research that produced the first useful and humane applications of this diabolical marvel. To master these terrifying forces of unlimited destruction, to have this stupendous invention metamorphose itself into a humane discovery through the filter of our national genius, this would bring honor to our country.

—*Raoul Dautry, first Administrator-General of the Commissariat à l'Energie Atomique, October 1945*[2]

In June of 1940, German troops marched into France for the second time in less than thirty years. On June 17, Marshal Pétain announced that he would seek peace with Hitler. Charles de Gaulle launched the Resistance the following day in a broadcast from London. Thus began four years of opprobrious occupation and fractured resistance. In June of 1944, Allied troops landed on the beaches of Normandy to liberate a nation humiliated by defeat, ravaged by war, disgraced by collaboration, and only partly redeemed by resistance.

France had lost nearly a million and a half people in the war. The industrial infrastructure was in shambles. Food was scarce and expensive. France had lost its self-respect. It had also lost its standing among world leaders—a loss made glaringly obvious by de Gaulle's absence at Potsdam and Yalta. The bombing of Hiroshima and Nagasaki highlighted the enormous technological gulf between France and the United States. The consequences of the war for the French empire remained unclear, but prospects already looked grim in Indochina. The embarrassed, destitute nation resigned itself to accepting American economic aid in the slow and painful task of reconstruction. To use Robert Frank's phrase, France entered the second half of the twentieth century "haunted by its decline."[3]

No wonder, then, that the nation expressed such enthusiasm when Zoé, its first experimental nuclear reactor, underwent a chain reaction in December 1948, only four years after the Liberation. This success, proclaimed one newspaper, was "a great achievement, French and peaceful, which strengthens our role in the defense of civilization."[4] The following year, scientists isolated France's first milligram of plutonium. President Vincent Auriol paid Zoé a visit and solemnly declared: "This achievement will add to the radiance of France."[5]

"The radiance of France"—a phrase usually interchangeable with "the grandeur of France"—appeared regularly in many realms of postwar discourse. These two notions referred back to France's glorious past, from the golden reign of Louis XIV to the "civilizing mission" of the empire.[6] France's radiance had taken a severe beating during the war, and decolonization threatened to hasten the decline.[7] How could the nation regain its former glory? What would radiance or grandeur mean in the radically reconfigured geopolitics of the postwar world?

Technical and scientific experts offered a solution to these dilemmas: technological prowess. In articles, lectures, and modernization plans, experts repeatedly linked technological achievement with French radiance. Industrial, scientific, and technological development would not only rebuild the nation's economy but also restore France to its place as a world leader. For the nascent nuclear program, "le rayonnement de la France" carried special punch: "rayonnement" means radiation as well as radiance.

The nuclear program epitomized the link between French radiance and technological prowess. Before World War II, Marie and Pierre Curie, Jean Perrin, and Frédéric and Irène Joliot-Curie had become national heroes thanks to their Nobel prizes in physics and chemistry. After the deadly explosions at Hiroshima and Nagasaki, nuclear technology became a quintessential symbol of modernity and national power.[8] France could claim a modest role in the Manhattan Project, thanks to a few researchers who had fled the occupation to work in Britain and Canada.[9] No other technology could better enhance French radiance. With this logic in mind, de Gaulle fostered the creation of an atomic energy commission in 1945. After building several experimental reactors, the Commissariat à l'Energie Atomique (CEA) began to work on plutonium-producing plants in the mid 1950s. Its scientists and engineers also collaborated with their colleagues at Electricité de France (EDF, the nationalized electric utility) in the construction of a series of power reactors. Despite the fact that similar reactors existed elsewhere (notably in

Britain), the gas-graphite design developed by the CEA and EDF became known as the *filière française*—the French system.

What was French about the French nuclear program? This question appeared vital to nuclear engineers and scientists during the 1950s and the 1960s. It apparently interested social scientists and humanists as well: I heard it repeatedly over the eight years I spent researching this book.

On one level the answer seems simple. French engineers, scientists, and technicians developed most of the designs and techniques for their gas-graphite reactors. The resulting system, therefore, was French. Or was it? After all, some of the CEA's most important scientists had learned a great deal from their Canadian experiences. The 1955 Atoms for Peace conference had made possible a slow but steady international flow of information. French nuclear engineers and scientists increasingly discussed technical matters with their colleagues abroad, officially and unofficially. Such circumstances make the Frenchness of the French nuclear program rather difficult to pinpoint.

The question raises more complex issues on a deeper level. Its very formulation presumes a stable notion of Frenchness: somewhere, it implies, exists an essential French identity that can provide not only a description for how things happen in France but also an explanation for why they happen that way. Yet French identity is not inherently stable.[10] The effects of World War II extended well beyond threats to French radiance. The war had called everything into question, from military and industrial structures to systems of government and cultural identities. What would be the essence of a renewed France in a world transformed by the atomic bomb and superpower geopolitics? Could a new social order regenerate the nation's identity? Of what would that identity consist? These questions did not have simple or immediate answers.

As a *guide* to historical inquiry, then, the question "What is French about the French nuclear program?" has little value. We cannot simply gesture toward the Napoleonic institutional heritage or the Colbertist tradition of state-directed industrialization in order to describe or explain nuclear development in France. There was no such thing as an essential French technological style. Engineers did not make the choices they did *because* they were French.[11]

Rather, I argue in this book, engineering choices must be understood as part of a struggle to define Frenchness in the postwar world. For this very reason, the question "What is French about the French nuclear program?" *is* valuable as an *object* of historical inquiry. How and why did the people who designed, built, worked in, wrote about, and lived near reactors forge and understand the relationship between nuclear technology and French national identity? What role did invoking Frenchness play in nuclear development? How did nuclear technology figure in changing notions of Frenchness?

The answers to these questions depend greatly on people's involvement with the nuclear program. Engineers wove links between nuclear technology and national identity into the fabric of reactor design and program development. Workers forged these links in both labor union ideology and workplace practices. Neighbors of reactor sites understood these links primarily as symbols that justified changes in local socioeconomic structures—symbols with which they had to contend in order to make sense of modernization.

The continuities among these three domains—nuclear engineering and program development, reactor work, and the communities around nuclear plants—were as significant as the discontinuities. For example, the fact that the "radiance of France" notion operated in all three domains demonstrates the strength and flexibility of the association between technological prowess and national identity. This notion, together with its constellation of symbols, provided the foundation for a vocabulary through which to imagine modernity and technological change. But the diverse meaning of the symbols it employed highlights the profound differences in how various groups imagined the new technological France and their places in it.

Important differences also existed *within* these three domains: the groups they incorporated were by no means homogeneous. For engineers, disagreements over reactor core design, construction materials, and industrial contracting were also debates about how to connect reactor development with France's political and economic future. Some nuclear workers felt themselves to be active participants in the making of an ultra-modern nation, while others viewed their workplace as the extension of an oppressive technocratic state. Some local residents perceived reactor sites as socioeconomic opportunities, while others experienced the sites as instruments of a suffocating modernity. Conceptions of technological France thus varied greatly. The fault lines for difference could be technological, political, institutional, professional, geographical, or cultural. Usually they were several at once. The stakes of any given dispute went far beyond the matter apparently under debate. The history of the French nuclear program, therefore, is both a history of technology and a history of France.

In this book I connect these three domains by tracing the multiple links between technological prowess and national identity. But the politics and culture of nuclear development did not revolve solely about redefining Frenchness. As an exploration of the complex relationships between technology and politics, my arguments have equal relevance for nations other than France. Let me first sketch out the book's terrain, and then discuss the theoretical considerations that frame my analysis.

From the very beginning of this project, my goal has been to trace the social, political, and cultural life of reactors as artifacts. This goal arose in part from my realization that much historical scholarship focuses on a single aspect of the life of an industry or a technology. This is particularly true of twentieth-century technologies. A single book might discuss design, development, and diffusion; or the organization of work in a system; or cultural representations of technology; or the social impact of an industry. But what kind of picture emerges from examining all these together? This question formed the foundation of my research strategy.

In the first three chapters of this book I explore the domain of engineering and state expertise. I begin not with the nuclear program but with a general consideration of the history and ideology of French state experts, whom I also refer to as "technologists." This overview provides a broad context for understanding how these "men of action" (as they thought of themselves) conceptualized the relationship between technology and politics. Debates about the nature of this relationship were contests for the power to shape the future of France and its identity. Technologists located French radiance not only within the technologies they built but also in their potential ability to export their expertise, thereby evoking the imperial connotations of the notion of "radiance." In chapters 2 and 3 I follow this theme into the nuclear program by examining the design and development of gas-graphite reactors in the 1950s and the 1960s. I focus on the CEA and EDF, the two state institutions that directed the program. These two agencies collaborated in designing nine reactors, but their administrators and engineers had different goals and espoused different design and development practices. I argue that the artifacts elaborated within each institution can be best understood as hybrids of technology and politics.[12] Engineers and administrators used these hybrids—along with invocations of the nation and of their public-service ideology—to define and implement military and industrial policy. In the process, they also made nuclear technology both French and indispensable to Frenchness.

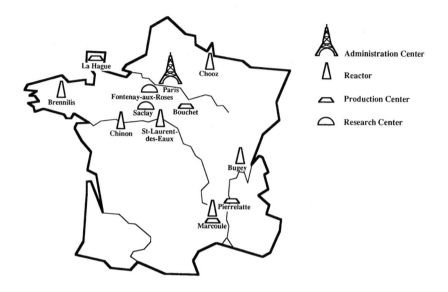

French nuclear sites in the late 1960s (not to scale). Drawing by Carlos Martín.

While state experts propelled large-scale technological development, their ideologies and conceptualizations did not dominate all levels of technological activity. Workers also played a role. The next two chapters shift to the domain of labor. In chapter 4 (which parallels chapter 1) I focus on labor union discourse about technological change in general. Again, I examine conceptualizations of the relationships between technology and politics. Labor militants reflected on the role of technological change in France's future, as well as on their own role in implementing such change. In one sense, their ideas *challenged* those of technologists. Militants envisioned an important position for technically trained workers in France's future social order. Two of the three labor unions actively criticized the development policies pursued by state institutions. In another sense, though, union militants and technologists were equally active as *participants* in the construction of a French technological identity. In chapter 5 I examine work inside two nuclear reactors: one operated by the CEA at its Marcoule site and the other by EDF at its Chinon site. Here I consider how the ideologies and technologies described in chapters 2 and 3 combined to produce two different workplaces. Marcoule and Chinon featured distinct labor organizations and work practices. Some of the issues raised by the labor unions at the national level were played out at these reactor sites: the relationship between technical training and authority, the role of workers in nationalized companies, the development of a high-tech workforce, and the relationship between workers' jobs and their place in the new technological nation.

In the next two chapters I explore a different kind of politics. In chapter 6, in parallel with chapters 1 and 4, I discuss articulations of the technological nation in the popular media. Moving beyond the national level, I examine how political and intellectual elites in the regions around Marcoule and Chinon presented the nuclear sites to their constituents. I argue that journalists, politicians, scholars, and technologists together (though not always in a concerted fashion) produced a drama of regional salvation and redemption in which large-scale technologies functioned as icons and actors. A survey of the critics of this spectacle demonstrates that not everyone rejoiced at the technological France imagined therein. In chapter 7 I discuss the reactions of the spectacle's audience. After a brief look at national public opinion polls, I explore the history and the memory of the two nuclear regions. The drama of regional salvation had promised residents around Marcoule a harmonious blending of tradition and modernity. Most residents, however, construed the arrival of the nuclear site and its employees as a wholesale invasion by the modern state. Residents around Chinon, meanwhile, had been promised a spectacular display in which reactors would function as modern châteaux. This was a far easier promise to fulfill. Though some tension did accompany the site's development, by and large the residents seemed to sublimate it and to concentrate instead on the economic benefits. Together, chapters 6 and 7 argue that the nuclear sites operated as a lens through which local communities re-imagined their political and cultural relationship with the nation.

In chapter 8 I unite the book's three domains in an extended consideration of the late 1960s' "war of the systems," which pitted the "French" gas-graphite reactor system against the "American" light-water reactor system. Each system had proponents in EDF, in the CEA, and in the French government. In this chapter I explore the debates among administrators, engineers, and labor unions in EDF and in the CEA. I discuss the 1969 strike in which CEA engineers and workers demonstrated in defense of the French system, examine the reactions of the residents around Marcoule to the abandonment of the gas-graphite system, and show how EDF workers dealt with the consequences of gas-graphite's demise during the cleanup of a reactor accident in 1969. During the course of this protracted "war," relations between the CEA and EDF were reshaped. The nexus of technology, ideology, expertise, and definitions of Frenchness shifted. Conceptions of the relationships between technology and politics were reconfigured. And workers and residents played out their imagined roles in the new technological France.

Inevitably, there is much I have not covered. My investigation stops in 1970, and even within this time frame I have set important limits. Readers hoping to find here an exhaustive scientific and technical history of the early decades of the French nuclear program will be disappointed. This study is limited to gas-graphite reactors; bombs, experimental reactors, fuel processing plants, waste disposal, and research programs remain peripheral. Further, I examine gas-graphite reactors primarily from the point of view of the two state establishments that developed the program; private industry enters my analysis only through their perspective. While these topics are important in their own right, I have omitted them in order to conduct a sustained examination of the multiple meanings of nuclear power and technological development for French politics and national identity.

Let me now turn to the methods and theories that inspired my analysis.

Technology, Politics, Culture, and National Identity

In researching and writing this book, I have combined methods and theories of technology studies with those of political and cultural history. When discussing issues of cultural or national difference, historians of technology frequently engage in cross-national comparisons. Thomas Parke Hughes masterfully demonstrated the fertility of this analytic tool in his study of electrification in the United States, Britain, and Germany, showing how distinctive approaches to system building emerged in response to particular political, geographical, and institutional conditions.[13] As other scholars have also shown, comparing technological systems and practices in different countries reveals national patterns that may remain hidden when countries are examined in isolation.[14] In contrast, French cultural historians focus on internal struggles over cultural forms and social relationships. They seek to understand how values, ideologies, and the language and symbols that constitute them arise and change. Culture thus provides not the explanation but the entity that demands explanation.[15] Understanding the significance of technological development across a broad range of sites within a single nation, as I attempt to do in this book, requires combining these two approaches and problematizing *both* technology and culture.[16]

The relationships of technology, politics, and culture have long pre-occupied the history and sociology of technology.[17] The major insight of this scholarship is that political, social, and cultural choices shape the design and growth of technical artifacts and systems.[18] Some scholars have sought to counter progress ideologies and other forms of technological determinism: the ideas that technology develops according to its own internal logic; that every technology has an inherently "best" design, which, left to market forces, will inevitably prevail; and that technological change clearly leads to social progress.[19] The battle against determinism has produced an impressive array of theoretical tools. Hughes's "seamless web" (a metaphor for the inseparable connections among technical, social, and economic aspects of large-scale technological systems) allows us to understand how those connections define and propel systems. John Law's "heterogeneous engineering" provides a way to talk about the inter-actions among the technical, social, political, and economic dimensions of engineering work.[20] A host of other concepts attempt to refine our understanding of these relationships.[21]

A loose consensus has developed around the notion that technology, politics, and culture are mutually constitutive, but by and large the history of technology and its disciplinary cousins have expended considerably more energy on the construction of technology than on the construction of culture or politics. Perhaps the fear of relapsing into technological determinism has led scholars to use culture primarily as an explanatory factor. Certainly the focus on constructivist approaches and on their attendant epistemological issues has induced many scholars to limit their research to technological design and construction, thereby avoiding anything that might resemble the "effects" of technology.

There have been a few attempts in technology studies to take the construction of society, culture, and politics seriously. Some sociologists and anthropologists have argued that "society" is itself an intellectual construct that cannot explain technology any more than the technical can explain the social.[22] Perhaps the most compelling sociological demonstration of the mutual shaping of technology and politics is Donald MacKenzie's account of nuclear missile guidance, which shows how socially constructed technology shaped policy decisions about nuclear strategy.[23] Recent significant historical attempts in this direction include Paul Edwards's analysis of computers and Cold War politics and culture and Ken Alder's study of gun manufacturing and revolutionary politics in eighteenth-century France.[24]

Only recently and sporadically, then, has technology studies begun to assemble a toolbox for examining the *mutual* construction of technology, politics, and culture. This is not really surprising: the history and sociology of technology have based their disciplinary strength on their ability to explain *technology*—not politics or culture, which are the province of many other disciplines. Recent efforts have shown, however, that seeking to explain politics and culture enriches our explanations of technology. Opening the black boxes of culture and technology *simultaneously* can (for example) give us insight into how technologies constitute a terrain for transforming, enacting, or protesting power relations within the social fabric. Taking politics and culture seriously as objects of analysis greatly deepens our understanding of technological change.

Of course, "politics" and "culture" are big, vague concepts. Before going any further, therefore, let me specify which pieces of these concepts I examine in this book.

The politics I investigate here consists of the constitution, assertion, and exertion of power through material and discursive practices. More concretely, I am interested in how technologists define their niches in national policy making and enact policy choices in technical practices and artifacts, how workers establish their place and assert agency in hierarchical structures, and how local communities situate themselves within a nation.

Culture is an even broader concept than politics. Here I limit myself to two manifestations of culture: national identity and social identity. By national identity, I mean the ways in which people imagine the distinctiveness of their country and define uniquely national ways of doing things. I explore social identity specifically with respect to the nation, and I define it as the set of ways in which groups understand and portray their relationships with one another and with the state. Particularly interesting to me are the aspects of identity that are related to politics. I thus focus on moments when statements about identity are acts aimed at asserting power or position within a sociopolitical order. Other, less deliberate and more affective forms and assertions of identity are also important, but they remain outside the scope of my analysis.

With these specifications in mind, I will now review how scholarship in political and cultural history has tackled some of these issues and suggest how the insights of technology studies might benefit these discussions.

Scholars have debated at considerable length the question of whether post-World War II France is a technocratic society.[25] Most agree that "technocrats" (a term that, in France, usually refers to high-level state administrators trained in the elite schools known as *grandes écoles*) make many of the nation's industrial and financial policy decisions. But the means through which this elite exerts power remain murky. Technologists themselves legitimate their power as meritocratic, arguing that only they are qualified to make certain decisions. Justifiably skeptical of such claims, many scholars have argued that technologists derive their power from a system of social privilege that enables them to create a closed community. The language of technical rationality and professional competence serves as a tool of exclusion and a cover for raw power. Yet these scholars have paid little attention to technologies, knowledge, and practices—perhaps out of a somewhat perverse combination of skepticism about technologists' claims and belief that technological knowledge is indeed hermetic and impenetrable.[26]

As the history and the sociology of technology have demonstrated, however, the construction of a technological system is not an impenetrable, apolitical act. In this book I argue that to understand how French technologists acted politically we must analyze their artifacts and their practices. Institutions certainly provided powerful support, but they did not, by themselves, constitute the means through which these men shaped national policy. Technologies gave this elite a unique vehicle for political action—one that cannot be dismissed lightly.

Although scholars in the cultural history of labor and in the history and sociology of technology almost never cite one another, they make similar arguments about the cultural shaping of the material world. Recently, for example, labor historians have sought to transcend the artificial opposition between experiential and linguistic approaches imposed by historiographic debates. Focusing on how language and culture mediate material experience in shaping identity or politics, they have observed, does not obviate the examination of experience. Their approach enables us to explore and explain experience, and the material world more generally, in fresh ways. The linguistic approach need not imply an anti-materialist position. Instead, it can show how the material world both derives meaning from culture and performs culture.[27] Although they rarely phrase it this way, historians and sociologists of technology also transcend stark oppositions between the material and the cultural world by showing that technical artifacts and practices (the supposed epitomes of the material world) are deeply social, cultural, and political.[28] Synthesizing these two literatures makes clear that, instead of asking whether workplace experience is prior to culture or whether culture is prior to experience, we should look for ways in which experience is cultural *and* culture is experiential.

I seek to do this by examining the relationship between workplace practices and the social identities of nuclear workers. The material practices in which workers engaged derived meaning from a constellation of sources, including the labor unions that represented them and the institutions that employed them. The men I write about did not have a priori identities as nuclear workers. Nor did they articulate or forge those identities solely or even primarily through language or union discourse. Instead, I argue, their identities as nuclear workers emerged as they performed the meaningful material practices of their jobs. Those identities, in turn, not only situated workers in the nuclear program but also defined their place in a national sociopolitical order. The identities of these men as nuclear workers both referred to existing ideas about national identity and reshaped those ideas to fit into the specific context of nuclear work.

What do I mean by "national identity" in this book? My conception of this notion is inspired by a broad range of scholarship on nationhood and nationalism. I ground my treatment in Benedict Anderson's classic formulation of the nation as an "imagined community." At the most basic level, this means that nations are not autochthonous social units but rather communities whose coherence is imagined through political and cultural practices. The content and function of these imaginings vary according to time and place. However stable a sense of nationhood may appear, national identity is in fact continually subject to negotiation and contestation. For Pierre Nora, this means that French national identity is "a reality that is entirely symbolic."[29] Ideas about national identity do not grow by themselves. They must be actively cultivated in order to persist. Further, articulating and rehearsing these ideas often reformulates them.[30]

Discussions of national identity typically refer back to the past. But ultimately national identity discourse is not about the past per se, or even about the present. Instead, it is about the future. National identity discourse constructs a bridge between a mythologized past and a coveted future.[31] Nations and their supposedly essential characteristics are imagined through a telos in which the future appears as the inevitable fulfillment of a historically legitimated destiny. This process naturalizes change; it makes proposed novelties appear to be the logical outgrowth of past achievements. In postwar France, the notion of radiance is precisely such a bridge: radiant through its empire before the war, France must maintain its radiance to maintain its Frenchness. This entails engaging in various political, cultural, and technological acts, many of which derive legitimacy by invoking the relationship between France and the rest of the world.[32] Similarly, proponents of large-scale technological systems justify modernization by placing the systems in direct historical lineage with past national achievements—for example, calling nuclear reactors the modern heirs of the Eiffel Tower and the Arc de Triomphe, nineteenth-century symbols of technological progress and military prowess. Such discursive moves give the nuclear program cultural legitimacy: they aim to make reactors French and to make a non-nuclear France impossible.

Invocations of national identity are thus not gratuitous acts, and this is one reason why historians of technology must take them seriously. Consciously or not, people usually invoke the nation to perform political, cultural, and sometimes even technological work. Anderson notes that the very concept of the nation conjures up the notion of disinterestedness: "For most ordinary people of whatever class the whole point of the nation is that it is interestless. Just for that reason, it can ask for sacrifices."[33] Here Anderson refers to the personal sacrifices entailed by warfare, but disinterestedness need not have such extreme ends. French state engineers cultivate an ideology of disinterested service to the nation that enables them to justify particular approaches to technological development. Invoking the nation thus creates a sense of objectivity, which in turn performs the work of legitimation.

Peter Sahlins discusses a different kind of work done by national identity discourse. Sahlins found that residents of communities on the border of France and Spain called upon national identity in their pursuit of local economic and political interests, thereby legitimating those interests and adjudicating among them. Through repeated invocation of the nation, locals in this borderland came to imagine themselves as national citizens.[34] This example frames another aspect of national identity discourse that will prove important in this book: that ideas about national identity are not simply imposed by the center on the periphery.[35] Provincial communities create their own ideas about national identity. As we shall see when we examine the reception of nuclear reactors in central and southern France, these ideas incorporate local interests, metaphors, and histories, and they are deployed in local contexts.

Explorations of French national identity have yielded rich analyses of how that identity is imagined in debates over issues such as Americanization, modernization, immigration, and colonization.[36] But technology (writ large) is glaringly absent from this literature, as though it were not a site for discourse about national identity. Indeed, in this scholarship technology is cultural only insofar as it becomes an icon or a consumer item; its construction and its attributes do not appear as outcomes of cultural processes. Yet in France technologists, workers, and provincial communities involved in large-scale technological development deployed national identity in a wide variety of circumstances and toward diverse ends. In so doing, they imagined not only a technological France but also their role in such a nation. Thus, if technological development is treated as a social, political, and cultural process, the history of technology can contribute to the historiography of national identity.

The reverse is also true: historians of technology can learn from the scholarship on national identity. One crucial point to take away from this literature regards the instability of culture. Considerable work goes into making culture, and into keeping it stable. Contests over culture are often political; debates about national identity are, at least in part, about who has the power not just to define the identity of a nation but also to shape the nation's sociopolitical order. In the case I examine here, this is not a merely symbolic matter. Debates about the identity of France were not so much about what France was in the present as about what France would become in the future.[37] Attempts to define a specifically French technological style were not frivolous gestures of nationalist fervor but interventions in a contest over the power to shape the future of the nation. Taking the instability of culture seriously means digging more deeply into the power dynamics involved in technological change.

Conceptual and Methodological Tools

In my search for a deeper understanding of the mutual construction of technology, politics, and culture, I have fashioned a set of conceptual and methodological tools that synthesize some of these scholarly insights.

The first tool consists of a question: How do the historical actors we study themselves conceptualize the relationship between technology and politics?

Historians have put great effort into examining the ontology of the relationships between technology and politics. Sociologists have probed these categories, arguing that we cannot decide ahead of time what counts as technology and what counts as society—that these categories emerge from, rather than precede, the construction of an artifact or a system.[38] But these scholarly efforts, and debates over technological determinism more generally, can overlook an important dimension of the story they seek to tell. Even if we do not or should not, historical actors *do* have a priori ideas about the nature and the relationship of technology and politics (or society, or culture). Their *beliefs*—be these beliefs in technological determinism, or more complex ideas about how technology and politics relate—shape their actions and decisions. We must therefore ask how engineers and workers *themselves* conceptualized such relationships, and explore what is at stake in those conceptualizations. Here, posing this question reveals that French state technologists did not conceive of technology as something radically separate from politics (or culture, for that matter). Quite the contrary: many of them saw technology as a thoroughly political entity. Comprehending the reasons behind and the manifestations of this view is crucial to understanding both the shape of the nuclear program and the political behavior of the technologists who built it. I do not mean to deny that we should seek our own understanding of these relationships. Of course we must. But in doing so, we cannot simply dismiss the conceptualizations of historical actors.

My second tool is an elaboration of the concept of *technopolitics*.[39] I use this term to refer to the strategic practice of designing or using technology to constitute, embody, or enact political goals. Here I define technology broadly to include artifacts as well as non-physical, systematic methods of making or doing things. Two examples of technopolitics in this book are nuclear reactors designed with the express goal of creating and implementing military atomic policy and optimization studies aimed at shaping industrial policy. From the very beginning, engineers and administrators consciously conceived of these reactors and these optimization studies as hybrids of technology and politics. Many of the criteria that shaped their technical choices were consciously political. Calling these hybrids "politically constructed technologies" is correct but insufficient, because technologists intended them as tools in political negotiations. At the same time, these technologies were not, in and of themselves, technopolitics; rather, the practice of using them in political processes and/or toward political aims constitutes technopolitics.

Why not just call that practice "politics"? The answer lies in the material reality of the technologies. These technologies cannot be *reduced* to politics. In deciding between fuel loading systems, engineers did not have infinite choices; they only had a few. Further, the effectiveness of these technologies as objects designed to accomplish real material purposes (such as producing plutonium, or calculating the energy efficiency of a reactor) *matters*. On the most basic level, it matters because, for example, this plutonium really did exist, and France really did develop a military nuclear capability, which it shared with other nations, including Israel and Iraq.[40] In addition, the material effectiveness of technologies can affect their political effectiveness. For example, the fact that the CEA's three Marcoule reactors generally worked well served to boost that institution's reputation in the eyes of Charles de Gaulle, while the fact that EDF suffered repeated technical setbacks in the construction of its Chinon reactors angered him; one result was that throughout the war of the systems de Gaulle staunchly backed the CEA over EDF. Finally, the technological aspect of these hybrids shapes the *kind* of political voice that technologists have. (Other factors shape that voice too, of course—especially educational background, institutional provenance, and sociopolitical hierarchies.) Technologists did not participate in French political life as members of a party, or thanks to their clever way with words (though some did have considerable rhetorical skills); they participated because they engaged in, or supervised, or organized the design of material artifacts. Their skills differentiated them from ordinary politicians and contributed greatly to their authority and influence. For all these reasons, the term "politics" captures neither the nature nor the power of these strategic practices.

The third and final tool I develop is the concept of *technopolitical regimes.* These regimes, grounded here in institutions, consist of linked sets of people, engineering and industrial practices, technological artifacts, political programs, and institutional ideologies, which act together to govern technological development and pursue technopolitics. This concept is anchored in the Hughesian notions of technological system and technological style. A technological system is a linked network of artifacts, knowledges, and institutions operating in a coordinated fashion toward a series of specified material goals.[41] Thus, the French nuclear program is a technological system whose components include state agencies, private companies, reactors, laboratories, uranium mines, university curricula, factories, and portions of the electricity distribution network. The technopolitical regimes that I examine operate within this system. They emanate from different institutions, and they have distinct (if sometimes overlapping) goals and ideologies. For the sake of convenience, I have labeled the regime based in the CEA the *nationalist* regime and the one based in EDF the *nationalized* regime. (These labels, however, are associated with institutional stereotypes, and I try to characterize the regimes more subtly in my analyses.) Both regimes seek to shape the French nuclear system. In this sense, one might say that they promote different styles of technological development. Yet "style," albeit an important concept for describing systems, elides the purposeful policies pursued by these regimes.

I have chosen the "regime" metaphor for three reasons. The first reason relates to the use of the term "regime" in political parlance to refer at once to the people who govern, to their ideologies, and to the various means through which they exert power. By analogy, "technopolitical regime" provides a good shorthand for the tight relationship among institutions, the people who run them, their guiding myths and ideologies, the artifacts they produce, and the technopolitics they pursue. The term aims both to evoke the similarity with political regimes and to convey the difference that technology makes. Second, "regime" conveys the idea of regimen, or prescription. The regimes I examine aim, through the pursuit of technopolitics, to prescribe not just policies and practices but also broader visions of the sociopolitical order. This is especially evident in regard to reactor operation: through artifacts and work practices, the workplaces in these regimes performed distinct visions of the sociopolitical order. Third and last, "regime" captures the contested nature of power. The two technopolitical regimes I examine aimed at governing nuclear development at a national level and at governing specific technological practices at an institutional level. But these regimes were not uncontested. Just as national political regimes (democratic or otherwise) must grapple with opposition, these technopolitical regimes had to contend with varying forms of dissent or resistance, both from outside and from within the institutions they governed. As we shall see, these regimes were neither static nor permanent: a technopolitical regime is easier to topple than the technological system within which it operates.

Research Stories and Oral Histories

I began this project with considerable trepidation. Several historians and political scientists had warned me that my ambitions might prove impossible to fulfill. France has fairly restrictive laws governing archival disclosure. Documents that might pertain to national industrial secrets are protected for thirty years, and those pertaining to national defense for sixty years. Waivers are sometimes granted, but I was warned that because of the sensitive nature of my topic I should expect no favors. Things looked even bleaker when I first tried to gain access to the official archives of the CEA and EDF. With exceedingly polite explanations ("We deeply regret, mademoiselle, that we have not yet catalogued our papers"), I was denied entry.

In desperation, I followed the advice of one historian who suggested that I interview old-timers in the nuclear industry. Perhaps, he speculated, some of them had kept private papers. My single experience with oral history was a series of interviews I had conducted with a computer scientist for an undergraduate paper four years earlier. Still, this route seemed my only hope. After a crash course in interviewing techniques, I nervously set out for my first appointment.

My luck changed immediately. The first man with whom I spoke, Claude Bienvenu, had been a project engineer at Marcoule, Chinon, and Saint-Laurent. He had kept a vast amount of documentation: blueprints, memoranda, reports, letters, meeting minutes, and more. Neatly arranged in chronological order in his office, the collection took up nearly 2 meters of shelf space. In a stunning act of generosity, Bienvenu not only allowed me to work in his office but even let me take the occasional folder home to read at my leisure. It took me several months to work through the entire collection.

Meanwhile, I had begun to develop a taste for interviewing. Most people seemed eager to share their memories, look for documents, and put me in touch with others who might help. A few devoted entire days to me. I benefited from extensive tours of Marcoule, Chinon, Saint-Laurent, Saclay, Fontenay-aux-Roses, and even a fuel rod manufacturing plant in Annecy. True, not all encounters went so well. One engineer sourly commented that my interests were outdated (*ringard*) and that I should really study light-water reactors. Another spent 45 minutes lecturing me on why Euratom would make a far better research topic. In a transparent effort to control the havoc they feared I would wreak in the official version of events, a few told me ludicrous, blatant lies: two researchers insisted that there had never been any conflict between the CEA and EDF, and another maintained that EDF had never produced plutonium for the CEA.

Still, most people I interviewed did not seem interested in lying to me. Of course, this does not mean that we can take their stories as faithful, transparent accounts. Everyone has a personal perspective on events, recent or distant. Memories reveal as much about the storyteller's relationship to his or her history and community as they do about the events themselves. Let me cite one striking example. In 1996, I spoke with a former director of EDF. This man had seen many journalists and scholars, and I had transcripts of some previous interviews. As much as I tried to steer the conversation elsewhere, he kept returning to the same stories. A skilled rhetorician and politician, he had no intention of revealing anything new. In fact, not only did he tell the same stories he had told fifteen years earlier; on occasion he even repeated the same sentences word for word. Clearly he had rehearsed these tales so often that they had become rote. I discuss other examples later in the book.

Thanks to a few key private collections, some treasure troves buried in the dusty closets of Chinon and Saint-Laurent, and the awesome documentation efforts of the Confédération Française Démocratique du Travail, I eventually found sufficient evidence to produce a version of this book that would not rely on interviews at all. Municipal and departmental archives contained plenty of information about local community responses to nuclear sites. And the CEA even granted me limited access to its archives during my last research trip. Why, then, have I made use of such a notoriously unreliable source as human memory?

One fairly simple reason is that some things conveyed in interviews are not in any document, accessible or not. These include accounts of how people related to one another, anecdotes about their reactions to particular events, stories about breaking safety regulations, criticisms of institutions, and so on. The accuracy of such tales cannot be verified, but they are all we have. I have made use of these memories in two ways. When the same story was recounted in two or more separate interviews, I have woven it directly into my narrative (signaling its source in the notes, of course). Stories I heard only once are generally quoted verbatim, sometimes along with comments on the nature of the conversation in which they occurred. Most people spoke to me on the condition that they would not be cited directly. I apologize to readers who find this frustrating. The bibliography offers a complete list of the people interviewed.

More importantly, I have included interviews because the tales I heard have shaped—perhaps in more ways than I realize—my understanding of life in today's technological France. From the engineers who decades later still express anger over the demise of the gas-graphite system to the neighbor of the Chinon nuclear plant who could not understand why anyone would care about her memories of its construction, these people have taught me a very personal lesson about how history shapes our understanding of the present, and how the present shapes our understanding of history. I have tried to convey a sense of this throughout the book, not only through the interviews but also through stories of my own research experiences.

1

A Technological Nation

We are a people who easily get enthusiastic over large industrial projects, perhaps owing to tradition—for last century we were among the promoters of modern technological civilization—and doubtless also due to a natural admiration for work well done and a certain pride in being able to compare national achievements with those of other countries. . . .[1]

—Marcel Bleustein-Blanchet, advertising executive

As a gesture of friendship at the end of my second visit to the Saint-Laurent-des-Eaux nuclear site in 1990, the public relations manager there gave me a heavy embossed medallion attached to a key ring. In the foreground stand the site's four reactors; in the background, the château de Chambord, the Loire Valley's "paradigmatic" castle.[2] The Loire River swirls around the two images. "EDF" is engraved along the bottom edge of the medallion, together with a lightning bolt to symbolize electric force; the name of the site appears along the top edge. This souvenir conveys a clear message: France's nuclear reactors are contemporary châteaux, symbols of national glory equivalent in scale and style to the grandest historic monuments.

This preoccupation with linking modern technology and historical monuments began in the context of a postwar national identity crisis centered on anxieties about wartime losses, reconstruction, decolonization, and American dominance. State engineers, planners, and other technologists proposed solutions to these problems through industrial development and engineering prowess. They offered up visions of a new technological France and claimed a central role in shaping this national identity. In so doing, they portrayed themselves as leaders, "men of action" with a deep sense of public service, men who would save the nation from a stagnation induced by politicians and aging industrialists. Historicizing their achievements provided a means of asserting their legitimacy: the past justified their leadership and affirmed their right to speak for the future of French national identity.

How did other political and cultural leaders respond to such claims? In the 1950s and the 1960s, social scientists, humanists, and some politicians expressed discomfort with sociopolitical visions offered by technologists. They feared that France would become a technocracy. Debates raged over the nature of technology and politics, and over the relationships between them.

I explore these debates here in order to provide a national context for what follows in subsequent chapters. In this chapter, I shall argue that much was at stake in debates about technology and politics—first and foremost, the question of who should have the power to define and construct France's future (and therefore its identity). An understanding of these stakes will help me make sense later of how nuclear technologists conceptualized and enacted the relationship between technology and politics. In the process of debating their role in the nation's sociopolitical order, technologists sought to create a hybrid notion of technology that deliberately and explicitly incorporated politics and culture. They did so by developing a trope (the notion of French technological radiance) and a set of practices (grouped under the general rubric of systems thinking) that would become common currency in the French industrial world. These would serve as important elements of the technopolitics of the French nuclear program. Examining the broader debates will also demonstrate that even though the nuclear program stood as the epitome of French technological radiance, it was by no means the only industrial effort that was subject to such interpretations in the postwar period.

Before focusing on the postwar period, though, I must provide a brief overview of state engineering in the nineteenth and early twentieth centuries. State engineers were a major subset of "technologists," a more general term for state experts of all kinds. Their history and ideology is important for understanding how they situated themselves after World War II, when all manner of technologists rose to prominence. I shall then turn to debates over the meaning of "technocrat" in this period. Next I shall look at how technologists' attempts to define (or redefine) their political and cultural role underlay their efforts to simultaneously describe a new technological France and define a specifically French technological style. Finally I shall look at how these efforts came together in the creation of multi-year plans aimed at modeling and shaping national development.

State Engineering before World War II[3]

Throughout modern French history, the engineers with the highest status have belonged to the state engineering corps, particularly the Corps des Mines and the Corps des Ponts et Chaussées (Bridges and Roads). In order to enter one of these corps, young men first had to attend the Ecole Polytechnique, the nation's most prestigious institution of higher learning, and then had to enroll in one of the more specialized engineering schools; the corps selected the top graduates of these schools for membership. Engineers who worked for private industry were known as *ingénieurs civils* and attended different schools. During the nineteenth century the two types of engineers increasingly came into conflict over issues ranging from design methods to professionalization mechanisms.[4] Despite the many professional gains achieved by civil engineers during these struggles, state engineers retained most of their social, technical, and political power. This power derived from both their institutional and their social backgrounds. Although the Ecole Polytechnique was a product of the Revolution, supposedly intended to provide meritocratic access to power, only those boys whose families could afford the school fees and the special schooling needed to prepare for the entrance exams could reasonably expect to attend. Then and now, the most common critique leveled at the school was that it provided merely another way for elites to justify and perpetuate their power.[5]

Deeply embedded in the system of state engineering lay an ideology of public service and leadership. The archetype of the state engineer was the *chef*, a strong, quasi-military leader. Thanks to his irreproachable morals and his irrefutably logical mind, he wore his considerable authority with ease and grace, and used it with fairness and restraint. This supremely male ideal emerged with particular force after World War I, when many lauded the heroic performance of state engineers.[6] In 1932, the civil engineer Georges Lamirand wrote:

'The engineer is a leader: he must 'serve and command' (p. 40), 'gain the sympathy of his men by his manner' (p. 50), 'look his men directly in the eye' (p. 55), and 'impress them . . . with the force of his mind and will' (p. 68). He must 'give the impression of physical superiority' (p. 55), which can be achieved through gymnastics and sports, and in short he must possess the virile qualities that make an officer: frankness, a firm sense of reality, courage, tenacity, and dedication to his work. . . . Finally, the engineer must 'know how to punish' (pp. 56–57).[7]

This archetype derived in no small measure from the figure of the *officier-ingénieur*, the young *polytechnicien* vigorously recruited by the military in the interwar period.[8] Though most civil engineers could not attain top leadership positions in which to exercise such talents, many state engineers did. This ideal persisted during Vichy, when the Ecole Polytechnique was removed from military purview and placed under the tutelage of Jean Berthelot and his Ministère des Communications. Defining the school's new mission, Berthelot (himself a member of the Corps des Mines) wrote:

> The general aim of the Ecole Polytechnique is to educate leaders [*chefs*] for every branch of national activity that requires extensive scientific knowledge coupled with an extensive general culture [*culture générale étendue*].
>
> By means of a strong moral, physical, and intellectual education, students are prepared to become leaders in the corps, services, and companies which need them.
>
> The moral and physical education aims at developing within them the qualities of a leader: vigor, character, decisiveness, sports mentality, command aptitude, a taste for effort and responsibility. It should develop their personality in a disciplined [manner], imprint on them team spirit and a sense of community; in a word, [it should] give them a high sense of national purpose.[9]

Polytechnique ideals thus blended national pride, public service, and masculinity. The "sense of national purpose" identified by Berthelot had always formed an important part of how state engineers defined themselves. Designing for private industry was considered demeaning.[10] Instead, state engineers' commitment to public service supposedly guaranteed that their canals and railways served the nation, not private profit—for example, by determining which transportation routes would best serve the largest number of citizens and ensuring that private companies did not skimp on construction.[11] After World War II some state engineers perpetuated this animosity toward private industry, resituating it in a new political and institutional context.

State engineers used this ideology of public service and leadership to cultivate a particular style of engineering knowledge, design, and practice. Good leadership required "polyvalent" knowledge: the state engineer had to be a generalist as well as an expert. This justified the Ecole Polytechnique's highly theoretical curriculum, since theory was supposed to give graduates the broad perspective required to lead properly. Once out of school, many state engineers conceived of their large-scale projects as monuments to the glory and eternity of the French nation. Engineers measured and quantified the value of public service. They plugged this value (rather than numbers generated by market-based exchanges) into the economic calculations of their state projects. For example, quantifying the usefulness of a stretch of rail to the industries and towns it served enabled Ponts engineers to argue against parochial or private interests in the name of the national interest. The Déclaration d'Utilité Publique, which officially certified the public usefulness of proposed projects, epitomized such calculations.[12]

State engineers sought not just to serve the nation, but also to build it. Theodore Porter notes that the Corps des Ponts et Chaussées "aimed to unify and administer the French territory, and even to civilize the French peasantry."[13] Ponts engineers expected to accomplish this mission through their designs and practices. The famous Legrand star, for example, had tracks that went from Paris to the six corners of France along straight lines, thereby trying to encompass and unify the French nation with the utmost spatial economy. Victor Legrand himself, meanwhile, saw the railroad as the "instrument of national civilization."[14]

Scholars[15] have observed that, historically, state engineers' understanding of their social role has bound them together more than their expertise: their group has identified less as a profession than as a social category. Consequently, some researchers have questioned whether state engineers really were (or are) engineers in the technical sense at all.[16] Terry Shinn has argued that the work of state engineers was "almost totally lacking in scientific and engineering content: the attributes that mattered were social and political."[17] Eda Kranakis, however, disputes such conclusions on two counts. First, she argues, the public works design and construction projects of nineteenth-century state engineers clearly included large amounts of technical work. Second, she notes, the emergence of large-scale technological systems increasingly required both state and civil engineers to perform a variety of different kinds of tasks, and defining engineering as a narrow set of technical tasks offers too simple a view of what it means to be an engineer of any kind.[18]

Ample work in the history and sociology of technology supports this argument.[19] Clearly, engineering is a heterogeneous activity. In this respect French engineers are no different from others. Consider Antoine Picon's description of Ponts engineers in the interwar period:

[They] refuse the opposition between technology and humanistic culture, technology and language, technology and society. For them, the ideal of a technique mastered by a fully efficient technology [is] often accompanied by an implicit or explicit social project, destined to reconcile man with his mechanical environment. In the minds of the boldest ones, such a reconciliation would strongly resemble an assimilation of the social to the technical. Why not make society function like a big, scientifically regulated factory?[20]

Yet this example also hints at a more unique characteristic of French state engineers that will be encountered repeatedly throughout this book: their frequent *willingness* to admit the social (or political, or cultural) nature of their work. Porter explains this by analyzing the nature and basis of state engineers' authority in the nineteenth century. Faced with pressures from local politicians concerning the location of roads and bridges, for example, Ponts men did not (as American engineers might have) devise a rigid set of rules by which to make decisions. Such rigidity would have been "inconceivable," because it would have undermined their status and power as an elite.[21] Instead, they derived power from their commitment to public service and their social standing, more than from their scientific training. In the words of one Ponts engineer, "There are many ingenious formulas to calculate the traffic volume on a planned route as a function of the population served; but to apply them with discernment requires taking account of the social, economic, and moral state of the population, and that is the greatest difficulty."[22] Only a broad education and a strong moral sense could possibly lead to such "discernment."

To put the matter somewhat differently, state engineers did not so much *derive* legitimacy from their technological achievements as the other way around. That is, their position within the state *conferred* legitimacy on their technologies. That position fluctuated over the course of the nineteenth and twentieth centuries, and so did the success of their technologies. State engineering flourished in the early nineteenth century. By the 1880s, however, economic liberals dominated crucial ministries; they curtailed deficit spending, and with it the ambitions of state engineers to plan and build the nation.[23] Between the two world wars, state engineers (and other technologists) became interested in national economic and industrial planning, but their projects met with limited success. During World War II such men also participated in the Vichy government and (less visibly) in the Resistance.[24] Many of them remained involved in state institutions after the war, when state engineering came into full bloom.

State Institutions after World War II

World War II led the French to question the foundations of their social, political, and economic life. Despite other profound differences, the dominant political groups of the Liberation period agreed on a few fundamental issues. They blamed France's abysmal military performance during the war on the economic "malthusians" who had run the nation and its industries throughout much of the Third Republic. Because these men had tried so hard to preserve the status quo, the economy had stagnated and French industries had fallen far behind their German counterparts. Leaders feared that France had declined—perhaps irreparably—from its former status as a great nation.[25] Anxiety about technological backwardness was one manifestation of this fear.

Most agreed that the remedy for these problems lay in rethinking the role of the state in directing the economy (in general) and in directing industrial, scientific, and technological development (in particular).[26] By engaging in and promoting investments aimed at modernizing and expanding industry, the state would accomplish the dual aim of resuscitating the economy and restoring France to its rightful place in the ranks of great nations. Of course, considerable disagreement existed among communists, socialists, centrists, and Gaullists over the long-term role of the state in French society. But enough consensus emerged among these diverse political groups to create or reform a number of new state institutions.[27]

These institutions covered a wide spectrum of activities. The Ecole Nationale d'Administration was founded to produce modern managers for the state. The Centre Nationale de la Recherche Scientifique, founded just before the war and stagnant during Vichy, was redesigned to cover a broad spectrum of scientific activities and expected to make recommendations on science policy.[28] The Commissariat Général au Plan (Planning Commission, often referred to simply as the Plan) was created to plan reconstruction and modernization through a program of nationwide production goals and coordinated industrial development.[29] The new Commissariat à l'Energie Atomique aimed to develop atomic research and technology. Finally, amid considerable political controversy, the electricity, coal, and gas industries were nationalized. The restructured state claimed to serve the French people rather than private interests. After decades of debate, the nation would finally modernize.

This profusion of new and reformed institutions proved congenial to a growing class of state experts. Issuing primarily from the *grandes écoles*, this group included state engineers, economists, and professional administrators. Their ideology of public service, polyvalent knowledge, and masculine meritocratic leadership dovetailed perfectly with the prevailing view that state institutions should take primary responsibility for directing national reconstruction and modernization. Joining the chorus denouncing the protectionist practices of traditional French businessmen, these men poured into the top ranks of ministries, the Plan, nationalized companies, and other state institutions. Experts could thus bolster their ideological claims to public service with the alleged disinterestedness of the nation-state: their background and their chosen institution would doubly qualify them to speak for the nation. From this position, they vigorously forged their visions of a French technological identity. During the Fourth Republic (1947–1959)[30]—a time of considerable political turmoil, marked by twenty changes of government—the most successful and visible of these efforts occurred primarily in nationalized companies such as Electricité de France and in state agencies such as the Commissariat à l'Energie Atomique. The Fifth Republic (1959–present) provided political support and leadership for such efforts, which gained correspondingly in momentum, visibility, and prestige.

What Is a Technocrat?

The important role played by state experts in shaping new ideas about the nation did not escape the attention of their contemporaries. Quite the contrary: as the euphoria of liberation gave way to the hardships of reconstruction, private industrialists and opponents of domestic policies from both ends of the political spectrum began to accuse state experts of undemocratically imposing their will on the nation. The word "technocrat," fairly neutral before the war, became a derogatory epithet—in part, probably, as a result of the specter of expert involvement in Vichy.[31] At the same time, "technocrats" attracted the attention of social scientists and humanists as an identifiable group to be analyzed and critiqued. Debates sprang up over the meanings of "technocrat" and "technocracy." At the root of these debates lay passionate defenses of a separation of technology and politics. For social scientists and for stronger opponents of state experts, "technocrat" designated someone who had breached a boundary, who had moved from his area of expertise into the domain of political decision making. The dangers inherent in breaching this boundary were considerable; first and foremost among them was the capitulation of democracy to technocracy.

Because "technocrat" became such a loaded term in this period, I will treat it as an *object* rather than a category of analysis. When I need to designate state experts as a group, I will use the term "technologist." This word is my translation of *technicien*, a more neutral and polite term used by all sides in these debates. Like its derogatory cousin, "technologist" often referred not just to state engineers but to any expert or high-level bureaucrat involved in state administration. (There were contests over this meaning too, but these were not as heated.) I will follow this usage, because—as we shall see when we look at the leaders of the CEA and EDF in chapters 2 and 3—it often does not make sense to distinguish between engineers and other types of technologists when discussing their positions in the upper levels of state institutions.

What was technocracy, and what relationship did it subsume between politics and technology? According to one organizer of a 1956 conference on "Politique et Technique," such questions had come to dominate problems in political science by the mid 1950s.[32] Technocracy, speakers at this conference said, entailed the replacement of politicians by experts—not just engineering experts, but also experts in finance and administration who engaged in quantifying practices. Some, like the eminent writer André Siegfried, saw this replacement process as a quasi-inevitable result of technological civilization:

> . . . is it not normal that we should be pulled toward technocracy? The primacy of our material preoccupations demands this, for the standard of living depends on technology: the quantifying spirit, the geometrical spirit in machine civilization has become dominant. With the expert replacing the politician, it has even invaded a domain where the spirit of finesse should continue to reign. These transformations were inevitable, but perhaps they pulled the State too far down the road of a potentially oppressive technocracy. The defense of the individual, of liberalism, must find new positions.[33]

Social scientists echoed these anxieties. Technologists were eroding traditional political power, and this erosion posed a grave danger to democracy. While in theory power remained in the hands of elected officials, in practice those officials no longer played any significant role in policy making. State planning offered a striking example: plans were devised by state experts, while elected officials, who had neither the time nor the qualifications to understand the calculations, merely approved the budget without questioning the plan. "Government by opinion," one speaker noted, "is giving way to the power of initiates who have the secrets of technology, or who simply know its rules."[34] Most conference participants did not think this change resulted from a deliberate takeover strategy on the part of technologists. Rather, they felt it emanated from the increasing complexity and scale of technological society. Nonetheless, the overall trend seemed clear. Democracy, the essence of the French republic, was threatened: crucial decisions were being made by an elite who had not been elected to do so.[35]

The technocrat figured as a villain across the spectrum of party politics. On the extreme right, the populist Poujadiste movement accused technocrats (as well as other intellectuals and mainstream political leaders) of oppressing ordinary French citizens with modernization, planification, large corporations, and the growing state. "You, Man of Science, for whom do you work?" Pierre Poujade railed in a typical display of inflammatory rhetoric. "If you are neither a nut or a sadist, does it make you happy to see your work crush men?"[36] The left too feared technocracy, though it used somewhat different language to describe its dangers. At the "Politique et Technique" conference, the Confédération Général du Travail militant Pierre Le Brun warned:

A State in the service of Technology that would itself serve Capital would be a technocratic State, a State that . . . would tend to *govern men as though they were things*. . . . Fascist corporatism, which we experienced in France from 1940 to 1944, is also, in its negation of class conflict, a form of technocracy and it is no coincidence that so many technologists and technocrats sat in the Vichy government. . . .[37]

Technocracy, continued Le Brun, could become a form of dictatorship that completely subjugated the wishes of the people. In political circles, castigating technocrats had become another way to criticize the government by playing on old fears about the encroaching power of the state.

Beyond the arena of party politics, discourse about technocracy also expressed anxiety about the apparently relentless advance of technological civilization. For example, many intellectuals feared that without any deliberate human agency, technological change and the lure of material goods had conspired to alter the very structure of social and political relationships. Would technology go so far as to make politics irrelevant? The fact that the Fourth Republic prospered economically despite the lack of strong political leadership suggested that politicians had indeed become unnecessary. Such a situation posed special dangers for political scientists, of course: one of them worried that on the day when technology could resolve all human problems their discipline would become superfluous.[38]

These anxieties did not fade after 1958, when Charles de Gaulle returned to power after a twelve-year hiatus. True, de Gaulle *announced* that he had no intention of leaving serious decision making to experts. Gaullists had lambasted "technocrats" on a number of occasions, blaming them for the disorder of the Fourth Republic: "Each ministerial crisis appealed to 'experts': a squad of interchangeable high functionaries in the ministry of finance. Always the same ones. Regardless of what color the Chancellor of the Exchequer was, they were there, always with a plan in their pocket: each time, the plan was identical to the preceding one and consisted of priming the financial pump."[39] Once Gaullists came to power in the Fifth Republic, however, such rhetoric appeared to have been a matter of expediency. De Gaulle believed in a strong state, national planning, and large-scale technological projects. Powerful leadership, he felt, had to rise above party politics. He accordingly appointed several state experts with no background in professional politics to his ministerial cabinets. His regime breathed new life into the flagging Commissariat Général au Plan by appointing Pierre Massé—*polytechnicien*, economist, and former EDF director—as High Commissioner of Planning. The general himself called upon the French people to consider the plan produced by the commission their solemn and urgent duty, dubbing it "the ardent obligation."[40]

Debates about the nature of technocracy intensified during the first decade of the Fifth Republic. In 1960 the political scientist Jean Meynaud published *Technocratie et politique*, in which he described the history of technocratic ideology and accused the French state of becoming increasingly technocratic. The technocratic mentality, for Meynaud, was "that state of mind which makes us conceive of technological achievements as the supreme evidence of humankind and that invites us to expect everything from scientific progress."[41] He offered a broad-ranging list of those who, by virtue of their technical competence and influence on public life, ranked as actual or potential technocrats: bureaucrats, mathematicians, military officers, professional consultants, scientists, and more.

Meynaud located the intellectual foundation of technocracy in an unlimited faith in the value of scientific analysis. The end result of technocratic action was the "abdication of the politician in favor of the technologist."[42] Numerous leftist critics claimed this had occurred twice already: in 1940 with the institution of the Vichy government, and again in 1958 with the Gaullist overthrow of the Fourth Republic. Ultimately, Meynaud agreed. He judged technocratic ideology to be little more than an apology for the technologist and a justification of the desire to reduce politics to technology.

Meynaud particularly suspected that ideology's exaltation of the technologist and his supposed ability to take a *vue d'ensemble*—a systems view. When paired with the alleged moral qualities of the technologist—such as a highly developed sense of public responsibility—this ability supposedly enabled men to make choices in the general interest. "In the extreme case," wrote Meynaud, "this is about possessing qualities that allow [technocrats] to direct valid social choices."[43] Meynaud found this a highly idealized portrait: men possessed of the technocratic mentality frequently did not have such an elevated moral sensibility. Uncertainties abounded in any kind of decision making, and technologists were no better qualified to deal with ambiguity than politicians. Furthermore, any kind of decision, no matter how "technical," necessarily incorporated political considerations. Most ridiculous of all were the cyberneticists who proclaimed the arrival of a "governing machine" that would ultimately mechanize political decision making. For Meynaud, the advance of technocracy heralded the weakening of democracy.[44]

Meynaud's analysis was representative of the writings of many social scientists. Popular representations painted the technocrat as a man steeped in abstract reasoning, focused only on Paris and evincing complete ignorance of the provinces, a productivity maniac, a bad Frenchman who drank whisky and Coca-Cola instead of wine, and a Jew. The portraits painted by social scientists and humanists were less racist and hysterical, but barely more flattering in tone. Technocrats were men who proclaimed the irrelevance of all ideology, called for the introduction of operations research in political decision making, argued that France had to find a middle ground between Russian rationality and American efficiency, and vehemently denied being technocrats.[45]

Social scientists and humanists found technocrats most threatening when they breached the boundary between technology and politics. These intellectuals deemed the maintenance of this boundary crucial to the proper and decent functioning of society. When Jean Meynaud acknowledged that technical decisions had political dimensions, he did not mean, as today's social scientists might, that technology and politics *could not* be separated (much less that the very process of design was political). Rather, he meant to make his own political statement: namely, that elected officials should be presented with a range of technical options so that *they* could make final decisions. Democracy and justice, in other words, demanded a clear demarcation between technology and politics.

Technologists defended themselves against their accusers in a great variety of forums. Some wrote books ostensibly aimed at the educated lay reader. Others made speeches to alumni associations. Still others wrote in the popular publications of the new ruling elites, such as *Le Monde* and *L'Express*. Taken together, these defenses had several overlapping aims: to assert the legitimacy of technologists' place within the ruling elite, to elucidate the specific nature of technologists' contributions to that elite and generally establish their identity, and to enroll the technological rank and file—including engineers and *cadres* (mid-level managers)—in their visions of a new sociopolitical order.[46]

Technologists adopted two strategies in their defenses, contradictory on one level but eminently compatible on another. They first denounced the negative connotations of "technocrat" and attempted to salvage a positive meaning for the term. This consisted of arguing that "technocratic" modes of action, while accused of being authoritarian, in fact fitted perfectly well within the democratic process. The second strategy involved articulating an identity for technologists that opposed the identity of politicians. On the surface, this consisted of affirming separate and opposing identities for the two social groups. But in arguing that technologists were superior to politicians, this approach ultimately militated for a conflation of technological and political means of action. Let us examine each strategy in turn.

Alfred Sauvy (a *polytechnicien* and the director of the Institut National d'Etudes Démographiques) had a chance to defend his kind at the 1956 "Politique et Technique" conference. The pejorative sense of the word "technocrat," he said, had emerged unjustly. All technologists did was propose ideas for reforming the nation, and ideas were part of the democratic process.[47] Sauvy vigorously defended the notion of disinterestedness, scornfully dismissing businessmen and lobbyists who denied its validity:

From the point of view of certain private interests, disinterestedness is a laughing stock, a monstrosity. At the very least it is an object of suspicion. When concern with the general interest opposes a private interest, the first reaction of the defenders of that interest is often, if not "who pays?" then at least "in the name of which private interests are these ideas proposed?" Then, if they do not receive a satisfying answer, the classic attitude is to object that these are the views of a theoretician, that they are abstract ideas. In the end, if these epithets do not suffice to condemn the adversary in the eyes of the audience, the last resort is to call him a "technocrat."[48]

While Sauvy admitted that the defenders of the general interest did not always communicate and negotiate as much as they should, this was largely because such men didn't have any forums where they could express themselves. Developing more and better channels of information could easily remedy this situation. How absurd, he exclaimed, that French citizens were more suspicious of the "technocrat" who defended the public interest than they were of industrialists who defended only their own private interests!

Sauvy thus argued that the ideals of the disinterested expert were perfectly compatible with democracy and dismissed the word "technocrat" as nothing but a petty insult. Others, however, attempted to re-appropriate the word and restore a positive meaning to it. In a book titled *Plaidoyer pour l'Avenir* (Plea for the Future), Louis Armand (*polytechnicien*, member of the Corps des Mines, onetime head of the Société Nationale des

Chemins de Fer, professor at the Ecole Nationale d'Administration) and his co-author concluded that "it is not being a 'technocrat' in the insulting sense of the term to want to base oneself on realistic data, to seek to understand [these data], and to finally attempt to synthesize them. This is being a man who loves life and wants to figure out what he can do to love it even more and to ensure that others love it."[49] Thus, far from inhuman or mechanical, the technocrat was full of passion. Technocracy, said the enthusiast Dominique Dubarle, was "exercising power inherent in scientific . . . and mathematical technology in order to [ensure] the good operation of society and the success of large social entities: large companies, nations, tomorrow perhaps all of humankind."[50] Technocracy contained within it the power to transcend the petty boundaries of national politics. But only true technocracy had such transformative powers. Polytechnique professor Maurice Roy attempted to define the "true" technocrat in expressing his desire that his school would continue to "produce authentic technocrats and at the same time avoid, under the banner of progress, inventing and training 'supertechnocrats' deprived of true and serious knowledge of the technologies in question."[51] For this *polytechnicien*, then, financial and administrative experts did *not* count as technocrats in the best sense.

The civil servant Jean-Louis Cottier also wanted to strip "technocracy" of its negative connotations. He found technocratic thinking eminently compatible with Christian humanist thinking, which in turn provided the moral basis for any technocracy. Technocracy was anything but anti-democratic—indeed, people from all social classes could be technocrats: ". . . engineers from the *grandes écoles*, industrialists who are self-taught or trained by family traditions, workers whose value makes them stand out from the ranks . . . , military [men] who are good at command. . . . These men participate in politics as technologists."[52] In order for this moral, democratic technocracy to function properly, technologists needed a sense of history and human society. Engineers had to study the influence of technology on history. By understanding the "laws" linking the evolution of science and that of civilization, technocrats could better construct the future. Thus informed by humanism, technologists *could*—contrary to the claims of their detractors—build a more human, moral, and democratic world.

Such attempts to salvage a positive meaning for "technocracy" did not work. Striking evidence for this failure appeared in a 1967 article published by Louis Armand in *La Jaune et La Rouge*, Polytechnique's alumni magazine. The article compared "technocrats" and "technologists." The former, said Armand, were graduates of the Ecole Nationale d'Administration and reigned over juridical and administrative domains; the latter were Polytechnique graduates and reigned over technological and industrial domains. Claiming that he wanted to explore the common ground between the two groups, Armand situated himself at their intersection by noting his affiliation with both institutions. But his description of the differences between the two groups made clear that his loyalty lay with the technologists. They were oriented toward the future. They might be idealists, but they were neither ideologues nor demagogues. While they believed in profits for their employers, they paid little attention to personal profit, and they were more attracted by professional recognition than by social rank or salary. Furthermore (and this held particularly significance for Armand, a strong supporter of the European union), technologists felt comfortable with their counterparts in other Western nations: they had a natural tendency toward international cooperation and wanted the whole world to benefit from their work. Technocrats, by contrast, focused only on immediate problems. They tended heavily toward *dirigisme*—a word that raised the specter of Vichy and total state control. They relentlessly pursued political power, constantly seeking to rise through the administrative ranks. Since their expertise consisted in elaborating legal and administrative texts whose purview was necessarily restricted to the nation, their outlook was narrow. Worse, they voiced automatic suspicion for anything European, which they felt threatened France's national sovereignty.[53]

This unflattering portrait of technocrats represented a clear attempt to distance those with technical training from that epithet. Technologists retained all the virtuous characteristics of previous definitions of technocrats, leaving administrators looking self-serving and power-hungry. But even though Armand had abandoned his efforts to rehabilitate the word "technocrat," he retained his convictions that technologists should occupy positions of power. Indeed, only this could save France from permanent second-rate status, for ultimately technology "is at the foundation of all that can ensure, if not total independence, which is no longer accessible for a country of our size, then at least sufficient means to enable us to play an active role in the federations of the future that will replace the Europe of borders."[54]

Armand's article brings us to the second, parallel strategy that technologists used to counter accusations against them: articulating an identity that explicitly opposed that of the politician—or, more accurately, the politician as seen by the technologist-engineer. In the pages of *La Jaune et La Rouge*, politicians emerged as corrupt, dishonest, and ineffective. One

Polytechnique alumnus defined the contrast with particular eloquence in a speech entitled "The Cardinal Virtues of the High-Class Engineer":

Do we want the best of our engineers to participate, through their acts, their pens, and their words, in making our economy healthy again? Well then! we must immunize them, from their very first steps, against the disease known as political thinking of which Machiavelli was the champion. . . . The Florentine did not hesitate to claim that the individual should sacrifice to the State not just his fortune and his life, which is very good, but also his honesty, which is despicable. . . . In the twentieth century, . . . a democracy cannot accommodate long-standing deception, and I cannot conceive that economic recovery could occur without the country being fully aware of the difficulties to surmount and having full confidence in the sincerity of its guides. [55]

Democracy—a cardinal virtue of the social order about which everyone could agree—could not be implemented by politicians and their political values. The heirs of Machiavelli would only lead the nation into further chaos. Instead, France needed heroes like Galileo and Pasteur. Such mentors could teach boys to defend truth and rigor under the direst of circumstances, turning them into fine, upstanding young men who would "restore our economy and give France the place it deserves in Europe." The alumnus continued: "Let us pray, my dear comrades that, in the expert hands of our imperturbably devoted teaching personnel, and with the help of its cadre of officers whose enthusiasm is infectious, our old establishment can still shape men who, thanks to their talents and their virtues, will have the Glory of accomplishing, thanks to their culture of Science, the task that the Nation demands of them!"[56] In contrast to the corruption and decadence of politicians, the eternal, disinterested values of the Ecole Polytechnique would thus guide the nation.

Not all technologists felt this need to attack the political class in order to articulate a distinctive identity that would naturalize their participation in running the nation. Some—like Armand and his co-author in *Plaidoyer pour l'Avenir*—defined the difference as one of method and thought process. The problem with politicians—elected officials as well as prefects or ministers—was their inability to think synthetically and systemically. This ability constituted the great strength of technologists, and their most important difference from politicians.[57] Allowing technologists to produce and direct organized systems would not render politicians superfluous, but it would render ideology "obsolete."[58] Thanks to the systems thinking and building of technologists, politicians would no longer need ideology to demonstrate the validity of a policy. Good policies would now emerge from rational rather than ideological choices. Thus systems thinking both defined and legitimated the participation of technologists in public life.

Another strand of efforts to elucidate a distinctive technological identity appeared in the portrayal of technologists as supremely masculine. The technologist was virile, decisive, and forward looking. He was, in one of Pierre Massé's favorite expressions, a "man of action."[59] Politicians, by contrast, were not men of action—or at least not very efficient ones. According to Jean Barets, "the political man of the Republics . . . , the product of chance, [is] badly prepared for the awesome task of a man of State, ignorant of international and economic problems, duped by his own ease of expression. He will be vanquished by facts."[60] It followed that only technologists, who had a true mastery of "facts," could be real "men of action." The striking durability of this masculine archetype shows in the title of a recent eulogy to a former head of the CEA: "Pierre Guillaumat, man of action." Citing Guillaumat's role in providing Israel with military nuclear technology, the writer compared him to James Bond: "You could see the shadow of a super 007, far away and inaccessible: Pierre Guillaumat."[61]

The virtues of the man of action included honesty and directness. In Massé's experience, some politicians did possess these attributes—men who laid their cards on the table and engaged in rational conversation. With them, he could talk "man to man."[62] But such politicians were rare. Technologists, however, as men of action, truly understood the merits of frank discussion.

Contrary to the image that political detractors painted of the cold, hard technocrat, men of action were not without heart. Indeed, another remark by Barets showed a passionate desire to procreate and nurture— actions that, indeed, could come directly from the male technologist, completely bypassing any female assistance: ". . . in the literary mind, the love of man is merely a platonic love, [a love] which does not create human life. The technologist loves man with a more carnal love and wants to continue to nurture the being whom he loves. He will therefore try to protect [this being]."[63] Technologists were thus the virile, passionate protectors of mankind (and presumably of womankind as well). Love spurred them to action. Sometime this action occurred behind the scenes, but it was always in a good cause. Politicians and writers did not act—they wrote, they talked, they waffled, but they did not act. Action made technologists masculine and therefore powerful.[64]

Clearly, technologists as well as humanists, social scientists, and political leaders found it vital to enact a boundary between technology and politics. But the boundary had different meanings and locations for the two groups. For non-technologists, the boundary upheld one of the foundations of democracy. Its transgression therefore signified the collapse of the social order. The boundary was thus located in the domain of practice: technologists should not behave like politicians, and technological methods should not be applied to political decision making. For technologists, the boundary was located primarily in the domain of identity. The relevant difference was, above all, one between themselves and politicians. In constructing this difference, technologists adopted extremely narrow definitions of politics. Politics could mean the implementation of classic ideological stances such as communism, socialism, or liberalism. Or it could refer to the activities of politicians, caricatured as corrupt, indecisive, irrational, manipulative, and Machiavellian.

"Politics" in the sense of classical ideology and corrupt machinations may have been "other" at one level of technologists' identity discourse.[65] But at a deeper level lay the implication that, because of their values and knowledge, technologists were ultimately better equipped to pursue at least some of the activities in which politicians engaged. Now stripped of connotations of ideology and corruption, politics took on a broader meaning, becoming part of what technologists did and should do. At this level, technologists sought to erode what they claimed was an outdated boundary between technology and politics.

The Future of France

One important site for the erosion of the alleged boundary between technology and politics was in the discourse about the role of technology in the future of France. This discourse attempted both to describe what a future technological France would look like and to define a specifically French form of technological and industrial development. General forms of these descriptions and definitions constituted a relatively mild erosion of the alleged boundary between technology and politics: they proposed that technological achievement replace more traditional measures of national power and prestige. Discussions of how to attain such futures, however, attacked the boundary much more aggressively. Shifting from goals to means meant searching for ways to shape the future and control destiny. Systems thinking—both qualitative and quantitative—loomed large in this drive to control destiny, constituting an important means for technologists to blur the technology-politics divide and define and defend their own role in shaping France's future and identity. In order to understand this function of systems thinking, though, we must first examine how technologists related technology and Frenchness.

The fundamental premise of discussions about a future technological France was that, in the postwar world, technological achievements defined geopolitical power. A typical article stated that "the possession of industry, especially heavy industry, appears to be a necessary element for respect and independence."[66] Technologists who had sometimes vainly insisted on this equation during the Fourth Republic found ample political support for it once de Gaulle returned to power: "We are in the epoch of technology," declared the general on one occasion. "A State does not count if it does not bring something to the world that contributes to the technological progress of the world."[67] For de Gaulle, technological prowess could be particularly important in helping France combat the crisis of grandeur brought on by the decolonization of its empire. Technologists happily agreed that technological development could provide the basis for a new relationship between France and its former colonies. Writing just a year after the Algerian crisis that had brought de Gaulle to power, Jean-Louis Cottier saw tremendous potential for this new way of conceiving geopolitical relationships: "In 1958, destiny knocked on the door. . . . In came the technocrats who would build the Franco-African industrial community. In them, the science of engineers is united with the will of captains. The new French strategy, French peace, will be brought to the world."[68] At the same time, France's former African colonies could remain in a sense French by providing the raw materials so essential to France's energy independence—particularly oil and uranium.[69]

Even worse than losing the empire, however, would be the economic and cultural colonization of France by the United States.[70] Indeed, technological achievement as the standard of geopolitical power did not mean that technological pursuits all over the world were identical. The loss of cultural specificity posed the greatest danger of adopting this standard. Even those French who found the United States fascinating dreaded the prospect of a thoroughly Americanized France.[71] The Groupe 1985, a collection of technologists convened by Pierre Massé in 1964 to think about the long-term future of the nation, issued a clear warning: "The first unexpected challenge is the intellectual and cultural survival of an original and individual France. Indeed this scientific civilization will increasingly tend to attenuate national specificities and deformities. From now on our presence in the world depends on our ability to imprint our mark on this civilization by means of significant contributions from French technology and French science."[72]

What made a technological or scientific project French? A difficult question to answer. Indeed, most technologists avoided addressing it directly, concentrating instead on listing and celebrating French achievements. At a 1959 press conference organized by the Conseil National des Ingénieurs Français, one prestigious engineer enumerated the accomplishments of French technology over the previous decade. French engineers had excelled in numerous domains: coal, electricity, steel, nuclear research, railways, aeronautics, building, and more. The ultimate proof of French prowess came when other nations consulted French engineers. For example, the Israelis had asked French engineers to help them design urban transportation systems; in gratitude, they named one of the terminals in Haifa "Paris."[73] Paris in Israel: what better evidence of the "radiance of France"? The pages of *La Jaune et La Rouge* were filled with praise for French technological achievements, succinctly expressed in titles such as "French aeronautics, a matter of pride and hope," "'The Caravelle': a national triumph," and "The radiance of France from the builder's scientific and economic and viewpoint."[74] In 1960 the explosion of France's first atomic bomb in the Algerian desert—triumphantly announced in a press conference at the Ecole Polytechnique—showed the "entire world the value of French technologists and considerably reinforc[ed] our country's position."[75] Two years later, the new terminal at Orly Airport filled this role: ". . . we deemed it indispensable that an undertaking of this size, destined to be seen by the entire world, should give everyone, inside as well as outside, an example of what we can do in France."[76]

Language provided another means of defining a French technological style.[77] The dominance of the United States seemed particularly challenging here, largely because of the "delay that French technology suffered with respect to American technology during the last war."[78] American words threatened to colonize French technical language. According to one group of *polytechniciens*, this posed several problems: these terms were difficult to pronounce, they sounded ugly in French, and they threatened the precision of the French language. In 1954, to guard against the wholesale invasion of American terminology, these men founded the Comité d'Etude des Termes Techniques Français, which was dominated by *polytechniciens* but which also included other engineers, linguists, university professors, and delegates from professional technical associations. This committee met monthly to find equivalents for foreign technical words—especially those that sounded particularly horrible in French. Monthly reports went to institutions and prominent industrialists for comment, after which the committee formalized its proposals and tried to get them adopted by engineering schools and the technical press. In essence, the committee saw itself as a kind of linguistic immigration officer: "Upon entering a country there is a service that sorts immigrants in order to ensure that only the useful ones enter; similarly, we must filter foreign words as soon as they mingle with French vocabulary."[79] Demoting American terms from colonizers to immigrants made them significantly more manageable. Meanwhile, it was felt, technologists should fight to reinstate French as the world's *lingua franca*. In a 1962 radio interview, the director of the CEA's research center in Saclay remarked that the English language was "not very rational," since the same word could have different meanings when used by the Americans or the British. Much better to use French, "which is a stable, . . . solid language, and which still allows for all the nuances needed to deal with the most modern science and technology. . . . We should teach the greatest possible number of foreigners [at least] a modest French."[80]

Technologists also attempted to elucidate what was—or should be—specifically French about French technology. Most of these efforts appealed to a sense of history or tradition. Tradition, it appeared, could define or describe Frenchness fairly unproblematically. Placing modern accomplishments in direct historical lineage with accepted traditions would therefore make them demonstrably French.

One traditionally French quality was a refined esthetic sensibility. "The beautiful," noted the Groupe 1985, "is a traditional export of France."[81] Technological achievements did not have to be ugly; modernity could be beautiful.[82] The time of hideous industrial landscapes had ended. Modern technology—especially in France—"engenders . . . its own beauty, [the beauty] of large dams and artificial lakes . . . , [the beauty] of large bridges. . . , [the beauty] of large buildings where lines, materials, and light play with each other . . . , and even [the beauty] of the metal towers of high-tension power lines."[83] Nor did this beauty have to come at extra expense: "Caravelle [the airplane] is both a technological success and an esthetic success, but its beauty comes as a surplus: it results from lines and materials, not from additional cost. . . . Similarly the beauty of large dams resides in the harmonious marriage of the object and its natural setting."[84] Cultivating an esthetic dimension to industrial projects would not only assert the Frenchness of French technology; it would also enhance the prestige that the nation could derive from its technological achievements. Even when beauty did add to the cost of a project, the supplement remained small in relation to its benefits: ". . . beauty brings income for tourism (it is important not to disfigure sites with inadequate equipment), it brings prestige because it represents a considerable attraction, even when there is no commercial profit. The CEA's installations, which receive numerous foreign visitors, would certainly gain nothing by being hideous, and the esthetic of certain nuclear reactors, whose cost is insignificant with respect to [the cost of the] equipment, does more for the radiance of France than would ten times as many millions spent on propaganda."[85]

Through beauty, tradition could legitimate modern technological achievements as being truly French. And the relationship worked both ways. France was a nation rich in tradition, but this tradition no longer sufficed to define the glory—indeed, the radiance—of the nation. Armand made a point of this in *Plaidoyer pour l'Avenir*: ". . . the wealth of the setting—churches, castles, rivers and their embankments, towns which each have their own personality . . . —should mean that France would continue to be a crucible of ideas. Yet all that subsists in France in the way of tradition—in the countryside as in the army—will only have real value and will only be able to radiate if the nation as a whole is solidly of our time."[86] Hence the other side of a symbiotic relationship: just as tradition was necessary in order to make French technology truly French, modernity was necessary in order to make France truly France.

The nation's nuclear achievements epitomized these dialectics between tradition and modernity and between national radiance and technological prowess. Consider, for example, a 1957 promotional film commissioned by the Ministry of Foreign Affairs and entitled "Le Grand Oeuvre: panorama de l'industrie française."[87] The film opens with a view of the sprawling Versailles castle and the words of Jean-Baptiste Colbert to Louis XIV. "Sire," Colbert declares, "the grandeur of a state rests on its arts and manufactures." The next 40 minutes recount, in epic style, French postwar technological development. Viewers see canals, coal mines, oil refineries, petrochemical plants, railroads, and airplanes. The narrator continually reminds his audience of the connection between technological prowess and national grandeur: "Airports are part of the infrastructure necessary to a great industrial nation." "What would France be without its railroads?" The country's industrial growth is attributable in part to thousands of heroic workers laboring "elbow to elbow in the mechanical fraternity," but mostly to the engineer—"the man of industry *par excellence*." Spiffy young *polytechniciens* talk energetically in the Jardin du Luxembourg, just outside the Ecole des Mines. The film continually affirms the connection between history and modernity: the Alsace, which once made dyes, now prints cloth; Lyon, which used to manufacture silk, now produces synthetic fabrics. At the end of the film, these themes coalesce in the depiction of the burgeoning nuclear program. We learn that the construction of Marcoule, France's first large-scale nuclear site, has "mobilized all of French industry." We see one of Marcoule's heat exchangers making its way by truck convoy through an old village filled with amazed peasants. Finally, we see two nearly completed reactors. The narrator intones: ". . . the latest great accomplishment of the century of the atom, the future's answer [at this point the film switches to a shot of the Eiffel Tower] to this other great symbol of French industrial grandeur, sketched in the Parisian sky." France's postwar industrial achievements thus fitted into the nation's historical teleology, nuclear technology its apotheosis.

Technologists envisioned a future France whose power would rest on technological prowess yet whose technological achievements would remain distinctly and identifiably French. The type and degree of nationalism in this discourse varied greatly. Some shared de Gaulle's vision of a strong, independent France; others argued that henceforth France could be strong only by associating itself with a larger European community. Such differences led to differences in the technopolitics pursued by technologists. But either way, the goal seemed clear: France had to become a technological nation. Its future depended on planning a wise route to this goal.

The Mentality of the Future

In order to attain this goal, most technologists argued that the French *mentalité* had to change. This refrain dated from the early postwar period. In those years, advocates of state planning had blamed French defeat on petty industrialists who had clung tenaciously to the status quo and had refused to invest in new technologies that could have made France strong. While the most egregious material problems had been corrected by the mid 1950s, advocates of state planning still found much to complain about in the French *esprit*. For example, one former planner reminisced about an interaction between Etienne Hirsch, Haut Commissaire du Plan in the 1950s, and Monsieur de Wendel, the elderly, well-respected head of a steelworks. Hirsch had invited de Wendel to sit on the steel production commission of the Second Plan. De Wendel was puzzled, and replied: "But Mr. High Commissioner, what is this about? After the war you explained to us that we had to make a big effort to modernize the steel industry. We listened to you, we did it, we took risks . . . now we are modern!

So what could you possibly want to discuss?" Exasperated, Hirsch replied: "But Mr. de Wendel, modernity is not a definitive state! You made an effort to make up for a delay and modernize certain installations, but this effort will never be exhausted once and for all!"[88] The point of this anecdote, which opposed the old-style industrialist (the very figure who had supposedly caused France's downfall and who represented private industry's alternative to and antithesis of the technologist) and to the modern planner, was clear: the listener was supposed to be amused that the old man did not realize that modernity was not a physical condition, but a state of mind.

This idea that modernity began with a change in attitude pervaded the discourse of technologists throughout the 1950s and the 1960s. Many reproached the French for not thinking big. Armand located the roots of the French obsession with smallness in the nation's revolutionary tradition, which he interpreted as the revolt of the small against the large— of the artisan and the bourgeois against the landowner. This theme of smallness, he argued, dominated the mentality of the non-industrial middle classes (and a big part of Poujadiste discourse): "A 'small job,' a small shop, a small house, a small garden . . . no worries, a small game of cards, and above all no complications. . . . (But one day a big defeat!)"[89] The biggest reproach of the technologists, however, centered around the French conception of and approach to the future. The French could no longer stumble blindly into their future; they had to learn how to control their destiny. This involved cultivating *une attitude prospective.*

The notion of *la prospective* originated with the formation of the Centre International de Prospective. Though most of those involved would have identified with the label "technologist," a few were also humanists.[90] The Centre was intended to provide a place and a publication (the journal *Prospective*) for systematic and systemic reflection and action oriented around three related poles: "human problems" such as employment and education, the relationship between Western and other civilizations, and the consequences of new developments in science and technology. The Centre forbade itself from conducting "any political activity"—meaning corrupt ideological machinations or affiliations with political parties— and made a point in its publications of lambasting "ideological" modes of reasoning.[91] Above all, the Centre aimed at cultivating *une attitude prospective.*

What was this attitude? It was one turned toward the future, especially the far future. It differed from short-term forecasting, and it had to be cultivated by more far-sighted individuals. Gaston Berger, the president of the Centre, explained this using a military analogy:

It would be dangerous for a combat officer to be associated with peace negotiations because his role is to fight even while peace is being discussed. But it would be unforgivable for [national] leaders not to dream of peace while making war. In the adversary of today they must already see the colleague, the client, the friend of tomorrow. . . . It even happens fairly frequently that short-term actions must be taken in a direction opposed to that revealed by a study of the long term. Those who implement such actions must pursue them with vigor, but at a higher level, responsible leaders must calculate the importance of these actions and situate their exact position in events as a whole.[92]

In other words, "responsible leaders"—reminiscent here of the *chef* or the "man of action"—had to take *une attitude prospective*. This attitude was essentially systemic: it involved defining a goal and figuring out how human, technical, and economic factors could be synthesized into a plan of action. Ultimately, its advocates argued, this attitude would enable men to "control their destiny" rather than "submit" to it.[93]

Many highly placed technologists—Louis Armand, Pierre Massé, and François Bloch-Lainé among them—advocated *l'attitude prospective*. They took it to mean applying qualitative systems thinking to problems that were at once technological and social. Armand sought to define *la prospective* for national transportation systems, a subject he knew intimately. The issue for the future of transportation, he argued, was no longer maintenance but coordination—not just within subsystems like the railways, but also between subsystems. Airlines, railways, and roads should be coordinated to provide optimal transportation routes for travelers and goods. The current chaotic state of affairs, in which these subsystems competed with one another, merely demonstrated the "need for governments to apply notions of political economy which are the domain of operations research."[94]

Like modernity, *la prospective* was above all a state of mind. Taking *une attitude prospective* involved seeing life as a "continual invention."[95] It demanded intimate knowledge of "large new technologies" [*grandes technique nouvelles*], of which the two most important were atomic energy and cybernetics.[96] It was an activity for an "elite"[97] composed of "men of action"—"men who not only have a taste for moral or philosophical meditation, but also a concrete knowledge of men and the experience of command and responsibility."[98] The action in which men should engage involved synthesizing "all the means at the disposal of modern society in order to know and to predict, to organize, and to decide."[99] Like the other men involved in the Centre International de Prospective, Berger placed a heavy emphasis on real-world experience: "We do not seek to operate a synthesis of knowledge and writings, but a synthesis of lived experience . . . only doctrinaires—inefficient but formidable—start with abstract ideas completely cut off from reality."[100] This experience provided a non-ideological foundation for action based on *la prospective*.

The central figure at the heart of *l'action prospective* was the engineer—not just any engineer, but an engineer who could take a *vue d'ensemble*. Such a systems thinker could master his destiny. He combated the defeatism of intellectuals: "If the myth of Sisyphus expressed our true condition, our engineers would have already discovered the means of using the regular fall of the boulder and Sisyphus, freed from repetitions, would devote himself to other tasks."[101] Invention provided the foundation for men to build their destiny; as such, the material world had spiritual value.[102] Berger evoked the myth of Faust to demonstrate this point. In the end, he said, Faust found fulfillment not through the gifts of the Devil but by working for other human beings. For this, God saved him from the Devil's clutches. "And what does this mean? This means that man had been a magician and became an engineer, but an engineer in the service of others. What is the magician? He is the one who uses spiritual forces for selfish goals. What is the technologist? He is the one who uses his work, his pain, his intelligence to bring to men and to others the things they need. Moving from magic to technology is not staying on the same level; it's substituting generosity for egotism."[103] Imbued with *une attitude prospective*, the engineer—leader, man of action, and systems thinker—could shape human destiny. This was "decidedly" not technocracy; it was simply good sense.[104]

What, besides a more rational, prosperous, and powerful nation, would the application of systems thinking produce? Armand put the answer very simply in an equation: "Technology + Organization = Culture."[105] While technological change could have disastrous effects, these could be averted by the application of heterogeneous, systemic organization. "Instead of dividing oneself to combat the noxious effects of technology through action inside companies, each firm, each industrial company must admit that progress supposes a larger discipline to which each must submit. This organization is indispensable to ensure the downfall of accusations levied against 'inhuman' technology."[106] The essence of all such action involved overstepping traditional boundaries, particularly those between technology and a certain kind of (presumably non-ideological) politics. In the words of another advocate of *la prospective*: "If the modern world demands an increasingly large number of specialized technologists and researchers, it is necessary that among and next to these a certain number of young men be able to dominate their technologies in order to participate in the definition and implementation of general industrial policy. For this, they must be or become more than just technologists."[107]

Ironically, the qualitative systems approach that Armand thought would combat accusations of inhuman technology was precisely what many non-technologists found threatening. Jean Meynaud ridiculed Armand's equation, and was even more outraged when it appeared that even the members of the literary-minded Académie Française admired and trusted Armand enough to elect him. Armand interpreted this election as "the entrance of technology, flags flying in the wind" into the Académie.[108] Clearly, the attempt of Armand and others to trespass into the territories of politics and culture by imagining and implementing heterogeneous systems was exactly what social scientists like Meynaud found threatening.

La prospective essentially consisted of a qualitative approach to systems thinking: by taking into account human and cultural factors in their inventive efforts, engineers and other "men of action" had the means to set goals for the future and to trace out trajectories for attaining those goals. Though some advocated specific methods for delineating those plans—particularly the techniques of operations research—even they kept their contributions to the journal *Prospective* quite general. This was not the case, however, for the Plan.

The Plan

The Commissariat Général au Plan produced the ultimate instrument for shaping the future and destiny of the nation, the ultimate effort to constitute and define a large-scale system: the plans.[109] In the plans and their planners, the various themes examined thus far come together. The planner epitomized the technocrat—or the broad and forward-thinking technologist, depending on one's perspective. Architects and advocates of the plans often held them up as a quintessentially French achievement—the ultimate marriage of certain select traditions and modernity. Making the inevitable reference to Descartes, one enthusiast noted that "the effort toward increased rationality that the French plan represents conforms to one of our best national traditions."[110] Planners themselves promoted the plans as instruments not only of national cohesion and internal economic development, but also of national power. Consider the introduction to the fourth plan:

. . . going beyond individual destinies, [the national goals of the fourth plan] define themselves as survival, progress, solidarity, [and] radiance. They consist of ensuring our defense by combining the modernization of the military with a reduction in its personnel, of giving research the material power necessary to ensure the full participation of the French spirit in the great scientific and technological enterprise of this century, of giving regions and less favored groups— be they the aged, repatriated soldiers, employees, or low-income farmers—concrete proof of a solidarity indispensable to national cohesion, and finally of pursuing our aid to the less-developed nations of the third world, especially those French-speaking African States which decided to keep special ties with our nation.[111]

Technological radiance and post-colonial geopolitics thus framed the fourth plan. Given such lofty goals, serving the Plan was not only an "ardent obligation" but potentially a quasi-religious experience: "inasmuch as the plans have become a necessity, all reticence is abnormal, even stupid. The only logical and efficient recipe is to make planification a psychological force of progress and solidarity. Serving the plans, participating in them at different levels, can feed the transcendence of each one [of us]."[112] Thus the plans could furnish a means for enacting the spiritual dimension of the material world. A quintessential manifestation of *la prospective*, the plans' very existence would turn the nation into a system.[113]

The plans would systematize the nation by providing information. Left to their own devices, private industries or organized sectors of the economy would, in the disastrous manner of the prewar period, pursue independent policies conceived from the sole point of view of the industry or sector in question. They would have no way of knowing how their actions affected other industries or sectors, nor would they know how the actions of others affected them. This ignorance might well lead them to make faulty decisions, not just in terms of the national interest, but potentially also in terms of their own interests. By providing decision makers with increasingly detailed maps of the infinite interconnections that bound the nation's economy together, successive plans would make possible a new kind of decision making: one that was not only more rational and efficient, but also more systemic. Indeed, it would ultimately be in everyone's best interest to work toward the national goals set by the plans (for the plans were not merely descriptive, but also prescriptive), even if in the short term these goals asked individuals to make decisions that appeared to go against their immediate interests. The whole would be bigger than the sum of the parts, and the parts would benefit in consequence. Hence describing and prescribing the system were also, in a sense, supposed to create the system (assuming, of course, that the actors operating within it paid attention to the maps and prescriptions provided by the planners; assuming, in other words, that they accepted the notion of themselves as actors in a system).

More complex and sustained systems thinking in the Commissariat Général au Plan began with the fourth plan. The methods for elaborating the first three plans had been primarily qualitative, and even the most fervent admirers of state planning conceded that those plans had mattered primarily for psycho-cultural reasons, serving to change the *mentalité* of the French by articulating a dynamic, modern future. In contrast, the fourth and fifth plans—under the impulse of linear programming enthusiast Pierre Massé—were developed using a combination of qualitative and quantitative methods. As such, they pursued national-scale systems thinking far more intensely.

The first three plans had used material indices (such as industrial equipment) and simple economic indices (such as productivity and efficiency) in order to set goals for national economic and industrial performance. In contrast, the fourth plan sought to develop dynamic models that would describe the economy—and the social and material relations that drove it—as a whole. The economy was composed of overlapping heterogeneous "subsystems." These included diverse sectors (such as agriculture, industry, commerce, and transportation), geographic regions of the nation, economic constructs (such as balance of trade, consumption, savings, and investment), and socio-economic relationships (such as employment and the labor market). These subsystems interacted in complex ways. With the help of the mathematical services of the Ministère des Finances and the Institut National des Statistiques et des Etudes Economiques, planners aimed to model these interactions as closely as possible—even while viewing such models as necessarily imperfect because they could only incorporate statistically describable relationships and interactions.

Though they fell short of building a single model for the entire nation, planners did produce a set of models that together described the national economy as a heterogeneous system composed of technological and economic artifacts, individual decision makers, and social relationships, and driven by the interactions among these various components. Using these models, they defined an "optimum" growth rate. (This process of using models to define optima would become important in the techno-politics of the nuclear program, where it would derive legitimacy from its use in national planning.) They then turned the growth rate and parts of the models over to the modernization commissions. Convened separately for each plan and not composed of expert planners, these commissions studied specific industrial, economic, geographic, or cultural sectors. On the basis of information they received from the expert planners, the commissions drew up tentative plans for their sectors. Finally, the expert planners collected these subplans and modified them in order to fit them into a system both described and constituted by the final plans.[114]

Intended to cover the period 1962–1965, the fourth plan outlined policy directions for a broad swath of French economic, industrial, social, and even cultural life. First and foremost came the need to develop scientific and technical research: ". . . the fate of a people is increasingly determined by the energy it deploys in opening new routes to knowledge, which is the very source of its radiance and the indispensable condition of the [continuous] renovation of its technologies."[115] Beyond this, the fourth plan set growth and development objectives for the sectors of the economy defined by its models, recommending for example that the nuclear industry seek to develop several different types of designs for power plants. Pointing to increasing international competition, it recommended that large industrial establishments pursue efforts to merge and to specialize. It set goals for urban development throughout the nation, recommending the destruction of dilapidated buildings, the construction of wide roads, and the creation of parks and sporting facilities in city centers. It introduced the idea of regional planning, and it outlined schemes for the modernization of rural areas.

In addition to outlining material development, the fourth plan promoted cultural harmony through technological and institutional means. A second television station would contribute to "increasing the information and cultural development of the population and to spreading the radiance of France beyond its frontiers."[116] Meanwhile, the construction of cultural centers throughout the nation would enable "culture" (which remained undefined) to "remedy that which often seems discordant and inhuman about technological civilization, . . . to penetrate the daily life of men and especially to become . . . as immediate a concern as hygiene and stable employment."[117]

The fourth plan thus represented an attempt to chart the future of the nation on every possible front. In theory, the national system that it conceived was open and unbounded. Even if some parts of the system could not be described quantitatively, and even if planners did not yet understand the precise mechanisms through which some of the subsystems or components interacted, ultimately no aspect of national life lay outside the whole. In greater or lesser detail, labor policy, industrial growth, urban development, technological change, investment, and cultural enrichment could all be related to one another and planned for the greater good of a new, modern France. The modernity of the nation could be described, in some instances measured, and in all cases enhanced. And, sweeping as the fourth plan was, the fifth plan covered even more ground (particularly in the domain of regional planning) and used even more elaborate quantitative methods.

Clearly, these plans were hybrids of technology and politics in the broadest sense of both words. Indeed, the architects of the plans saw matters in precisely these terms. In a report written to promote the diffusion of the fifth plan's programming methods, one technologist wrote of the "simultaneously technical and political nature" of the plan's "elaboration process":

On the political front this process is a mechanism for determining preferences, that is to say [a means of developing] social awareness and [making] political choices about goals and means. On the technical front, the goal of the process is to establish coherence. . . .[118]

Throughout its entire course, the planning process wove together technical and political methods: ". . . the kind of variables involved in the technical work [and] the determination of their values or of the relationships between them are themselves tied to explicit, or more often implicit, political choices. This web of political choices and technical work is woven during the preparation of the [fifth] plan. . . ."[119] While no one went so far as to argue, as we might, that the technical and political aspects of the process were indistinguishable, it seemed clear to all those involved that they were at least closely related, and furthermore that a close relationship was necessary for the success of the plans.

In a sense, taking a systems approach constituted an attempt to naturalize the erosion of the alleged boundary between technology and politics. Conceiving of the nation as a system and arguing that all its heterogeneous components were interrelated implied that all these components could be planned. In other words, all components—technological, regional, cultural, economic—fell within the purview of the planners.

The enrollment of an extremely heterogeneous group of people into the planning process provided a key means of enacting this erosion. While the group of expert planners remained small and select, the modernization commissions included a great variety of people: industrialists, labor union representatives, bankers, regional administrators, architects, urban planners, even a few artists and writers. Their participation was meant to ensure the democratic character of the plans. It was also intended to promote compliance by increasing the investment of different social groups in the plans.

In a sense, the Plan as a whole attempted to enroll the entire nation in a broad program of sociotechnical development. Because the plans were not coercive, they could not impose an agenda on industrialists, workers, regions, banks, or flows of money. But they could attempt to persuade the nation to follow their lead. Their very existence was one form of persuasion, and the High Commissioners for Planning engaged in other forms by visiting politicians, regions, and industries to promote the plans. Witness, for example, how Massé described his job two decades after retiring:

My profession was to send a message that would not falsify the truth, but that would be accessible to labor union members, politicians, [and] public opinion. I had to convince the Government to adopt my plan project, convince the Economic and Social Council to emit a favorable overall opinion, [and] convince Parliament to vote it [into effect]. I repeat the word 'convince' three time because this was an essential part of my job, carried out for seven years with a respectable measure of success. . . . In sum, I had a responsibility of a political nature that went beyond the mission of the experts.[120]

Here, politics was not ideological but persuasive, an essential part of making a technological nation.

*

In debating and enacting the relationship between technology and politics, technologists used a central trope—the radiance of France—to articulate and legitimate their place in postwar French society. This trope exuded historical referents, from the shining monuments of the Sun King to the glorious days of the French empire. By the same token, it encapsulated the crisis of national identity faced by France in the postwar period. Radiance was what France had lost in its wartime occupation and defeat and would continue to lose through the process of decolonization. And so radiance, that quintessential quality of Frenchness, was what technologists offered the nation.

Language, tradition, and esthetics would make technologies truly French, thereby performing the double operation of legitimating technology as an expression of national identity and preserving French uniqueness through the painful process of modernization. Thus embodying Frenchness, these technologies in turn would restore and enact French radiance throughout the world. Technology and Frenchness would shape each other.

Evoking the radiance of France in discussions over modernization represented an effort to generate agreement over technological development. Almost everyone could agree on the desirability of French radiance. In appropriating this trope, technologists hoped to generate similar support for technological prowess. They also sought to become the legitimate leaders of modernization more generally, portraying themselves as decisive "men of action." In the process, they blurred not only the boundaries between technology and (French) culture but also those between technology and politics.

For many social scientists, humanists, and politicians, the erosion of the boundary between technology and politics threatened the very foundations of democracy. Technologists, however, presented it as necessary to social and economic progress on all fronts. Systems thinking provided an ideal means for naturalizing and enacting this erosion. If heterogeneous elements were related to each other in identifiable, describable, controllable ways, and if technologists could predict and control these systems, then they could engage in politics better and more reasonably than could politicians. Such systems thinking was composed of a series of qualitative and quantitative practices—*la prospective*, planning, optimization—which came to constitute the basic toolbox of state technologists.

When arguing with social scientists, pleading for a modern future, grandstanding in *La Jaune et La Rouge*, or sitting on planning commissions, technologists generated a set of concepts and practices that they could mostly agree about. Matters grew considerably more complex when it came to enacting the technopolitics they advocated so enthusiastically. They agreed on the ideal of a technologically radiant France, but they did not necessarily agree on the best route toward that ideal.

2

Technopolitical Regimes

Before I began interviewing engineers about their involvement in the development of nuclear power, I expected our conversations to be dry, technical affairs in which these men would describe their small corner of reactor design and indignantly deny that their work had political or social components. My expectation arose from two sources. First, much scholarship argues that scientists and engineers expend a good deal of energy denying the political, social, or cultural, dimensions of their activities. Donald MacKenzie demonstrates this point particularly forcefully in regard to the engineers involved in the development of nuclear missile guidance in the United States.[1] Second, many American commentators argue that nuclear technology has been "depoliticized" in France. By this they mean that parties across the political spectrum agree that the nation should pursue both nuclear power and atomic weapons, and that there is little or no public debate about these choices.[2] Anticipating, then, that direct attempts to address the political aspects of technological work would induce suspicion and mistrust, I resolved to begin my interviews by asking about the "scientific and technical" decisions in which my informants had participated. I hoped that a discussion of technical details would lead, discreetly and indirectly, to comments about the political and social aspects of nuclear engineering.

My first appointment was with a man who had been a project engineer for six gas-graphite reactors. By the time I met him, he was a high-level manager with an enormous, sumptuously furnished office in EDF's headquarters. I tried hard not to feel intimidated as I sat down in front of his vast, polished wood desk. In hopes that challenging gender stereotypes would counterbalance the disadvantage I felt as a young woman interviewing a much older male expert, I had worn a suit and tie. I quickly established my technical credentials, reaffirming (as I had already explained in the first paragraph of my query letter) that I had a degree in physics from MIT, the "Polytechnique de l'Amérique."[3] Thus armed, I began my introductory spiel: I wanted (I explained) Monsieur le directeur to tell me about the scientific and technical decisions in which he had taken part during the 1950s and the 1960s. Imagine my surprise when Monsieur le directeur slapped his hand on the desk, leaned toward me, glaring, and roared: "But Mademoiselle! These were not scientific or technical decisions! They were economic decisions! Political decisions!"

Over the next few years, this encounter became the paradigmatic story of my research. In one respect, it was unique: this particular man turned out to be an unusually colorful character who enjoyed emitting shocking statements. None of the other men I interviewed made such bald declarations. A few did deny that their work had political dimensions. Yet most of them appeared to assume that politics was a normal part of their job. In another respect, then, this incident provided only the most explicit example of a widespread belief among these engineers in the necessary interweaving of technology and politics.

In the previous chapter I showed how this belief formed the central premise of elite state technologists' efforts to shape national discourse about France's identity and future. But Monsieur le directeur's exclamation suggests that the deliberate interweaving of technology and politics ran much deeper than this public discourse. In the nuclear program at least, it permeated all levels of development, from the interactions between nuclear leaders and government officials to the artifacts and practices of reactor design.

In this chapter, in order to explore the multiple facets of this interweaving, I develop the notion of *technopolitical regimes*. The two regimes I discuss were grounded in state institutions, one at the Commisariat à l'Energie Atomique (CEA), and the other at Electricité de France (EDF). They consisted of linked sets of individuals, engineering and industrial practices, technological artifacts, political programs, and institutional ideologies acting together to govern technological development and pursue *technopolitics* (a term that describes the strategic practice of designing or using technology to constitute, embody, or enact political goals). As I noted in the introduction, the "regime" metaphor is meant to evoke the tight relationship among institutions, the people who run them, their guiding myths and ideologies, the artifacts they produce, and the technopolitics they pursue. It also conveys the notion of prescription: as I will show, each regime aimed to prescribe policies, practices, and visions of France's future. Finally, "regime" captures the dynamics of power: even within the institutions that housed them, these regimes were subject to negotiation and constestation.

These analytic points will become clearer as we see the two technopolitical regimes in question take shape. In the present chapter I will discuss each regime in turn. From the mid 1940s to the mid 1950s, the nuclear program belonged to the CEA. I will begin with this institution, describing the emerging dominance of a nationalist ideology within it. This was simultaneous with, and closely linked to, the choice of the gas-graphite design for the CEA's first industrial-scale reactors. A nationalist technopolitical regime developed within the institution. As the CEA began to build its gas-graphite reactors, this regime grew stronger and more established, articulating increasingly ambitious goals not just for the nuclear program, but for French industrial development more generally.

Next I turn to EDF. I briefly situate its creation in the wave of postwar nationalizations and discusses competing views of the meaning of nationalization in the utility. Because EDF devoted its first decade primarily to building ordinary power plants, I skip over that period and move straight to its first involvements with the CEA and nuclear power. The nationalized technopolitical regime that emerged to govern nuclear development within EDF situated itself at the intersection of two emerging technological systems: the fast-growing electric power network and the budding nuclear program.

Each regime developed its own reactor site, but limited resources forced them to collaborate on both. This collaboration was fraught with conflict, because each regime sought to mold its reactors into components of its technopolitics. Comparing the design and the industrial contracting process of the reactors built by the two regimes reveals two dimensions of the technopolitics. The first involved French nuclear policy. The CEA's reactors produced weapons-grade plutonium at a time when official government policy had not yet decided in favor of a French atomic bomb; they thus constituted the nation's de facto military nuclear policy. EDF's regime, meanwhile, positioned its reactors in deliberate counterpoint to the CEA's technopolitics: it wanted its first reactor at Chinon to constitute the first step toward an economically viable nuclear *energy* program. The second dimension of these technopolitics involved French industrial policy. Should the state promote national "champions" in different industrial sectors, which ultimately might mean creating consortia of private companies within the same sector? Or should it, on the contrary, promote competition among companies in order to force them toward higher standards of technical excellence and economic efficiency?

And which route would best enable French companies to export their technologies and compete successfully in the international market? In the mid 1950s the CEA's regime chose the first route when organizing industrial contracting for reactor construction, while EDF's regime chose the second.

The Creation of the CEA

The Commissariat à l'Energie Atomique began its life as the ultimate post-Resistance institution, the product of a common vision of communist wartime resistants and Charles de Gaulle.[4] The communist physicist Frédéric Joliot-Curie had spent the war in France as a member of the Resistance, attempting to hide the results of French nuclear science from the German occupiers and helping to smuggle crucial nuclear materials out. In these efforts he had the support of Raoul Dautry, the minister of armaments before the German invasion. At the war's end, the two men easily convinced de Gaulle that a nuclear program would both elevate France's stature in international politics and accelerate its industrial and economic recovery. Following de Gaulle's recommendation, the National Assembly quietly approved the creation of the CEA in October 1945. The agency's stated mission was to "pursue scientific and technical research in the view of using atomic energy in the various domains of science, industry, and national defense."[5] Joliot-Curie and Dautry both argued—and de Gaulle agreed—that the CEA should be protected from the whims of ministerial politics. Yet it needed to remain "very close to the government because the fate or the role of the country might be affected by the development of [atomic energy]."[6] The statutes therefore specified that the CEA was accountable only to the prime minister, and that it would not be subject to the same financial controls as other state institutions.

The institution's leadership structure reflected an ambiguous marriage of science and politics: it was a dyarchy headed by a scientist who carried the title of High Commissioner and a professional administrator who carried the title of Administrator General. The two men would share power equally. Not surprisingly, de Gaulle appointed Joliot-Curie and Dautry to these two posts. The CEA's steering committee included several of the institution's top scientists, a number of government-appointed administrators from a variety of ministries and other state institutions, and a military general. Despite this military presence and the mention of "national defense" in the CEA's creation ordinance, official government policy stated that France would limit its atomic endeavors to peaceful ends.

After de Gaulle stepped down from power in 1946, the government lost interest in the CEA. Parliament passed a new constitution, and the Fourth Republic began. The new government was headed by a rapid succession of prime ministers (twenty different men over thirteen years), who initially focused on the pressing concerns of national reconstruction. Left to their own devices, scientists dominated the CEA's operation during its first five years. They concentrated on conducting fundamental research in nuclear physics and chemistry, developing large-scale experimental equipment such as reactors and accelerators, and prospecting and mining uranium. The steering committee agreed that these activities constituted the basic building blocks of any nuclear program, especially since the Americans and the British displayed no intention of sharing their research results or their raw materials. This agreement persisted as long as no one evoked more ambitious goals.

As the Cold War intensified, however, successive governments found Joliot-Curie's communist affiliations increasingly embarrassing. In 1949 the Soviets successfully tested their first atomic bomb. Shortly afterward, the British convicted the scientist Klaus Fuchs of passing nuclear secrets to the Soviets. Tensions were already high, therefore, when Joliot-Curie declared in April 1950 that he would never build an atomic bomb, because such a weapon could only be aimed at the Soviet Union and would therefore help to precipitate another world war. Speaking out against nuclear weapons per se did not conflict with the government's position: spokesmen repeatedly declared that France cared solely about peaceful applications of atomic energy. But Joliot-Curie's reference to the Soviet Union was unforgivable. The United States protested the presence of a communist at the head of such a strategically sensitive institution. Prime Minister Georges Bidault dismissed Joliot-Curie in late April. Nearly a year passed before Joliot-Curie was replaced by another eminent (but less vocal) scientist, Francis Perrin.[7]

The interval gave the non-scientists on the CEA's steering committee the opportunity to increase their influence within the institution. René Lescop, a *polytechnicien* whom Dautry had appointed to the position of secretary-general, seized the moment forcefully. In January 1951 he and Dautry spearheaded a formal institutional reorganization which subordinated scientific authority to administrative authority. Both men had privately expressed enthusiasm for a French bomb, so the scientists recognized that this move could have national political implications. In August of that year, Dautry died, leaving Lescop as the highest-ranking administrator on the steering committee.

The Emergence of a Nationalist Technopolitical Regime

That same month, the CEA acquired an important political ally: the young parliamentary deputy Félix Gaillard, newly appointed as state secretary for atomic energy.[8] Gaillard's position placed him on the CEA's steering committee, and it quickly became clear that he intended to push an ambitious program for the agency, one that fit better with Lescop's ideas than with those of the scientists. Convinced that France's future lay in the strength of its nuclear program, Gaillard urged the committee to draft an ambitious five-year plan for the development of atomic energy—one that would seduce Parliament by promising material benefits in the near future. It would be easier, he said, to justify a 20 billion franc plan that included developing atomic energy on an industrial scale than a 3 billion franc[9] plan devoted only to basic research. Atomic technology appeared vital to France and its dwindling empire. "The use of atomic energy can command the future of France (and of the French Union)," Gaillard declared confidently. "Our country's lack of industrial capacity is increasingly dramatic, and inasmuch as atomic energy can provide a solution in a few years, the CEA's future budget is a national issue."[10]

Perrin and other scientists expressed doubts about whether the CEA had the scientific and technical ability to carry out an extensive program. They would have preferred to concentrate on education in order to build a solid base of trained scientists and engineers.[11] But François de Rose of the ministry of foreign affairs, and other administrators, supported Gaillard. The CEA should aim high, he said. France currently led the second-tier nuclear nations, but (he argued) this might change if Germany decided to start a large-scale nuclear power program. France's future leaders would thank the CEA for having the foresight to plan extensive nuclear development.[12]

The scientists eventually agreed to the principle of a large-scale reactor program. This raised an important question: What sort of system should the CEA choose? "Primary" reactors, such as those built by the British, ran on natural uranium, of which France had plenty, largely thanks to its colonial territories. "Secondary" reactors, developed in the United States, ran on enriched uranium.[13] But enriched uranium was not for sale anywhere, so the CEA would need to build an enrichment plant—something it had neither the time nor the knowledge to do. Primary reactors, meanwhile, could produce both plutonium and electricity. And Bertrand Goldschmidt, who had spent the war in Canada working on the

Manhattan Project, already knew how to extract plutonium from irradiated uranium.[14]

In principle, plutonium could be used in secondary reactors—though no one had yet managed to do so. The metal had, however, proven highly suitable as bomb fuel. Some scientists worried that building plutonium-producing reactors would effectively constitute the first step toward a French bomb. They feared that producing the fuel would, at the bare minimum, whet the military's appetite.[15] The military representative on the steering committee had already expressed his personal interest in the atom bomb.[16] Further, asked the scientists, wouldn't building plutonium reactors alarm the United States?[17] Gaillard's supporters dismissed such objections. They reiterated that plutonium could be used as reactor fuel, and they reminded the committee of France's urgent need for energy sources. Without further ado, the committee settled on primary reactors. And without specifying the end use, it set a production goal of 15 kilograms of plutonium within five years.[18]

Having made this decision, the committee next had to pick a moderator for its reactor. The choice was between graphite and heavy water.

Natural uranium contains two isotopes of uranium: U_{238} and U_{235}. Fission occurs when a neutron hits a U_{235} atom, causing its nucleus to split and liberating a great deal of energy as well as more neutrons. Some of these additional neutrons are absorbed by other U_{235} atoms, causing more fission. With enough uranium piled up (what is known as critical mass), this fission reaction will "go critical" and be self-sustaining. Other neutrons, absorbed by U_{238} atoms, will not cause fission. Rather, upon absorbing a neutron, a U_{238} atom becomes U_{239}, which eventually changes into Pu_{239}—weapons-grade plutonium.

The committee members knew that, in order to split a U_{235} atom successfully, a neutron must be traveling at a speed lower than that at which it was released. Therefore a moderator was required to slow down the neutrons. The ideal moderator would not itself absorb any neutrons. Finally, a coolant was needed in order to extract the heat from the reactor core.

At the time of the steering committee's September 1951 meeting, the CEA had already built experimental heavy water reactors.[19] Physicists preferred heavy water as a moderator because it absorbed fewer neutrons. But heavy water could be made only by electrolysis, which itself required electricity. A heavy water plant seemed complicated and expensive to build, while the French company Péchiney already manufactured graphite. Such were the official reasons for choosing graphite over heavy water as a moderator.

An additional reason, however, was suggested by an engineer who worked on the early gas-graphite designs. Many of those who had worked on the experimental heavy water reactors were communists. Some had been dismissed along with Joliot-Curie. Since the plutonium produced by the first industrial-scale reactors might go into a future French bomb, some committee members wanted an easy way of excluding communist scientists and technicians from the new projects. Not picking the technology in which they had experience greatly facilitated this task.[20]

The CEA steering committee thus settled on a five-year plan for 1952–1957 that committed the CEA to building two reactors, powered by natural uranium and moderated by graphite. The plan also included a factory to extract plutonium from the spent uranium fuel that would emerge from the reactors.

Pleased with these goals, Gaillard took the plan to Parliament for approval. Because it conferred prestige and glory, he argued, France needed nuclear energy. In one radio broadcast, he warned that "those nations which did not follow a clear path of atomic development would be, 25 years hence, as backward relative to the nuclear nations of that time as the primitive peoples of Africa were to the industrialized nations of today."[21] Without nuclear technology, France's global position might move from that of a world empire to that of a backward, colonized nation. Gaillard reminded his fellow deputies of the nation's weakness in energy resources, and noted that expanding the nuclear program meant developing France's industrial base and ensuring its future energy supply.

Deputies on both ends of the political spectrum were clearly persuaded of the symbolic significance of nuclear technology. They used this symbolic value to reenact an increasingly familiar Cold War debate over the position of French communists. One right-wing deputy expostulated that the persistence of communists in the CEA was "scandalous, for it subordinates the work of an organ where the atomic future of our country is being worked out to Moscow's control."[22] Still reeling from Joliot-Curie's dismissal, meanwhile, communist deputies suspected Gaillard of concealing military goals. They demanded that the plan explicitly state that France would never build an atomic bomb. But the other deputies refused to vote for this amendment—not because they were prepared to approve a French bomb, but because they did not want to make any concessions to the communists. Someday, one right-wing delegate argued, France might need a bomb to "safeguard her independence and security."[23] He worried that on that day the communists remaining in the CEA might prove more loyal to their former leader (Joliot-Curie) than to their nation. Might not the plutonium eventually be used in a bomb, and shouldn't all communists be removed from the CEA as a precaution against that day? To calm anxieties, Gaillard played both ends against the middle. He agreed that France should not rule out a bomb a priori. He reassured the right that plutonium production would take place in a separate division of the CEA and hence be subject to special security measures. At the same time, he insisted that his plan was in no way directed toward military ends: the cost of a single bomb, he claimed, represented ten times the funds he had requested.[24] (This cleverly avoided the question of whether the proposed development would contribute to a bomb.) Appeased, Parliament approved a budget of 37.7 million francs for the CEA in July 1952.[25]

Plutonium was thus represented, not as bomb material, but as a life-saving fuel for the energy-starved nation. Even Antoine Pinay, the French president, understood the element in this way.[26] But the CEA steering committee had chosen the gas-graphite design knowing that it could yield weapons-grade plutonium. Beginning in November 1951, the agency's new Administrator General, Pierre Guillaumat, ensured that the reactors would do exactly that.

Guillaumat had graduated from the Ecole Polytechnique in 1931 and joined the Corps des Mines. He had begun his career in the far reaches of the French empire, first in Indochina and then in Tunisia. He had become a friend and ally of de Gaulle during the war, serving the Resistance as a secret agent.[27] After the liberation he had been appointed to direct France's energy policy and fuel supply.

Guillaumat was the quintessential "man of action." He held deep convictions about the necessary relationship between technological prowess and national radiance. According to one *polytechnicien*, Guillaumat's "extraordinary force of conviction, his charisma, his incomparable talent for building those modern cathedrals that were the great projects of national independence, won him the admiration and respect of the entire community of state engineers."[28] His wartime record proved his capacity to make tough choices under dire circumstances and gave him useful talents. His experience in the secret service, for example, convinced him that "actions taken behind the scenes [were] often more effective than those taken on stage."[29] He applied this lesson extensively during his directorship of the CEA. While successive government leaders continued to proclaim France's interest in a solely peaceful atom, within the CEA Guillaumat aggressively pushed the production of weapons-grade plutonium as well as other technologies essential for making atomic bombs.

As one of his first moves, Guillaumat created a new division—the Direction Industrielle—to direct the reactor construction projects. He placed Pierre Taranger, another *polytechnicien*, at its head. Guillaumat and Taranger made it clear to their top engineers that they had to build a plutonium production facility as quickly as possible. In less than five years, both the first reactor, G1 (heavily modeled on the American Brookhaven reactor, which Taranger had visited[30]), and the plutonium extraction factory were operating at the CEA's new Marcoule site in southern France. Studies for G2—a second, larger, more innovative reactor—were underway.

Parliament had approved the Gaillard plan in part because it supposedly represented the first step in a more extensive nuclear *energy* program: the plutonium produced in these reactors would fuel other, future reactors, and in the meantime Marcoule could serve as a prototype. Yet no one even mentioned extracting electricity from G1 until its design was almost finalized. Then Pierre Ailleret, the head of EDF's research division and a member of the CEA's steering committee since 1950, suggested appending a 5-megawatt plant to G1.[31] Questions had already arisen over which institution would provide France with nuclear energy: EDF, the nation's designated electricity supplier, or the CEA, the official guardian of all things nuclear. For Ailleret, G1 provided the perfect opportunity to involve an EDF team in the nuclear adventure. The rest of the steering committee consented. Perrin and other scientists apparently hoped that an alliance with EDF would veer the CEA away from the military atom.[32] Generating electricity at Marcoule suited Guillaumat and his allies because it strengthened their public claims that these reactors represented the first step in producing nuclear electricity; in house, however, they stressed that electricity generation should not interfere with plutonium production. With this same caveat, the steering committee also let EDF build a 25-megawatt plant for G2.[33]

The decision to develop gas-graphite reactors together and the arrival of Guillaumat signaled the beginning of a new technopolitical regime in the atomic energy commission. This regime strengthened the ideological principles upon which the CEA had drawn since its inception: the importance of nuclear technology in ensuring French grandeur, the significance of independent energy sources for national autonomy, and the primacy of nuclear expertise. Under Guillaumat's regime, however, this ideology had added dimensions. Communists became increasingly unwelcome in the CEA, and in early 1952 they fell victim to another wave of dismissals when Guillaumat reorganized the agency. In the unstable political climate of the mid 1950s, Guillaumat took full advantage of the CEA's vaunted autonomy from ministerial control, using more and more of the institution's resources to pursue the military atom behind the scenes. He thereby embraced the spirit of the CEA's original statutes as expressed by their author:

—if it is a question of mining research in the colonies, the Commissariat . . . *is* the Ministry of Overseas France;

—if it is a question of mining concessions in France, it is the Ministry of Industry;

—and, if it were a question of manufacturing atomic weapons, it would be the Ministry of National Defense.[34]

The CEA's new regime expressed an ideology that saw national grandeur first and foremost in terms of military technological prowess. It valued institutional autonomy and nuclear expertise, and it upheld a vision of nationalism that excluded communists. Its primary goal, articulated by Guillaumat, had become to make a French atomic bomb.

Had the CEA merely lobbied in favor of a French bomb, I would not be justified in identifying this constellation of ideologies and people as a technopolitical regime. But Guillaumat and his men did not stop at lobbying. Instead, they directed the design of reactors that would effectively enact their policy goals. In the process of constructing these reactors, they also developed additional goals pertaining to French industrial policy. The scientists and engineers who worked on the gas-graphite program explicitly and consciously used political as well as technical criteria to make design and contracting choices. Their reactors emerged as hybrids of technology and politics. To understand how this process worked, let me examine G2, Marcoule's second and most innovative reactor.

The G2 Reactor: Developing a Nationalist Technopolitical Regime

One of the most important issues faced by Taranger and Guillaumat in directing the G2 project was the relationship between the CEA and private industry. Until Marcoule, the CEA had been primarily a research institution. It had no experience with large-scale construction, nor did it have the knowledge, the personnel, or the mandate to engage in construction projects directly. It therefore had to contract reactor construction to private companies. How should this contracting relationship be structured?

Taranger and Guillaumat believed that the relationship between the CEA and its contractors had prescriptive potential: it could set an example for other large-scale industrial developments. More was at stake than just the reactor project, important though that project was. They believed that the prestige of nuclear development gave them the opportunity to shape the structure of French industry (and hence the nation's future) more generally. French companies, they argued, should not waste time or resources competing against one another. In order to stand up to increasingly large foreign companies, French industry needed to consolidate its resources and develop its strengths. Accordingly, the CEA leaders espoused what came to be known as the "policy of champions."[35] This involved hand-picking a single company to design each major reactor component, without issuing a request for bids. Initially they took this approach out of necessity as much as out of conviction: many companies did not initially want to become involved in nuclear development. The CEA engineer in charge of coordinating the construction of G1 and G2 remembers:

> . . . as soon as Taranger became the industrial director, he made the full rounds of French industries . . . and asked them, "are you interested?" When he finished his rounds, during which he must have met with something like fifty large industrialists—all the big names of French industry, in every sector—he found four big firms who wanted to work. These were Schneider, Alsthom, l'Alsacienne, and Rateau. That's it, the end. The others had said "no, maybe, okay but." One can't say that they were all rushing toward the door at the beginning.[36]

Taranger and Guillaumat felt that the "policy of champions" made it easier to convince private companies to participate in a venture that would not yield large immediate profits. Building G2 wouldn't make companies rich, but it would give them know-how, confidence, and prestige that they would be able to use in the future to export technology. Ultimately, then, this policy would enhance France's industrial base in the short term and its economy in the long term. The "policy of champions" thus constituted one prescriptive dimension of the CEA's technopolitical regime.

Under Taranger's guidance, the CEA's Direction Industrielle grouped the chosen companies into a consortium and placed the Société Alsacienne des Constructions Mécaniques (SACM), itself a conglomerate of electrical and mechanical engineering companies, at the head. The CEA signed a contract with the SACM, which subcontracted to the other companies and coordinated the overall design and construction process of G2. The design process was a cooperative effort: after CEA engineers defined the function of a reactor component, industrialists would propose an initial design, which would then be discussed in a series of meetings. Two sorts of such meetings occurred: meetings between a CEA team and a single company to talk about a specific component, and large monthly meetings grouping together representatives of all the CEA teams, all the companies, and EDF when relevant.[37] Guillaumat and Taranger wanted the reactor to be built quickly so that it could start producing plutonium.[38] This decision-making process had the short-term advantage of producing solutions that industry was capable of building. Left to their own devices, said an EDF engineer present at these meetings, CEA teams would have envisioned complex solutions beyond the means of French industry, and the deadline would never have been met. Even so, he added, the solutions chosen were often costly and cumbersome.[39]

Compared to time, however, cost was a secondary consideration for the CEA, and in the case of the G2 reactor EDF had little input. From the beginning, CEA engineers made it clear that EDF had to play a subordinate role. It had also signed the main contract with the SACM, but its "energy recuperation installation" was considered an auxiliary device to the reactor. Thus, although EDF engineers sat in on the monthly meetings with industry, they were not expected to voice concerns over the design of their installation. Industry had to give the CEA contracts priority over those signed with EDF.[40] Furthermore, EDF engineers did not always know about design changes that had a direct impact on their work.[41] This meant that no part of the reactor itself was designed to optimize the production of electricity.

The "nuclear" part of the reactor, then, took priority over the "classical" part.[42] The "best" companies were chosen to build the trickiest, "most nuclear" parts, with little regard for cost. Contracts signed for such parts were contracts of principle: the company agreed to build a device that would perform certain functions, but the specifications were not fixed ahead of time. In contrast, the "less nuclear" parts of the reactor, such as the "energy recuperation installation" or the prestressed concrete vessel, were covered by contracts detailing both specifications and cost.[43]

What sort of reactor emerged from all this?

Figure 2.1 depicts the design of G2. Most of the reactor was housed in a large building designed to protect its contents from the vagaries of the weather. (The Marcoule reactors did not have containment buildings.) The core, contained in the large cylinder, was made up of a stack of graphite bars piled in horizontal layers. Distributed through this pile were 1200 channels, into which the uranium fuel was loaded. The uranium came in small cylindrical rods hermetically encased in aluminum cladding. Each channel could hold as many as 28 of these uranium fuel rods (figure 2.2). When enough rods were loaded into the reactor, it went critical, setting off a self-sustaining fission reaction. The fission taking place inside the rods liberated a great deal of heat, which the

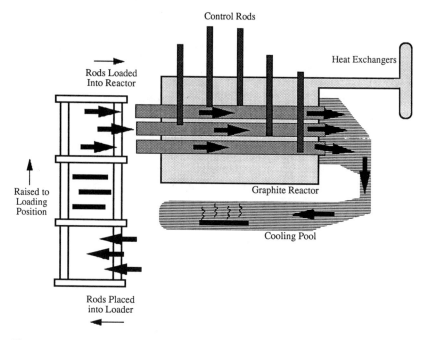

Figure 2.1
Schematic diagram of G2 (not to scale). Source: *Bulletin d'Informations Scientifiques et Techniques du CEA*, no. 20 (1958). Drawing by Carlos Martín.

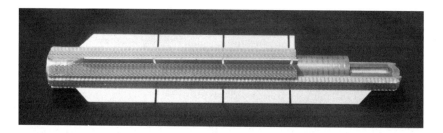

Figure 2.2
A uranium fuel rod with aluminum cladding. This particular rod is a 1968 model that was used in EDF's Chinon reactors, but its design is close to that of the G2 rods. The uranium is cast in a cylindrical mold in the center. The aluminum cladding has grooves on the outside to facilitate the even circulation of cooling gas. Photograph by F. Roux, 1968. Source: EDF Photothèque.

Figure 2.3
G2 and its twin, G3, in 1960. The four heat exchangers are lined up along the back wall of each building. Source: CEA/MAH/Jahan.

cladding absorbed. Carbon dioxide gas, entering the channels through openings on the back face, flowed around the rods and cooled the reactor by absorbing this heat. Upon leaving the core, the coolant traveled to the "energy recuperation installation" where the heat was converted into electricity.

Even this quick overview of G2 reveals that generating electricity was secondary. The "energy recuperation installation" stood outside the building that housed the reactor, both physically and symbolically removed from the fission reaction. In order to show in greater detail how the political agenda of plutonium production took precedence, I will concentrate on two aspects of G2's design: the loading and unloading of fuel in the reactor and the "energy recuperation installation" itself.

For the CEA, the key point in making weapons-grade plutonium was to obtain as much Pu_{239} as possible with as few "poisonous" isotopes of plutonium as possible. The Pu_{239} produced when a U_{238} atom absorbed a neutron was not a stable isotope: with time, it absorbed more neutrons and changed into Pu_{240} and Pu_{241}. A bomb containing too large a proportion of these isotopes might "fizzle" or detonate unpredictably. The CEA team working on Marcoule's plutonium extraction factory had already settled on a chemically based process to separate the plutonium from the spent uranium fuel—a process that did not distinguish among different isotopes of plutonium. The G2 teams had little knowledge and

Figure 2.4
The graphite block of G2 under construction in 1957. The men climbing on the block's face are construction workers from private industry; the men in bright white suits are probably CEA engineers supervising the construction. This photograph captures the "artisanal" nature of early reactor construction. Source: CEA/MAH/Jahan.

a severe time constraint to work with; under these conditions, the only solution they could devise that would minimize the "poison" involved removing the fuel rods before too much Pu_{240} or Pu_{241} appeared. The shorter the time that each fuel rod was irradiated (that is, allowed to undergo fission), the less "poison" was produced. CEA engineers calculated that, at the optimal irradiation for producing the right balance of isotopes, any given fuel rod should not stay in the reactor longer than 250 days.[44] Had G2 been designed to produce electricity, this short irradiation period would have represented an extremely inefficient use of fuel, since it involved removing the rods before they had yielded maximum heat.

These considerations led CEA engineers to impose a technopolitical constraint on the SACM, the company in charge of designing and building this system. With 28 fuel rods in each of 1200 channels, stopping the reactor, then unloading and reloading the core channel by channel, and restarting the reactor every 250 days would have wasted far too much time.[45] And saving time was crucial to the CEA, both technologically (to avoid getting "poisonous" isotopes of plutonium) and politically (since they wanted the maximum amount of weapons-grade plutonium as quickly as possible). CEA engineers hence asked for a loading system that could function while the reactor was operating.[46]

SACM engineers chose a costly solution that fulfilled the CEA requirements perfectly (figure 2.5). A cement block containing tubular holes was built flush against the northern face of the cylindrical vessel. This block contained one tube for each channel. On the far left side, the tube connected with the loading device, which traveled on a crane built on a platform adjacent to this block. The device itself consisted of two lock chambers side by side. By maneuvering the crane up and down and from side to side along the cement block, an operator sitting on top of the crane could couple these lock chambers with any of the channels of the core. Because the operation took place while the reactor was on line, these chambers were constantly exposed to radioactivity. They were therefore encased in 56 tons of metal and concrete.[47]

The lock chamber linked up to a storage chamber containing the new fuel rods. New fuel rods would be loaded onto an elevator and brought up to the storage chamber. The lock chamber, moving back and forth on a track, would pick up the rods and bring them over to the channel; a mechanical arm would then reach into the tube and undo the plug. The new rods would be loaded into the channel, pushing the irradiated rods out. Because this entire procedure took place while the reactor was under pressure, a complex system of locks and sensors was used to create a

Figure 2.5
The loading machine for the G2 reactor sat flush alongside the reactor core. It gained access to the core through the canals' ends, arranged in a hexagon at one end of the core. The operator sat in the glass-encased booth at the top of the machine. Source: CEA/MAH/Jahan.

perfectly hermetic seal every time the device coupled with the storage chamber or a channel.[48]

Once pushed out the back face of the reactor, the irradiated rods would fall down a chute. They would travel onto a toboggan which dropped them into the disactivation pool, where they would cool down for several weeks before being removed and sent to the plutonium factory for decladding and processing. This system, considerably more expeditious than the one EDF would use for removing its rods, also manifested the CEA's eagerness to obtain plutonium rapidly.[49]

Fuel loading and handling was by no means the only aspect of G2's design shaped by its plutonium production goal. Another example can be found in the CO_2 cooling circuit and in the "energy recuperation installation." The "installation" contained four heat exchangers, one turbo-generator, and auxiliary equipment. The hot CO_2 gas exited the reactor core into the heat exchangers, where it cooled by transferring its heat to water. In the process, the water turned into steam. After passing through a series of pressure stages, the steam would arrive in the turbo-generator, where its heat would be converted into electricity.

Had the main purpose of the reactor been electricity production, EDF engineers would have calculated the pressure, temperature, and flow of CO_2 that would have yielded the most efficient energy retrieval. G2's plutonium priority, however, imposed severe constraints on this energy recuperation cycle. First, the reactor had to operate continuously to avoid thermal shock to the fuel rods. Second, because they had little interest in energy efficiency, CEA engineers had not designed the aluminum cladding surrounding the fuel rods to withstand high temperatures. These two constraints led the CEA to determine specific values for the pressure and temperature of the CO_2—values that did not correspond to those for optimal electricity production.[50] A third, more significant constraint was that the CEA wanted to operate the reactor at maximum power all the time. Maximum power meant U_{238} would be converted into Pu_{239} more quickly. Combined with the rapid unloading of the fuel rods, this meant that a maximum quantity of Pu_{239} could appear and be removed before too much of it decayed into poisonous isotopes. Running the reactor continuously at maximum power, however, could not be handled by the electrical network to which the heat generator was hooked up; because of variations in energy consumption, the network could not always absorb all that energy. So all these constraints forced EDF engineers to add a "desuperheater" to the circuit, placed just before the steam generator, to absorb excess heat. Furthermore, the fourth heat exchanger existed only as a safeguard in case of breakdown; in fact, three exchangers would have sufficed to run the reactor and the plant. Finally, so that the reactor could be run at maximum power and low temperature, the CO_2 had to flow through the core at a very high rate. This did not favor energy efficiency: the high rate required more electricity to power the blower, and the exiting CO_2 was at a lower temperature and therefore contained less energy.[51]

CEA engineers thus translated Guillaumat's enthusiasm for a French atomic bomb into a reactor design whose ideal function was producing weapons-grade plutonium. That G2 was something beyond a prototype for an electricity generating reactor became even clearer in 1955, when Guillaumat negotiated a secret agreement in which the Ministry of Defense agreed to finance its twin, G3.[52] While the French government waffled over whether to build a bomb, Guillaumat and his engineers took the crucial first step toward that bomb. They had almost finished building G2 in April 1958, when Prime Minister Félix Gaillard signed the order to have a bomb ready in early 1960. Without these Marcoule reactors, France could have never exploded its first bomb so quickly.

To the engineers and technicians at Marcoule, the military aspect of their work was no secret. A sense of excitement and urgency pervaded the offices in which engineers struggled over design problems and the construction sites where the huge reactors took shape. They were creating a brand new technology, and one that was of singular importance to their nation. Although they had indirect knowledge of some nuclear work that had been done in America, Britain, and Canada, they apparently did not have access to many of the technical solutions worked out by researchers in those countries.[53] They thus relied on their limited experience with experimental reactors and their ingenuity. Sometimes they favored solutions because they had heard that the British were working on something similar. More often, they favored ideas that appeared to provide the quickest, if not always the most elegant, route to completion. Frequently they had no idea whether a device would work until it had been built and attached to the reactor. The uncertainty that thus dominated their work created what many later referred to as a "pioneering atmosphere" on the job. The excitement of this atmosphere saw engineers through the 60- or 70-hour work weeks that prevailed throughout construction. One engineer has said that on February 13, 1960, after the first French atomic bomb—loaded with plutonium produced at Marcoule—exploded, he and his colleagues had been so proud of their country and the part that they had played in this achievement that they had "shed tears of joy."[54]

Much as this internal sense of pride mattered for the CEA's success, it seemed even more important to ensure that the rest of the nation's science and engineering community saw the Marcoule reactors as the epitome of French technical prowess, even if their military dimensions had to remain hidden. A few relatively minor delays in G2's construction had led to rumors that Marcoule might not live up to expectations and to complaints about the size of the CEA's budget. The CEA's public relations officer urged division heads to counteract these rumors. Guillaumat concurred, saying that "engineers who read a technical journal [had] to keep a good opinion of what we do at Marcoule" and that "for this reason, it [was] very important that stories about Marcoule multiply throughout all the branches of the technical press."[55] Accordingly, Marcoule's site director held a press conference to extol the virtues of his reactors.[56] Similarly, CEA division heads published in most of the major science and engineering journals articles that characterized Marcoule as a uniquely French achievement, piloted by the CEA and heroically implemented by French industry.[57] Francis Perrin wrote:

[The CEA] deemed it necessary to associate French industry with these great achievements which prepare the way for the development of the industrial use of nuclear energy. French industry answered this appeal, despite deadline and supply constraints that were often severe. . . . Above all, the result of this collaboration is apparent through the two massive buildings that dominate, by their 50 meters, the banks of the Rhône. [They are] a modern replica of the ancient wall of Orange that faces them. . . .[58]

Marcoule was not only French by birth; it was also French by association and heritage. According to the site's director, "the Arc de Triomphe of the Etoile would easily fit in the vast metallic structure that shelters . . . G2."[59] The prestige that Marcoule derived from being associated with historical monuments was transitive: as a great and uniquely French achievement, Marcoule in turn would embody and strengthen French greatness. Thus the CEA's technopolitical regime sought to propagate its ideology throughout French industrial and engineering circles.

Such articles also aimed to present the Marcoule reactors as producers of reactor fuel and prototypes for power plants. That G2 did in fact generate a modicum of electricity (25 megawatts under the best of circumstances) enabled engineers and managers to describe it as a successful prototype worthy of the tremendous investments made in the nuclear program. A 1957 article in a French civil engineering journal said:

The first stage of the plan called for the construction of two nuclear reactors (G1 and G2) that were supposed only to produce plutonium *destined to fuel the secondary reactors of the future.* . . . But during the study, we were led to envisage using the heat released by these reactors to produce electric energy. . . .
Currently, the predicted total investment amounts to 60 billion francs. *This financial effort is justified by the necessity to develop, on the industrial scale, the production of electrical energy of nuclear origin,* due to the insufficiency of European resources in fossil fuel. The role of Marcoule is essentially to allow this development, to train teams of operators at different levels, and to promote technical and industrial progress in this field.

Investments should thus not be measured against the power of the installations, but against the development potential that they bring. In fact, the [amount of these investments] is quite in proportion to the increase in our energy needs. . . .[60]

The CEA thus actively sought to control the political meanings of its reactors. The ambiguity of its gas-graphite design enabled engineers to do so easily. Unable until 1958 to promote Marcoule as a linchpin of French military security, they portrayed G2 and G3 as power plant prototypes, sources of precious fuel (and therefore independence), and exemplars of French engineering prowess. These representations, in turn, enhanced the technopolitical flexibility of the reactors' design.

This flexibility made G2 a powerful strategic tool for the CEA's technopolitical regime. We can see a striking example by briefly examining the first ministerial-level discussions on the French bomb. These occurred in late 1954 in a series of meetings presided over by Pierre Mendès-France, then prime minister. Those present included Pierre Guillaumat and Francis Perrin of the CEA, the Minister of Finance, the Minister of National Defense, the Secretary of State for Research, and various ministerial cabinet members. Guillaumat, the Minister of National Defense, and others in favor of building a French bomb tried to push Mendès-France into making an official decision to that effect, arguing among other things that a bomb effort would have advantageous fallout for the civilian sector.[61] Mendès-France later recalled the meeting this way:

I remember asking which part of the research under way was of economic interest, and which was only of military interest. They retired to a corner of my office to discuss matters in a low voice, and several moments later, they came back and told me, "for another three years, we won't be able to distinguish the military from the civilian; only after three years will we reach a branching point when we can say: this is purely military, and that holds a purely economic interest." Under those conditions, I said, there's no problem: we must continue to do research. . . . There was no question of amputating the positive aspects of such research work from the French economy.[62]

Thus Mendès-France chose not to decide.[63] His government lasted only two more months, so he never reached that fateful "branching point." His response, meanwhile, enabled Guillaumat to continue pursuing his military agenda. The CEA Administrator General later recalled: ". . . each one interpreted [the meeting] the way he wanted to. . . . Without lying too much, [I] understood that Mendès had given us the go-ahead."[64] Successive ministers also flirted with making a firm decision, repeatedly falling back on the versatility of the Marcoule design as a means of avoiding a potentially unpopular choice.[65]

Meanwhile, scientists within the CEA who opposed the idea of a French bomb also attempted to capitalize on the versatility of the gas-graphite design. Eager to experiment with other types of reactor design, they wanted to use Marcoule's plutonium in breeder reactors (and in other types). Their proposals for the CEA's future included scenarios in which Marcoule's plutonium would go straight into secondary reactors—first to augment impoverished uranium fuel, then as the main fuel for breeder reactors.[66]

Attempts to co-opt Marcoule's plutonium proved fruitless, however, and by 1956 most scientists appeared resigned to—though in some cases resentful of—the CEA's military mission. Guillaumat later recalled: "Perrin always resented, kind of legitimately, the artisans of this first atomic plan for having suddenly produced a mass of plutonium that really only had one use. Sure, the possibility existed of making plutonium-fueled reactors, but that was extremely chancy."[67] Some proposals for experimenting with other types of reactor design did go through, but it was clear that Marcoule's plutonium would fuel bombs for the foreseeable future.[68] Indeed, the CEA had begun to pursue other aspects of weapons technology, through covert agreements with the military.[69] Techno-political versatility was an important strategy in all these efforts. Publicly, for example, the creation of a military division within the CEA in 1956 was aimed at building a nuclear submarine; in addition to a submarine, however, researchers in that division also investigated aspects of bomb design. That year the CEA's steering committee also began seeking approval to build a uranium enrichment plant.[70] Ostensibly, this plant would enrich uranium to fuel future secondary reactors—indeed, several scientists wanted the plant for precisely this purpose.[71] The plant designs under consideration, however, all included technologies that would enrich uranium to weapons-grade concentrations, and the final plant design was clearly geared to this end.[72]

Thus, by April 1958, when Gaillard signed the bomb order, a complex technopolitical regime governed the CEA. Gas-graphite reactors, industrial contracting according to a "policy of champions," and the uranium enrichment plant were more than mere outcomes of that regime's choices. They were the means through which CEA technologists expressed and enacted their commitment to a French atomic bomb, and to French technological prowess more generally. As hybrids of technology and politics, they solidified the CEA's regime and were key components in its technopolitics. Precisely because of their ambiguous and hybrid nature, G2 and G3 extended the power of the CEA's leaders beyond the confines of the institution into the national political arena. Their ambiguity—and their power—derived simultaneously from the versatility of their design and the ways in which CEA administrators and engineers capitalized on that versatility. Depending on the audience and the political climate, the Marcoule reactors could be presented as purely civilian, purely military, or somewhere in between. This flexibility ensured their continued development; it also enabled the de facto pursuit of a nationalist military nuclear policy well before the government was willing to commit to any such thing.

EDF: The Emergence of a Nationalized Regime

Unlike the CEA, which had emerged from backstage negotiations, Electricité de France was formed after protracted debate among the multitude of political parties that vied for power after World War II. The main technical idea behind EDF—to unify the production, transmission, and distribution of electricity in a single, enormous utility—was not new. Before the war, several members of the Corps des Ponts et Chaussées had begun to design such a system, trying to make sense of the plethora of smaller networks that ran on different frequencies and voltages and attempting to merge some of the nation's private utilities. But political backing for this plan did not exist until after the war. Only then was there a widespread consensus that these private utilities epitomized the problem with French industry: they were represented as "Malthusian" companies that shunned innovation and privileged short-term profit making over reliable public service. State engineers, labor unions, and politicians of most stripes agreed that the new France should be on a single, standardized electrical network run by a single, public utility.

Even so, nationalization meant different things to different groups. On the left, some saw it as the first step toward a socialist system, while others viewed it as a simple improvement of conditions within the capitalist system. Centrist parties saw nationalization as merely a practical step toward economic modernization, with no ulterior political meanings or implications. The right denounced it as the first step toward totalitarian statism. Debates ensued on the autonomy of the new utility, the source of its capital, the hierarchical structure of management in the company, and the labor contract for its workers. In April 1946, a nationalization law was passed that regrouped the private companies into a single electric utility, EDF, accountable for its expenditures to the Ministry of Finance and for its development program to the Ministry of Industry.[73] The men appointed to the upper echelons of EDF management included numerous *polytechniciens*, especially members of the Corps des Ponts et Chaussées. Like the state engineers who dominated the CEA, these men brought a strong ideology of public service to EDF.

The left-wing coalition (which included the major labor unions) had the strongest voice in structuring the new utility and in defining its symbolic meaning. The left viewed EDF as the model and prescription for a redefinition of the relationship between the French worker and the French state. The utility's labor contract embodied this model by guaranteeing paid leave, making pay scales public, and incorporating representatives of labor unions in the company's managerial structure (albeit in a subordinate position).[74] The Confédération Générale du Travail (the communist labor union) dominated EDF (and has continued to do so ever since). At a time when labor strife permeated French industry, this contract, therefore, had tremendous symbolic value.[75] Pierre Simon, the utility's first president, articulated the meaning of worker participation as follows:

In contrast to the spirit of routine, we must have a revolutionary spirit. Without a doubt, the workers can be very intelligent and can even escape their roles to become governmental ministers. Until now, there was a widespread tendency toward a separation between the roles of workers, who were to perform mechanical tasks, and managers, whose role was to define the methods of work. Today, that distinction is being suppressed. Under the old system, those who performed direct work were excluded from its conceptualization; today, we seek to associate the workers in that conceptualization.[76]

In other words, EDF promoted the active and valued participation of workers in the reconstruction and modernization of the nation. As such, it functioned as a potent public symbol of a new social order.[77]

Nationalization certainly represented a victory for the left, but it did not signal a unanimous consensus on the structure and role of the public utility. The onset of the Cold War in 1947 provided an occasion for attacks on communist strongholds, including EDF. Although these attacks weakened the formal power of the communist labor union, they did not succeed in purging communists from EDF. Unlike the CEA, EDF remained a bastion of the left. Attacks on the utility's managerial and financial practices did, however, mean that issues such as the rate structure, the choice between hydroelectric and coal power plants,[78] and the relations with private industry continued to be points of conflict within EDF and also among EDF, the government, and capitalist companies. The struggle over the meaning of nationalization for EDF would thus continue long after the utility's creation.[79]

Still, even diverging factions within EDF could agree on a few basic issues. First and foremost among these was EDF's mission: namely, to make France energy independent by producing and distributing the most electricity at the least cost. More generally, everyone agreed that the amount of electricity generated and consumed by a nation directly reflected its modernity (as did the level of ownership of electrical appliances).[80] By extension, the network for distributing electricity united and defined France symbolically as well as technologically: complete electrification would enable all French citizens to participate in the modernization of their nation. Though the implementation of worker participation provoked some disagreement, it would have been impolitic for anyone to deny its value in principle. Nationalization thus made room for everyone at EDF to embrace an ethos of public service. Working for EDF meant apprehending and serving the entire nation through the production and distribution of electricity.

These basic ideological principles underlay EDF's efforts to establish its own technopolitical regime within the nuclear program. The utility's interest in nuclear power began in the early 1950s with Pierre Ailleret, the director of its research division. Ailleret, also a member of the CEA's steering committee, had persuaded the CEA to add small power generating units to the Marcoule reactors. He simultaneously sought to drum up enthusiasm for nuclear power within the utility. Gas-graphite reactors had the potential to fit well within EDF's ideological scheme: designed and manufactured in France, they could provide an additional path to energy independence. Not all of EDF's top administrators shared Ailleret's burning enthusiasm for the technology itself, but the opportunity was too good to pass up. EDF had neither the time, nor the money, nor the expertise to launch an independent nuclear program—but it could collaborate with the CEA. Beginning in 1954, EDF's Director General Roger Gaspard signed a series of protocols with Guillaumat that specified the distribution of technical and financial responsibilities not only for the Marcoule reactors but also for the utility's first plant at Chinon. Financially, the CEA would be responsible for Marcoule and EDF for Chinon. Technologically, the two institutions would work together to develop EDF's reactor, whose main purpose would be to produce electricity (not plutonium).[81] The two institutions would draft development plans together and take them to the Commission Consultative pour la Production d'Electricité d'Origine Nucléaire (known informally as PEON), which had been formed in April 1955 to advise the government on matters of nuclear power development. PEON was composed primarily of high-level engineers and managers from the CEA and EDF; thus, provided the two institutions could agree on plans beforehand, commission approval would be little more than a formality.[82] And indeed the first gas-graphite development plan went through easily enough. EDF's first reactor would be a 60-megawatt plant. A succession of increasingly powerful reactors would follow, which by 1965 would total 800 megawatts.

This apparently amiable arrangement soon gave way to a series of conflicts between technologists at the two institutions. During the course of these conflicts, EDF engineers and managers established a nationalized technopolitical regime, through which they sought to prescribe and enact their own vision of France's industrial policy and their own ideas about how gas-graphite reactors should look. In order to understand the technopolitical regime that EDF upheld within the nuclear program, let us now turn to its first reactor project: EDF1.

The EDF1 Reactor: Developing a Nationalized Technopolitical Regime

The two regimes needed each other, both technologically and politically. Despite the fact that the Marcoule reactors did not produce energy in an optimal fashion, they did at least have an electricity generation unit. EDF engineers were learning valuable lessons by working on G2 and G3 (and also by taking CEA courses in nuclear engineering).[83] Politically, EDF's participation in Marcoule had buttressed the CEA's claims that the reactors there were prototypes for power reactors. Conversely, Marcoule's success strengthened the case for building separate EDF reactors. In many ways, the partnership seemed ideal.

In the course of working out the collaboration, however, each regime was also eager to establish its role in defining the future of the nuclear program and of French industrial development more generally. In theory, the terms of cooperation for EDF1 were clear: CEA teams would design the "nuclear" parts of the reactor (the core and the fuel rods). EDF teams would design the "classical" parts (electricity generation). Ultimately, EDF headed the project, and thus it would make the final decisions. In practice, though, the EDF1 project was fraught with tension between engineers in the two institutions. This tension centered around two issues: the role of private industry in the project, and the actual design of the reactor. Conflicts did not emerge because the CEA didn't want to build an electricity-producing reactor. Part of the CEA's mission was to develop nuclear technology in any form. But the distinct design and contracting practices developed at Marcoule had become integral to the

CEA's regime, and its engineers wanted EDF to follow the same practices. Notably, they wanted to preserve the dual nature of gas-graphite reactors: just as EDF had gotten some electricity out of Marcoule, CEA engineers hoped to get some plutonium out of EDF reactors. They had come to regard this versatility as an integral technopolitical part of gas-graphite design: on a technical level, gas-graphite reactors inevitably produced at least some plutonium, and this might as well be put to a good political purpose.[84] And engineers in each institution jealously guarded their expertise: CEA engineers held that their intimacy with nuclear matters gave them the edge, while EDF engineers maintained that their experience with conventional power plants gave them the upper hand.[85]

More was at stake than EDF1 alone. It was still far from clear that the nuclear program would receive long-term support. Precisely what would receive support (a military program, a civilian program, or both), and to what extent, was also uncertain, particularly before 1958. Furthermore, project participants expected that the set of working methods and expertise that prevailed in the EDF1 project would dominate, or at least influence, future reactor projects. They therefore felt that they were conceiving not one but a whole series of French nuclear reactors.[86] An examination of these conflicts will demonstrate how EDF engineers developed their own technopolitical regime by inscribing their political, economic, and technological agendas into the project and by making EDF1 into their own instrument of technopolitics.

Tensions between the two institutions first became manifest in the organization of the project. As with G2, Guillaumat and Taranger wanted private industry to coordinate the design and construction of EDF1. But EDF's nuclear team wanted to follow the contracting practices the utility had used for its conventional power plants. They held that EDF should fill the dual role of project coordinator[87] and general contractor—the dual function that the SACM had fulfilled for G2. Team members espoused the anti-capitalist sentiment that had spawned nationalization. By building EDF1, they were providing a public service. The best way to do this was to optimize the cost and efficiency of the reactor.[88] Ailleret argued that EDF, not private industry, should conduct the optimization studies "in order to be sure that we are not influenced by the industrialist's tendency to develop certain types of materials rather than others."[89] Another team leader commented disparagingly that Guillaumat and Taranger were "oil men" who "dreamed only of private industry."[90] Furthermore, the team argued, EDF should coordinate the overall design and building as well. The best way to keep costs down was to divide the reactor into parts and request bids for each part. EDF would thus retain greater control over both the knowledge needed to build the reactor and the cost of the project. To top it all off, this method of working was "politically correct": in the ironic words of a high-level EDF manager, "the pure and white EDF, a nationalized company, would acquire the know-how while leaving the builders, the capitalist companies, with the banal task of supplier."[91] In EDF's technopolitical regime, the utility would direct the development of nuclear power in the best interests of the state, with private industry merely following orders.

Overriding a furious Taranger (who maintained that the CEA's method was better for the overall industrial health of the nation), the EDF team proceeded according to its intial plan.[92] The first step now was to draft preliminary blueprints. Jean-Pierre Roux, the head of EDF's design team, had asked the CEA to do so in July 1955. But his team found this proposal, heavily based on G2's design, unacceptable: the team intended to generate electricity "optimally," something G2 did not do.[93]

The EDF team sought to change practically everything in the CEA's proposal.[94] In order to optimize the reactor for electricity generation, they wanted to control the definition of almost all the components and parameters, including components such as the uranium-graphite pile and the devices for loading and unloading the fuel, as well as parameters such as the pressure of the CO_2 cooling gas and the operating power of the reactor.[95]

The finished design of EDF1 (figure 2.6) looked quite different from that of G2. The most noticeable and perhaps the most symbolic modification was the location of the heat exchangers: right next to the pressure vessel that contained the core (rather than many meters of energy-losing pipes away), and inside the reactor building (rather than outside). The reactor still ran on natural uranium, encased in fuel rods similar to those of G2, and it was still moderated by graphite and cooled by CO_2. Just about everything else, however, had been changed.

The EDF team insisted on changing the operating pressure of the reactor and the pressure vessel containing the core. The CEA team had suggested a prestressed concrete vessel like the one at Marcoule. It argued that, in addition to being a tested technique, prestressed concrete was a domain in which France had outdistanced other nations. Adhering to the "policy of champions," CEA officials felt a responsibility to encourage French industry to reinforce its areas of excellence.[96] But EDF engineers found this vessel too expensive. They feared that prestressed concrete could not withstand the temperatures at which they planned to operate

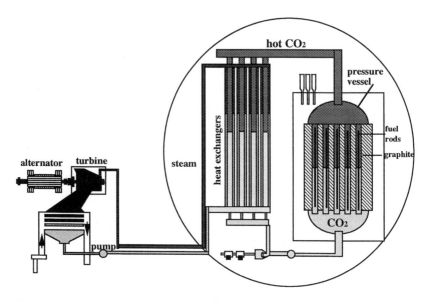

Figure 2.6
A schematic diagram of EDF1 (not to scale). Source: EDF, Rapport de sûreté Chinon A1, 1980. Drawing by Jay Slagle.

the reactor, and that it would require its own special cooling circuit. This would increase overall operating costs and lower the reactor's energetic efficiency: the blowers needed to pump the CO_2 through the special circuit would use up 10 percent of the electricity generated by the reactor.[97] Instead, EDF engineers chose a steel vessel, cylindrical in shape, capped by a steel hemisphere on either end.[98] Steel could withstand higher temperatures and pressures. In addition, the fact that the United States and Britain had built steel reactor vessels gave the French engineers confidence that they too could build a working steel vessel. For them, national pride would derive from the production of cheap, reliable electricity, not from promoting minor, if uniquely French, technologies.[99]

Early on, EDF engineers decided that EDF1 should function at a higher pressure than G2: 25 bars instead of 15. Lower-pressure reactors were easier and faster to build, and speed had mattered politically to CEA technologists in the G2 project. But a lower operating pressure meant that a higher flow of CO_2 was needed to extract the heat, which required more powerful blowers, thereby lowering the reactor's efficiency.[100] EDF engineers had also decided that the loading and unloading of the fuel would take place while the reactor was stopped. Unlike the CEA, EDF wanted to burn up the fuel rods as much as possible in order to extract the maximum amount of heat. In 1955, engineers designing EDF1 could imagine little use for a device that could move rods in and out of the reactor very quickly. They hoped that a loading device that could only work off line would limit how much weapons-grade plutonium the CEA could demand from EDF1.[101]

Having decided on this loading principle, the EDF team then decided to orient the channels containing the rods vertically, rather than horizontally as in G2. In a vertical configuration, the CO_2 could be pumped in at the bottom. It would thus follow the natural convection of heat, growing hotter as it rose. This meant that less pumping power was required for the CO_2, and it made the overall design safer in case of blower failure. A vertical pile also required fewer openings in the pressure vessel, thereby making it easier to ensure that the core was hermetically sealed. It also meant that the reactor could be loaded and unloaded from the bottom. Bottom loading involved using "a single loading arm capable of reaching all the channels and requiring only one opening in the shell, although clearly a large one."[102] EDF engineers found this system simpler and cheaper than the G2 design,[103] which had separate openings in the vessel for each channel and a huge machine designed to have access to every channel.

EDF engineers thus advocated a design that they felt would make most efficient use of both fuel rods and investments and that would be as simple as possible so as to provide a good basis for future reactors.[104] Both through the design itself and through the industrial contracting process, EDF engineers sought to redefine what a reactor was, how it should be built, and what it should be used for. By modifying pressure and temperature, EDF engineers had designed a reactor whose performance and capabilities matched their regime. They hoped that in future collaborations the CEA would have to work with these new parameters.

EDF engineers found that the process of designing a power reactor involved a great deal of guesswork and intuition. In the mid 1950s, they had no more access to foreign technology than did their CEA colleagues. They therefore chose technical solutions that they thought would further their political, economic, or industrial goals. Sometimes they simply favored options that would differentiate their regime from the CEA's. By the mid 1960s, when they began designing EDF4, they had revisited several such solutions and had found ways to make prestressed concrete vessels and continuous fuel loading suit their purposes. What mattered in the mid 1950s, though, was that technologists within the utility (and in

Figure 2.7
EDF1 under construction in 1962. Photograph by H. Baranger. Source: EDF
Photothèque.

the relevant ministries) believed that EDF1 engineers had designed the
project that best suited the utility's regime.

How the EDF engineers viewed their work is evident from the way they
promoted their achievements to other French engineers—inside as well
as outside their institution, for not everyone at EDF believed that nuclear
energy would ever compete with conventional power plants.[105] Some of
their prose paralleled that of CEA engineers, explaining that "anguish-
ing" shortages in energy resources justified the huge "financial sacrifices"
made for the nuclear program—sacrifices that, in any event, would soon
pay off, since nuclear energy increasingly seemed like a "providential solution."[106] At the same time, their rhetoric frequently sought to differentiate their regime. For example, one engineer contemplated the day when
EDF would no longer have to use natural uranium—a choice in which it
had played no part, although its engineers had accepted the choice without complaint: "The inferiority of natural uranium piles is less economic
than it is energetic. Later, when we move to another kind of reactor that
allows us to use enriched fuel, it will be less to lower the cost of the kilowatt-hour than to reduce the specific consumption of fuel and increase,

in considerable proportions, the amount of energy that can be drawn from natural reserves."[107]

Certainly EDF engineers had reason to be preoccupied with the overall "efficiency"—both energetic and economic—of their electricity-generating technologies. In order to get France's energy sector back on its feet, EDF had built as many conventional power plants as it could, as quickly as possible, in the first ten years after the war. The resulting hydro-electric program had paid less attention to cost than to speed and reliability. In the face of sharp criticism in the early to mid 1950s, EDF had adopted an institution-wide policy of *rentabilité* (best translated here as economic viability), which coincided with the priorities of the second plan elaborated by the nationwide Planning Commission.[108] Engineers hence had to show that their designs would not lose money and would make efficient use of fuel. Already, the engineers who had built the hydro-electric plants were fighting with those in charge of coal-fired thermal plants over whose work best fulfilled these requirements.[109] EDF's nuclear team therefore aimed its arguments about the benefits of nuclear energy at the world outside its regime, whether that be within the utility or beyond it.

Utility engineers compared their achievements with those of other nations, especially Britain. Jean-Pierre Roux compared EDF1 with the Calder Hall reactor and concluded "that this French project holds up under comparison with the English projects."[110] Especially, he continued, when one considered that the British took five to six years between reactors, whereas the French were only taking two. Waxing eloquent on the benefits of nuclear energy, other engineers emphasized that building nuclear reactors fulfilled their mission of public service to the French state and the French people:

> The path taken in giant steps during the past few years in the four large atomic countries, and especially in France, allows the highest hopes.
> It is not chimerical to think that the moment of massive realizations approaches rapidly.
> Placed at the disposal of all, in the workshop and in the home, nuclear energy will allow economic and social progress to continue everywhere in the world, and in the European community in particular.
> France must reap the moral and material benefits that she has the right to expect from a technology so often fertilized by her scientists and already so widely developed by her engineers.[111]

Just as CEA engineers had sought to shape the meanings of Marcoule to enhance its technological and political versatility, EDF engineers sought to demonstrate how Chinon upheld—indeed constituted—the utility's technopolitical regime within the nuclear program. Like the utility itself, nuclear energy would promote "social progress." It would promote democratic values by being "at the disposal of all." Under EDF's guidance, France's development of nuclear power would proceed rapidly and efficiently, easily competing with Britain's and thereby bringing the nation the "moral and material benefits" that were its due. EDF took just as much pride in French achievements as did the CEA, but it located the source of pride in the practices of a nationalized institution.

*

Both G2 and EDF1 were hybrids of technology and politics. There existed no single best way to build these reactors; they were not the inevitable products of some progressive logic inherent in the technology. Nor were they the infinitely malleable products of political negotiation. Rather, each reactor resulted from a seamless blend of political and technological goals and practices.

For both regimes, building these reactors entailed the pursuit of technopolitics. French military nuclear policy in the 1950s was not made by government officials contemplating their nation's place in the postwar world and firmly deciding to build a bomb. The political chaos of the Fourth Republic precluded any deep consideration of nuclear policy. Heads of state, ministers, and elected officials gladly allowed state technologists to make nuclear policy. In the absence of a traditional political formulation of nuclear military policy, the Marcoule reactors *were* that policy, containing both the ambiguities and ambivalences of Fourth Republic governments and the goals of men like Pierre Guillaumat and Pierre Taranger. As a counterpoint to G2 and G3, EDF1 was also policy and politics. EDF engineers seized on the energy-producing potential of the gas-graphite design to direct nuclear policy more firmly toward energy production. They used their technological choices in the EDF1 project—and the fact that those choices differed from the CEA's—to convince others in their institution, as well as bureaucrats and ministers who might fund their program, that nuclear energy could present a viable economic alternative to conventional power sources.

Thus CEA and EDF technologists deliberately—even proudly—sought to make their technologies into instruments and embodiments of politics. CEA engineers may have obscured the full extent of their political aims for some audiences, but they never tried to hide the fact that they had political aims. Nor did their EDF counterparts. *Politics and policy making gave the reactor projects significance,* both within the each regimes and in the interactions each had with its surroundings. For example, EDF1 was important not because it would produce economically viable electricity, but rather because it represented the first step in a nationalized nuclear program that would enact and strengthen the utility's ideology and its industrial contracting practices. At the same time, the *technological form* of their politics gave technologists power and influence. For example, Pierre Mendès-France displaced his decision onto the shoulders of CEA leaders, whose authoritative assurances about the flexibility of their technologies enabled him to abstain from deciding about a bomb. Meanwhile, this same flexibility allowed CEA technologists to persist in their pursuit of the military atom.

CEA and EDF engineers had a common interest in promoting a nuclear program and a shared heritage of public service. Working together, they sought to establish the nuclear program as an arena in which to play out issues of great significance to the French nation and its identity. For both groups, developing a nuclear program provided a means of making France a technologically powerful nation—of recasting the symbols of French identity in technological form.

But the precise nature of that form differed. Engineers and managers in the two institutions had diverging visions of the public interest and of the nation's future. Their efforts to translate these visions into techno-logical practices and artifacts resulted in two distinct technopolitical regimes. The CEA's nationalist technopolitical regime found form in its Marcoule reactors and in its "policy of champions." EDF's nationalized technopolitical regime found form in its Chinon reactor and in its efforts to micromanage industrial contracting. Both regimes sought to develop prescriptions for governing nuclear development within their institutions and for directing nuclear and industrial policy on the national stage. Embedding these prescriptions in artifacts and prac-tices constituted a strategic move in which technology and politics were deliberately conflated. Thus the reactors at Marcoule and Chinon func-tioned as strategies through which the two regimes aimed to retain power over both the technological and the political dimensions of nuclear development.

We can understand this strategic practice of embedding policies in reactors as technopolitics—that is, politics conducted through specifically technological means. Technopolitics differ from regular politics in two important respects. First, technopolitics is conducted not by elected offi-cials but by technologists (in the broad sense defined in chapter 1).

Second, its power derives from its grounding in expert knowledge and its expression in material artifacts or practices.

These regimes were neither uncontested nor static. Shortly after his arrival, Guillaumat dismissed several communists from the CEA under the guise of a reorganization. Employees widely interpreted this move as signifying that the CEA would engage in military activities, and a series of protest strikes ensued that lasted, on and off, for two years.[112] The strikes did not alter Guillaumat's course, but they did show that the nationalist technopolitical regime he had spawned would need to remain vigilant in order to retain control of the institution. As we have seen, a subtler form of resistance came from some of the CEA's scientific leaders, who sought to redirect gas-graphite technology toward more peaceful ends by promoting the development of breeder reactors. These efforts met with partial success, largely because the breeder promoters fitted their project proposal within the framework and prescriptions of the nationalist technopolitical regime. Although Marcoule's plutonium remained (at least in the short run) destined for weapons, the CEA also launched an experimental breeder reactor project. This project would provide important technopolitical support for that regime in the late 1960s, when the gas-graphite program would be threatened.

EDF's earliest efforts to establish a technopolitical regime in the nuclear program did not engender much opposition inside the institution. As the program grew, however, so did the stakes. The 1960s witnessed growing struggles, both inside the utility and between EDF and the CEA, over the methods and practices according to which reactors should be designed and built. In the course of these struggles, both technological and ideological components of EDF's nationalized regime underwent a series of shifts.

3

Technopolitics in the Fifth Republic

As the strategic practice of designing or using technology to constitute, embody, or enact political goals, technopolitics is a distinctive form of political action. Its effectiveness, however, depends at least partially on the broader political framework. The success of the CEA's technopolitical regime in the 1950s was due in good measure to the ministerial instability of the Fourth Republic and its leaders' collective unwillingness to engage in more conventional forms of nuclear policy making. These conditions changed in 1958, when Charles de Gaulle returned as head of government after a twelve-year absence.

De Gaulle and his allies attached tremendous symbolic importance to French nuclear achievement. The ensuing centrality of the nuclear program to government politics had double-sided and somewhat ironic results for its developers. The program acquired greater importance, but the actions of nuclear leaders underwent greater scrutiny. Gaullists fully embraced the notion of technological radiance, which gave technologists greater visibility and respect. This in turn meant that technopolitics became a more important and powerful form of political action. It also meant, however, that technopolitics became more complex and contested. Conducting nuclear technopolitics now involved more than embedding preexisting political goals into technological artifacts. Increasingly, technologists had to shape their agendas and practices in ways that would be compatible with Gaullist discourse on national identity and industrial development. The technopolitical regimes of the CEA and EDF no longer had exclusive control over the terms in which debates about France's nuclear future were conducted.

In this chapter, in order to understand these changes, I return to the debates between the CEA and EDF over the development of the gas-graphite reactor program. Here I concentrate less on final reactor design and more on design practices, project organization, and program development. After surveying the design battles over the EDF2 and EDF3 reactors, I focus on three topics that arose repeatedly in battles between the two regimes from the mid 1950s to the mid 1960s: optimization techniques and the nuclear kilowatt-hour, the control and pricing of plutonium, and industrial competitiveness. Utility engineers began to include economic models and other optimization techniques into their design practices, not only because these forms of systems analysis facilitated reactor design but also because they were useful defensive strategies against attacks on EDF expertise. EDF later used those changed practices to negotiate with the CEA over plutonium production. Meanwhile, the issue of industrial competitiveness on foreign markets became increasingly important in national politics. This issue was a focus of debates between the two regimes over how to export reactors, which in turn led to conflicts over contracting, project organization, program development, and the national interest.

My analysis shows how engineers and other technologists used technopolitics not just to solve the problems at hand, but also to extend their influence beyond their regimes, articulating and enacting their vision of the nation. At the same time, I examine instances in which the technological foundations of these technopolitical activities shaped their political effectiveness or potential. Notably, I discuss the political implications of the cracking of EDF1's containment vessel, and of the fact that all gas-graphite reactors (even those designed to produce electricity) produced at least some plutonium as a by-product. These examples highlight the distinctiveness of technopolitics as a strategic practice and emphasize the importance of taking the physical attributes of technologies seriously even when discussing their political dimensions.

Before I return to the subject of the nuclear program, however, I must discuss the role of technological development in Gaullist discourse on national identity.

Technology and Gaullism

In 1958, the mounting colonial crisis in Algeria induced Charles de Gaulle to return as head of government. The following year marked the official beginning of the Fifth Republic. The new constitution gave the state a stronger role in directing the economy and rendered the executive branch of government less vulnerable to political upheaval. De Gaulle hoped that strong leadership would help him heal his nation from the hardships of reconstruction, the political turmoil of the Fourth Republic, and the rifts caused by the wars in Indochina and Algeria. Above and beyond all this, he wanted to restore France to its former glory. Among other things, this meant steering an independent course in the escalating Cold War—a course clearly separate from that of the communist Soviet Union and also from that of the United States. (De Gaulle saw the United States as economically and culturally imperialistic.) Technological development was central to this independence.

In making a case for French technological radiance, de Gaulle used language and images that strongly resembled those used by the technologists we encountered in chapter 1. At the end of the war, he had declared: "Vanquished today by mechanical force, we can vanquish tomorrow with a superior mechanical force."[1] Like the technologists and the planners, he associated France's political and economic weakness with scientific and technological backwardness. As Olivier Wieviorka has argued, de Gaulle believed that only intensive scientific research could solve this problem. One of his first acts as president was to promulgate a decree restructuring the organization of state-sponsored scientific research.[2] For de Gaulle, scientific research led directly to technological development. And he linked technological prowess to political status in no uncertain terms: "We are in the epoch of technology. A state does not count if it does not bring something to the world that contributes to the technological progress of the world."[3] Technological prowess could therefore serve as an important foundation for international diplomacy. Losing so many former colonies had seriously diminished France's worldwide radiance, but the nation could recuperate much of its lost prestige by offering technological assistance to developing nations. Indeed, doing so would help break the hegemony of the United States and the Soviet Union over the rest of the globe.[4]

Of course, for technologies to function effectively in these symbolic and diplomatic roles, they had to be French. Here too de Gaulle's rhetoric paralleled that of technologists. "Being the French people, we must reach the rank of a great industrial state or resign ourselves to decline. Our choice is made. Our development is in progress," he declared in 1960. But achieving a properly French course was somewhat trickier. According to Wieviorka, de Gaulle viewed technology as a double-edge sword: it had the power to wreak social and cultural havoc, but human choice could also turn it into a tremendously useful political, cultural, and economic tool. De Gaulle particularly feared the homogenizing power of widespread technological development. He wanted to ensure that France would remain unique. In a 1965 memo he asked his advisors: "In the scientific, technological, [and] economic competition in which the world is engaged, 1) by which means, in which areas, and to what extent is our national character threatened (in particular by the United States)? 2) in which directions (research, technology, economic sectors) should we direct our principal efforts in order to maintain, and, if possible, develop our national character?"[5] For de Gaulle, technology was ultimately malleable. A supremely French technology—one that would both express and develop French identity—constituted an obvious and attainable goal. Again, this goal meshed well with that of many technologists to develop a specifically French technological style. The Gaullist regime provided an ideal framework within which to pursue the technopolitics of national identity and radiance. The question was thus not whether to engage in technological *grands projets* but how to do so.

De Gaulle considered the nuclear program to be the jewel in France's technological crown. He attached special importance to the development of a nuclear *force de frappe* (strike force). He did not harbor the slightest doubts on this score. In 1963 he declared: "The question [in the 1950s] was . . . whether we ourselves would possess these means of dissuasion and these new ferments of economic activity, as we easily could, or if we would hand over to the Anglo-Saxons our chances of life and . . . death on the one hand, and . . . our industrial potential on the other. This question is settled."[6]

De Gaulle's return to power could not have come at a better time for the CEA's nationalist technopolitical regime. The agency as a whole was a favorite of the general's, in part because he had helped create it, and in part because it could fulfill his most cherished dreams of French radiance. The regime's principal mission officially became the production of a nuclear arsenal. No longer needed as a backstage negotiator, Pierre Guillaumat joined de Gaulle's ministerial cabinet in 1958 as Minister of the Army. Later he was a special advisor to the prime minister. It would have been difficult for any institution to have more government support. Not surprisingly, the CEA regime's persistently Gaullist, nationalist outlook would continue to attract important political backing throughout the 1960s.

For EDF's nationalized technopolitical regime, the advent of the Fifth Republic had more complex implications. Many engineers and managers continued to espouse the tenets of their regime, including the dominance of nationalized companies over private industry. But this tenet conflicted on several fronts with those of the government. The Gaullists favored a version of the "policy of champions." They sought to reduce

Table 3.1
French reactors of the 1950s and the 1960s. Values for power represent the maximum potential operating power of the reactors; these do not always correspond to original predictions or average yearly power production. Source: Lamiral 1988.

Reactor	Design type	Power (MW)	Project decided[a]	Ground broken	Operation started
G1 (CEA)	Gas-graphite	7	1952	1955	1956
G2 (CEA)	Gas-graphite	40	1955	1956	1959
G3 (CEA)	Gas-graphite	40	1955	1956	1960
EDF1	Gas-graphite	70	1956	1957	1963
EDF2	Gas-graphite	210	1957	1958	1965
EDF3	Gas-graphite	400	1959	1961	1966
Chooz A (Franco-Belgian project)	Light water	305	1960	1962	1967
Brennilis (CEA)	Heavy water	70	1961	1962	1967
Phenix (CEA)	Breeder	233	1961	1968	1973
EDF4 (later SL1)	Gas-graphite	460	1963	1963	1969
Bugey 1 (EDF)	Gas-graphite	540	1965	1965	1972
EDF5 (later SL2)	Gas-graphite	515	1966	1966	1971
Vandellos (built in Spain by EDF)	Gas-graphite	480	1966	1967	1972
Tihange (Franco-Belgian project)	Light water	870	1968	1969	1975

a. Year of decision to build.

domestic competition between companies and to help the strongest private companies become more powerful presences in international markets. Such ideas gained considerable ground over the course of the 1960s, not just in the private sector but also in the utility. Within EDF itself a new group of economist-managers rose to power, slowly in the 1950s and more rapidly in the 1960s.[7] These men sought to recast EDF's regime in a mold that would be more favorable to capitalist industry. They advocated a somewhat modified version of the "policy of champions" toward such ends. While they continued to promote the concept of nationalization, they had little use for that concept's social(ist) implications. Their rise led to intense debates over the meaning of nationalization, and to a political and professional split within the utility between economists and engineers.

One of these economist-managers was Pierre Massé, who had served as the utility's associate director general for more than ten years when he was named Plan commissioner in 1959. In the early years of his reign, de Gaulle repeatedly declared his faith in state planning. By appointing Massé, he sought to restore the significance and prestige of the Plan, the influence of which had waned considerably by the end of the Fourth Republic. Under Massé, its mission shifted from reconstruction and modernization to broad social, economic, and industrial development.[8] Massé's appointment gave EDF's remaining economist-managers a powerful ally within the state. He promoted at a national level the design practices and industrial policies they pursued at the institutional level, thereby supporting and legitimating shifts in the utility's technopolitical regime. EDF nuclear engineers changed their language and techniques to accommodate these shifts, but in the process they lost a great deal of their influence over programmatic issues.

How did the technologies of the program (including artifacts, practices, and forms of organization) come to constitute technopolitics? And how did these technopolitics reflect and shape broader debates about France's identity and future? Between 1955 and 1969, EDF and the CEA collaborated in designing five more gas-graphite reactors (table 3.1). Tracing the shifting terms of debates throughout this collaboration will help me address these questions.

Technopolitics from the Fourth to the Fifth Republic: EDF2 and EDF3

For better or worse, the EDF and CEA regimes had to continue collaborating on the gas-graphite program. The EDF1 project established a rocky but manageable working relationship. As the foundations for this reactor were being laid at Chinon, engineers began contemplating the next two reactors destined for the same site. Once again, the differences in the two technopolitical regimes manifested themselves. This time, the central object of dispute was reactor power. Engineers imbued the number of megawatts that future reactors would produce with a variety of political and industrial meanings.

Engineers at EDF's Direction de l'Equipement (the division in charge of plant design and construction) continued to concentrate on producing as much electricity as possible. They therefore focused on *rendement* (output). From this perspective, it made sense to make EDF2 more powerful than its predecessor. The utility's strategy for developing conventional power—which involved increasing the output of successive plants—supported this tactic. Furthermore, EDF engineers saw themselves as competing with their British counterparts, who also pursued this strategy in nuclear development. Operating in a technopolitical regime directed at producing energy, Equipement engineers measured technological prowess by power output. Industrial habit, national competition, and institutional pride all clearly indicated, therefore, that France should steadily increase the power of its reactors. In a design hastily drafted in September 1956, EDF engineers proposed a 100-megawatt reactor that would use 150 tons of uranium encased in a spherical metal pressure vessel and cooled by carbon dioxide flowing through the core at a pressure of 35 kilograms per square centimeter.[9] They presented these parameters to their CEA counterparts for review.

Though CEA engineers conceded that EDF reactors had to produce electricity, they also wanted more plutonium. This goal remained tacit—though not exactly secret—until the French bomb was announced; thereafter it became quite explicit. The CEA engineers hoped to design a reactor that would fulfill both purposes simultaneously.[10] To this end, they revised EDF's proposal by adding more than 100 tons of uranium to the core and halving the pressure of the cooling gas. The revised design would produce 14 percent more electricity than the original one, using nearly 70 percent more fuel.[11] The extra uranium, presumably, would be converted into plutonium.

This new design contradicted the very essence of what EDF's regime held up as good engineering practice: the idea of *not* trying to extract as much energy as possible out of fuel seemed scandalous. If the CEA could actually supply over 100 extra tons of uranium, EDF should get more than a meager 14 extra megawatts in return. Furthermore, since drafting their preliminary design, utility engineers had held several discussions with their British counterparts. The British had developed a fuel rod for their gas-graphite reactors that could stay in the reactor longer, producing more energy and leading to even greater fuel efficiency. Upon hearing this news, EDF engineers thought that the French should at least match (if not exceed) the British in this domain, and hoped that their CEA colleagues could design an equally impressive fuel rod. Working furiously to devise a counter-proposal before the next meeting with the CEA, EDF engineers triumphantly produced a new reactor project. This version would use only 2 tons more uranium than the CEA had proposed, yet it could produce 167 MW of electricity. This performance, however, was predicated on the CEA's ability to design a fuel rod similar to the British one.[12]

CEA designers disliked this counter-proposal for several reasons. They too wanted to match or outdo the British, but their regime had a different criterion for prowess: reliability, not power. Britain had just experienced a serious accident at its Windscale reactor, so the French would outdo the British even if they simply managed to build a power reactor that ran without failure. CEA engineers feared that the much larger reactor proposed by EDF would increase the chance of problems in construction and operation. They especially feared that they might have difficulties designing the fuel rods posited by EDF. If their fuel rods were inadequate, the reactor would have to be stopped frequently for reloading. This, in turn, would result in a drop in availability, which would certainly undermine the image of France's nuclear program, and for which CEA engineers did not want the blame.[13] Most of all, though, a more powerful reactor would make extracting weapons-grade plutonium from the reactor's spent fuel extremely difficult, since running a reactor at higher power meant that the plutonium produced by the uranium fissioning inside the core would itself turn into other, non-weapons-grade isotopes.[14]

Debates between the two groups of engineers continued for several months as they countered each other's proposals and attempted to reach a compromise. The EDF team kept pushing the power threshold higher, while the CEA team grew more explicit and forceful in demanding plutonium. The final design, settled in April 1958, represented both a technical and a rhetorical compromise between the two institutions. The reactor core, encased in a spherical metal pressure vessel, would contain 251 tons of uranium, cooled by CO_2 at a pressure of $27 \ kg/cm^2$. The reactor would run two alternators of 125 MW each, but instead of publicizing all 250 MW, designers would bill EDF2 as a 175-MW reactor. Hence, EDF could get the prestige of running a powerful reactor and retain the option of eventually extracting even more power from it. The CEA, meanwhile, had more leeway in designing new fuel rods. And, as later became clear, since EDF had only committed itself to extracting 175 MW from EDF2, the CEA would have an easier time getting some plutonium out of it. Thus, EDF2's design features were inextricably bound to the politics of national prestige and of industrial and military production. As a hybrid not only of technology and politics, but also of two technopolitical regimes, EDF2 itself would become part of the technopolitics of ensuing battles over the program's future.

The compromise achieved with EDF2 had in no way tempered the goals of either team. Disagreements resumed during the CEA-EDF meeting held in May to discuss the next reactor. The EDF team wanted to

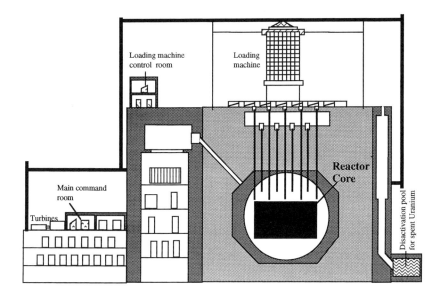

Figure 3.1
A schematic diagram of EDF2 (not to scale). Source: C. Bienvenu et al., "Les centrales nucléaires EDF 2, EDF3, EDF4," Conférence de Genève, 1964. Drawing by Jay Slagle.

make another big power leap with EDF3 by using two 250-MW alternators (the latest innovation in conventional power production) to run the reactor at 500 MW. Raising the same objections as they had in the previous round, CEA designers resisted and argued that a 250-MW reactor would suffice.[15] EDF engineers countered that if EDF3 ran at 500 MW it would be the world's most powerful reactor—a highly significant national achievement that would fit well within the CEA's regime. Once again, invoking the nation proved compelling. The CEA offered a compromise at 375 MW: this way, EDF3 would still achieve the distinction of being the most powerful of reactors, but instead of two new alternators it would use three of the more tried and true 125-MW alternators.

A technological mishap introduced a new hurdle into the discussions before anyone could settle the matter, however. Early one morning in February 1959, an explosion ripped through the Chinon site as a huge crack suddenly appeared in EDF1's spherical steel containment vessel, which builders had almost finished welding together. The cause of the problem—inadequate thermal treatment—soon became clear.[16] The solution, however, eluded both the private company building the containment structure (Levivier) and EDF. Of course, the incident also greatly

Figure 3.2
An aerial view of the Chinon site in 1966. EDF1 (the sphere), EDF2, and EDF 3
are lined up along the bank of the Loire. This image and others like it were used
in EDF's annual reports, in brochures for the Chinon site, and on postcards
sold in the region. Source: EDF Photothèque.

embarrassed both outfits. It received extensive press coverage (in which
much was made of the fact that the accident had occurred on Friday the
thirteenth), and for several weeks EDF and Levivier were the laughing-
stock of the CEA.[17]

The Chinon incident soon became technopolitical ammunition for
the CEA and others to use against EDF. Some engineers at the CEA
immediately blamed the crack on EDF's stubbornness in choosing to
build the containment vessel out of steel rather than prestressed concrete.
Indeed, the material of the vessel had been another bone of contention
during EDF1 negotiations: the CEA (promoting, as usual, its "policy of
champions") had argued that using prestressed concrete would enable
French industry to develop its strength in that area, and EDF had countered that steel was cheaper and better. Not only was the vessel material
already part of the technopolitics pursued by both institutions; the incident also enabled the CEA to reopen the issue of industrial contracting.
The problem would have never occurred, some argued, had the utility

not insisted on being its own general contractor rather than appointing a private company for that job.[18]

Perhaps most significantly, however, the accident shifted debates about nuclear technology into a more public arena. EDF as a whole had faced plenty of public criticism since its creation, but this was the first real mishap that the *nuclear* program had encountered.[19] Significantly, it represented the first time that the state criticized the nuclear branch of EDF's technopolitical regime. Planning commissioner Pierre Massé questioned the judgment of this regime and suggested that it rethink its relationship with private industry. Massé and several private industrialists told the Direction de l'Equipement that they did not like the idea of building increasingly powerful reactors, as engineers proposed to do with EDF2 and EDF3. Better, they said, to follow the CEA's advice and build a larger number of small reactors. Their reasons, however, did not emphasize reliability (the criterion proposed by the CEA's regime) so much as the need to give private companies more experience in the nuclear sector—again, in the interest of the nation (specifically, its economy).[20] In effect, Massé and the private companies proposed a revision of EDF's technopolitical regime away from a leftist version of nationalization. EDF's nuclear engineers[21] mistrusted this proposal, suspecting private industry of trying to seize technical terrain and make commercial gains without regard for the long-term future of the nation and its technological development.[22] Capitalism, in other words, was trying to overrule nationalized industry so that it could profit at the expense of the public and the national good.

The EDF1 incident reveals another aspect of the mechanisms of technopolitics. A technological event—a crack in a steel containment vessel— became a tool in a broader debate over industrial policy. This could happen precisely because the technology already had multiple political meanings. Had this not been the case, the failure of an EDF1 component would not have made good technopolitics. Under the circumstances, however, the cracked vessel provided an event that served to make and justify a critique of EDF's technopolitical regime. The undeniable materiality of the event dramatically strengthened this critique. Previous arguments for or against this regime had rested almost entirely on ideology and theory. The crack provided material evidence—which carried superior weight in both regimes. EDF's nuclear engineers could not ignore this critique. Nor could they deal with it effectively simply by pointing once again to their special sense of public responsibility as state engineers and members of a nationalized institution. They too would need other evidence to defend themselves.

Optimization and the Competitive Kilowatt-Hour

To defend their technopolitical regime and the France it sought to produce, EDF nuclear engineers introduced modeling techniques into their design practices. At first they used economic studies and models mainly as rhetoric, to justify decisions they had already made. As they became more proficient with economic analysis, however, its techniques began to permeate their decision-making processes. Soon they began to use models that incorporated both technical and economic data not only to justify but also to make choices about reactor design and program development. What started as a rhetorical defense weapon ended up as an integral part of EDF's nuclear technopolitical regime. Throughout the 1960s, it enabled this regime not only to make and defend choices, but also to reshape the terms of the nuclear debate.

As we saw in chapter 1, these modeling techniques derived their technopolitical credibility and visibility from their role in developing the fourth and fifth plans. Their origins in France, however, lay within EDF itself. According to his own account, Pierre Massé was largely responsible for combining the techniques of econometrics[23] and systems analysis[24] and introducing them to the French industrial world. As EDF's Associate Director General, he had elaborated economic optimization models and employed systems analysis to help design EDF's distribution network and regulate its overall system of energy production.[25] Massé's models had won the respect of engineers and managers throughout the utility, and recognition in private industry as well as abroad. When he moved from EDF to the Plan, he left a solid legacy of systems analysis and economic modeling behind, mainly concentrated in a division specially devoted to studying the economics of energy supply: the Service des Etudes Economiques Générales (SEEG), run by the brilliant young economist Marcel Boiteux.[26] The SEEG's main tasks were to forecast the nation's electricity demand, to analyze external factors that would influence the cost and pricing of electricity production, and to prepare management and rationalization tools to help "optimize" the electricity production and distribution system.[27] Initially, most of these economic studies were directed less toward developing accurate forecasts of energy demand than toward convincing those outside EDF that demand would in fact rise. As Boiteux told one of his economists, "the important thing is to convince."[28] With Massé directing the Plan, the SEEG certainly provided EDF with the tools it needed to persuade the government to support its development programs.

By the early 1960s, systems analysis and economic modeling were well established as powerful means of technopolitical persuasion. With the rising prestige of the Plan, the mode of reasoning represented by these techniques gained more and more respect and credibility throughout French industry. Its power derived from several synchronous and related developments: the increased prestige of numerical analysis, the growth and mathematization of economics, the ability of models to provide apparently objective solutions to otherwise unmanageable problems, and the increasing numbers of econometrically trained men in state institutions.[29] At least one of the high-level managers in charge of nuclear affairs at EDF—Pierre Ailleret—had intimate knowledge of these techniques. In the 1950s he convinced the SEEG to change how it predicted the growth of energy demand. Uses for electric power were developing so rapidly, he argued, that consumption would increase geometrically rather than linearly. He predicted that it would double every ten years, and what started out as a simple forecast became, in Ailleret's own word, a "doctrine"—a basis for demand forecast and therefore for programmatic development.[30]

Ailleret differed from Massé in his opinion of the nuclear program. Massé had a lukewarm attitude toward nuclear energy in the 1950s and the 1960s: he thought it had a future, but a distant one, and he doubted the wisdom of pursuing the gas-graphite line (or so he claimed retrospectively).[31] Ailleret, as we have seen, had nothing but enthusiasm for nuclear technology—so much so that nuclear reactors were originally known at EDF as "Ailleret's playthings."[32] Recall that he had persuaded the CEA to include "energy recuperation installations" in the Marcoule reactors. He had been closely involved in EDF's nuclear activities ever since, most notably heading the utility's Nuclear Energy Committee, where the top nuclear engineers and managers made design and development decisions. He therefore had a tremendous stake in the success of EDF's gas-graphite reactors.

When EDF's nuclear regime began to face serious criticism after the Chinon mishap, Ailleret suggested conducting a few simple "economic studies" to compare the capital costs (i.e., the amount of investment required) of a second reactor based on EDF1 with those of EDF2's proposed design—in other words, to compare the policy advocated by the CEA, private industry, and the Plan with that advocated by EDF's nuclear team. Regardless of the results, Ailleret said, "it is easy to justify our present policy by saying that we are building nuclear plants larger than those originally planned, but further apart [in time]."[33] Completed a few months later, the study showed that the capital costs of EDF2 (30 billion francs) were less than twice those of EDF1 (16 billion francs). Considering that EDF2 would produce nearly three times as much electricity, Ailleret and his nuclear engineers concluded that copying the first reactor hardly seemed worthwhile.[34] EDF2 went ahead as planned.

The favorable outcome of this study encouraged the Nuclear Energy Committee members. They promptly launched a more sophisticated study aimed at settling the EDF3 quarrel. This study analyzed the costs of both proposed designs. To carry this out, EDF asked private companies to bid on both designs. Incorporating these bids into the models showed the CEA and the Planning Commission that EDF's solution, a reactor with two 250-MW alternators, was cheaper. Utility engineers marshaled additional arguments to bolster their position. A smaller reactor would still not reach the "industrial stage," and the more reactors they had to build to get there, the more money they would have to spend on research and development. Furthermore, nuclear agencies outside France had begun to treat 500 MW as the appropriate power threshold for gas-graphite reactors. The British had just announced that they would build a plant of that size, and the International Atomic Energy Agency in Vienna was using a 500-MW reactor as the gas-graphite prototype in its cost analysis of different reactor types. Clearly, 500 MW was the goal to shoot for, and the sooner EDF got there, the better for the French nuclear program and the nation. In the end, the two institutions reached a compromise: they would adopt EDF's technical solution, but they would announce a figure of 375 MW. This would give the CEA a sufficient margin of error (both technically, in terms of designing the appropriate fuel rod, and politically, in terms of not losing face for France should its engineers fail in this endeavor). At the same time, it would give EDF the experience of building a sufficiently large reactor.[35] Meanwhile, EDF engineers assured one another in a closed meeting that they would later do everything possible to try to run the new reactor at its full power capacity of 500 MW.[36]

This study of EDF3's cost greatly aided the utility's nuclear team to obtain its desired design. The analysis was still fairly rudimentary, though, providing only the roughest estimate of the costs of the reactor's largest components. Its main value to engineers lay in its rhetorical powers of persuasion. Yet together with the EDF2 study, it also had another effect: that of introducing nuclear engineers to the general ideas behind systems

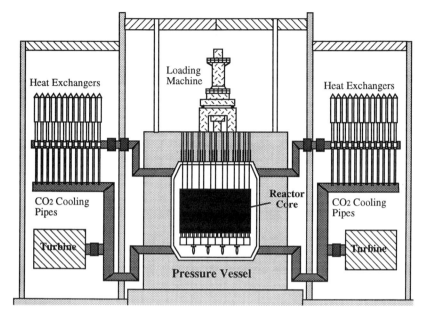

Figure 3.3
A schematic diagram of EDF3 (not to scale). Source: C. Bienvenu et al., "Les centrales nucléaires EDF 2, EDF3, EDF4," Conférence de Genève, 1964. Drawing by Jay Slagle.

analysis. In fact, one kind of systems analysis, known to the utility as "optimization studies," had been used for several years elsewhere at EDF in designing conventional power plants. These studies (which were somewhat more sophisticated than the one engineers had run for EDF3) broke down the cost of a power plant into the cost of its individual components. They then minimized the overall cost, either by finding ways to lower the costs of specific components or by redesigning certain components so that the whole plant would produce more power. They thus helped engineers find an "optimum" relationship between a plant's cost and its power output. These methods were not applied to reactor design until 1960, in part because so many variables were involved in reactor design that calculations were extremely difficult to perform with mere adding machines. The arrival of computers in EDF's research facilities changed this state of affairs.[37]

Indeed, just when the use of more complex forms of systems analysis became politically popular through Massé's Plan, computers made it possible for engineers to conduct optimization studies for their reactors.

Figure 3.4
Workers building the graphite pile for EDF3. The graphite casings are stacked on top of one another. The fuel rods will go in the casing holes. Photograph: Michel Brigaud, 1965. Source: EDF Photothèque.

Running the studies led to substantial changes in the design practices of EDF's nuclear engineers. Years later, one engineer described the design process before and after the use of optimization. Before, he said, "the whole trick . . . was to find the best compromise possible, without much economic data. . . . It was a mixture of common sense [and] intuition." Computers and optimization studies changed everything. "Until then these calculations had been done by hand. I'd had a young woman engineer with me [who ran most of these numbers]. . . . At the time, the people behind the computers wore white coats. [They] took your calculations, a bit like a doctor would see you for a visit. . . . The machine put out for me in one run what the young woman engineer would have taken two years to do. . . . We could 'play' in a much more sophisticated way. . . ."[38] This increased sophistication meant that engineers could plug different design options into models in order to test which option would best suit their purposes.[39]

Figure 3.5
Workers hooking up part of the cooling system for EDF3. Photograph: H. Baranger, 1964. Source: EDF Photothèque.

The first optimization study done for reactors covered the design of EDF3. Its guiding principle was simultaneously technological and economic: to maximize the reactor's power while minimizing the volume of its core. This reflected the design traditions of utility engineers as well as their interest in getting as much energy as possible out of their fuel. Using reference costs provided by the SEEG and relationships among various core dimensions, engineers tried out different core configurations to calculate how to derive the most power from the least uranium. They could hence "prove" their assertion that increasing the unit power of a nuclear plant decreased the overall price per kilowatt of electricity.[40] Such proof greatly weakened the arguments of private industry and the CEA in favor of smaller reactors.[41]

Optimization studies thus refined how EDF engineers conducted their technopolitics. What started as a rhetorical device had now become an integral part of design practice. The means helped justify the ends. This

Figure 3.6
Workers mounting the concrete blocks for EDF3's pressure vessel. Photograph:
H. Baranger, 1964. Source: EDF Photothèque.

shift helped EDF engineers present a reactor design optimized to fulfill
their own goals as *the* optimal design, with those goals safely buried in the
model. Nothing in the *models* prevented them from being used to optimize reactors for maximum plutonium yield or any number of other possible criteria. But the CEA did not seize upon these models. Thus, only
EDF designs carried the prestigious label "optimum." Robert Frost and
others have described the battles fought by EDF economists beginning in
the 1950s to make their pricing schemes politically acceptable and the
role played by the notion and techniques of the "optimum" in those battles.[42] By the early 1960s, those battles had been won, and the idea of the
"optimum" had emerged victorious. Indeed, it had acquired an almost
magical aura, again helped by Pierre Massé's position in the Plan. Even
the finance ministry had been persuaded by the power of EDF's models:
a high-ranking official of the Ministry of Industry later recalled that "EDF
was . . . one of the first big enterprises to have done in-depth techno-
economic studies. It is important to underscore this point, as [these stud-
ies] were very much appreciated by [the Ministry of] Finance. . . ."[43]

Everyone agreed that "optimal" technologies outranked others—and the only way to get an optimal technology was to run a model.

In one sense, these models functioned as microcosmic laboratories. They provided a way of testing different reactor designs without having to go through the extreme expense and effort of building them. Bruno Latour has argued that laboratories give scientists a resource that politicians can never have: the ability to make mistakes without suffering public humiliation. It is precisely in this specificity of science, Latour notes, that its power lies.[44] A similar argument holds for these optimization models. In the early 1960s, there were no hard operational data about the economics of nuclear power—sizable nuclear power plants had only just begun to come on line elsewhere in the world. Statements about the relative economic merits of different designs were, at best, educated speculations. The models, however, provided a widely respected method of "experimentation." By constructing different scenarios and extrapolating data, the models gave these speculations technical heft. They made calculations on paper acceptable substitutes for industrial experience. In the absence of operational data, then, the models gave EDF engineers a way to demonstrate the superiority of their designs: by transforming judgment calls into matters of fact,[45] they constituted an ideal means of technopolitics.

Learning the language and techniques of optimization helped utility engineers navigate the changing waters of the early 1960s. As the nation's priorities shifted from reconstruction and maximum production to economic efficiency and market-oriented production, EDF's overarching mission shifted from *rendement* (producing as much electricity as possible) to *rentabilité* (producing the cheapest possible electricity).[46] For nuclear engineers, this meant that the cost of the kilowatt-hour (i.e., the unit cost of the electricity actually generated by a plant in real time once it goes on line) became more important. Correspondingly, the cost of the kilowatt (i.e., the total capital cost of building the plant divided by the maximum number of kilowatts it can produce at any given time) diminished in importance. Increasingly, the power of a plant or how much it cost to build mattered less than the price of the electricity it produced, which combined these two quantities with others. This posed a problem, because none of EDF's reactors had actually begun to produce electricity (the first one, EDF1, wouldn't do so until 1963). At best, engineers could offer predictions.

Optimization studies offered a solution by helping EDF engineers turn the calculation of the nuclear kilowatt-hour into a "doable" problem. In the process, however, the EDF engineers had to reshape their technopolitical goals. Instead of aiming at a physical definition of efficiency (producing the maximum power in the most thermodynamically efficient manner), they began to search for a more economic concept of efficiency.[47] Optimization studies could not, of course, yield the actual cost of producing a nuclear kilowatt-hour. They did, however, give engineers a pseudo-experimental technique that enabled them to credibly predict that cost for different design options. Through this process, EDF engineers developed a new technopolitical goal: producing nuclear electricity that would compete economically with conventional power. Very quickly, comparisons between the two forms of power began to dominate their work. One participant said: "We lived in economic comparisons, in comparisons of the cost of the kilowatt-hour."[48]

Comparing nuclear power with conventional power by means of optimization studies gave nuclear engineers new ways of both discussing and practicing their work. The change became evident in 1961 as they began to contemplate EDF4. Until then, they had aimed to make each plant more powerful than its predecessor. With EDF4, however, they decided to stay at the same power level. Not only was 500 MW an internationally accepted threshold for gas-graphite reactors (which allayed fears of lagging behind other countries); in addition, the engineers' priorities had changed. Now they would try to make the reactor "competitive."[49] They saw two possible routes to this goal: either they could copy EDF3, improving each component as much as possible without changing the general parameters, or they could produce a radically different design that would place the heat exchangers inside the pressure vessel, underneath the reactor core. In June 1961 they expected their choice to depend not only on the price of each design but also on the future of the energy market. They did not expect copying EDF3 to save much money in the long run, but they felt it would provide an easier and faster short-term solution should an oil shortage develop in the next few years. If, however, prospectors discovered a significant reservoir of oil or natural gas in the Sahara, then the second option, which they expected to yield considerable cost savings in the long run, would be a better solution.[50] Aiming to compete with conventional power plants thus added the art of prediction to the art of designing a nuclear plant and drew on the forecasting techniques of the Plan. And the price of conventional fuel and power was not all that required forecasting. Designers also had to include predictions for fluctuations in interest rates and amortization periods. How long could they expect their reactors to produce electricity? Some predicted 20 years, others 35, but no one really knew. The most important thing, given these uncertainties, was to devise a credible, persuasive model. Showing that new reactors would compete economically with conventional power plants would go a long way toward acquiring political support for those reactors. Optimization studies thus made technopolitics more complex by inextricably linking the goals and the practices of EDF's nuclear engineers. Having proved themselves politically efficacious in attaining predefined goals, these models also provided the impetus and means for engineers to reshape their goals.[51] The models did not invent the entity of the "competitive nuclear kilowatt-hour"; however, they made its existence possible by enabling engineers to produce the number that would define that entity. The existence of this entity, in turn, helped EDF's nuclear engineers garner political support outside the utility. No one—not the Plan, not the CEA, not even de Gaulle—would deny that a competitive French nuclear kilowatt-hour constituted a desirable goal.

Optimization technopolitics also served EDF engineers in the ongoing quarrels between EDF and the CEA over fuel and plutonium production. De Gaulle's inexorable pursuit of the military atom helped the CEA to impose its plutonium requests on EDF. But EDF's design practices, politically supported as they were by the priorities and interests of the Plan and the Ministry of Finance, and entwined as they were with the goal of "competitive" nuclear power, enabled the utility to shift the terms of debates about reactor fuel and plutonium from production to cost.

Controlling Fuel and Pricing Plutonium

"So I hear you've been asking about plutonium," said one CEA engineer as we shook hands at our first meeting. "You know, EDF made some plutonium too." I was too stunned to reply at first. Then I smiled weakly and assured him that I wanted to hear all about it. I sat down and tried to calm myself as I set up the tape recorder. What had shocked me was not the revelation about EDF plutonium. This was something that every gas-graphite engineer knew, and treated either as a dirty public secret ("This may surprise you, mademoiselle, but they made plutonium at Chinon too") or as a completely uninteresting fact ("Oh yes, sure, the Chinon reactors made some plutonium—but that was technologically inevitable, you know"). What had shocked me was the fact that my interview subjects had evidently been talking about me and my questions. Clearly they were watching me as closely as I was questioning them.

Reactor fuel had always been a source of contention between EDF and the CEA. In the early to mid 1960s, as questions that had plagued the relationship between the two institutions became increasingly important, it was one of the main foci of dispute. Who would control the fuel at which point in the cycle? Would the CEA be willing and able to design fuel rods that would perform according to EDF's wishes? How much fuel would each reactor use? CEA technologists strongly felt that their institution had "exclusive responsibility" for providing fuel for all French nuclear plants. The director of the CEA's fuel division put this in no uncertain terms:

> There is no question of . . . establishing a [fuel] supply program in common. It is only a matter of noting the size of the order by type of fuel [rod]; it is the CEA's business to deal with the rest. . . . We have noticed that [EDF] would be ready to claim that it can ensure its [fuel] supply itself. I think that it would be most unfortunate if this inclination were to develop further, because it would remove one of the CEA's primary responsibilities and means of action in this joint effort.[52]

Controlling the fuel cycle, in other words, constituted the most important means by which the CEA's regime participated in the development of nuclear power. Relinquishing technical control over any part of that cycle would also mean relinquishing political influence within the program.

How much fuel would go exclusively toward electricity production, and how much would the CEA remove from the reactors for treatment at Marcoule's plutonium facility? This question caused the most acrimony between engineers in the two regimes: every demand made by the CEA for plutonium from EDF reactors felt like an invasion to EDF engineers, while every resistance on the part of the latter felt like a betrayal to the CEA engineers. Matters were not helped by personality clashes: the heads of EDF's engineering teams, including Claude Bienvenu and Boris Saitcesvsky (who had been involved in EDF's reactor projects from the earliest days at Marcoule), could not abide Jules Horowitz, the *polytechnicien* head of the CEA's Direction des Piles Atomiques. The feeling was entirely mutual.

Though EDF reactors were optimized for producing electricity, the fission reaction in their cores would inevitably yield some plutonium as a byproduct: such was the nature of natural uranium reactors. This technological fact opened up a political possibility for CEA engineers. Perhaps they could persuade EDF to remove some fuel rods before they were fully exhausted (by the standards of power production) in order to extract weapons-grade plutonium? EDF engineers were neither surprised nor thrilled when, in 1960, their CEA colleagues made the first official request to this effect.[53] They would have liked to refuse the request, but de Gaulle's enthusiasm for the *force de frappe* made that impossible. It was now the CEA's turn to invoke the nation: refusing to supply plutonium would have been positively unpatriotic. After extensive discussions with the Ministry of the Armies, the Ministry of Atomic and Space Affairs, the Ministry of Industry, and the CEA, EDF gave in.[54] Once again the technological versatility of the gas-graphite design had served the CEA's regime well. But this agreement of principle did not dictate the *terms* of EDF's cooperation. In negotiating these terms, EDF's nuclear engineers used their newly developed practices and goals to redefine the economic and political implications of plutonium production.

As of February 1961, the official arrangement was that (barring an unforeseen problem at the Marcoule reactors) no more than one-sixth of Chinon's fuel would go toward producing military plutonium, and this only as of 1966. Two months later, CEA engineers asked for more: they wanted to use one-fourth of Chinon's fuel capacity. Both EDF and its protectors at the Ministry of Industry protested vigorously, arguing that this quantity would seriously impair EDF's ability to derive adequate operational experience from its own reactors. Already, EDF's Nuclear Energy Committee had begun thinking about various forms of compensation. One option involved calculating how much energy would be lost by removing fuel rods before full irradiation and pricing that energy on the basis of the cost of producing the equivalent amount in a coal plant.[55] This kind of formula seemed straightforward enough, and the CEA had been willing from the beginning to contemplate financial compensation in the form of a rather vaguely conceived and ill-defined "plutonium credit."[56] In April 1962, the two institutions redrew their agreement. The CEA would give EDF a set number of specially designed fuel rods reserved for plutonium production; it would pay for changes that EDF had to make in the fuel loading machines of both EDF1 and EDF3 to facilitate this production; the two institutions would set a limit on how many rods of each type would go into the reactors; and both institutions would evaluate the "inconveniences" caused in the operation of the reactors by plutonium production and establish compensatory measures.[57]

These measures remained undefined, however, and the escalation in the dispute became almost humorous as each engineering team thought of more and more factors that simply *had* to enter the calculations. Soon after the second agreement, for example, the CEA resurrected the possibility of civilian uses for plutonium. It had begun to work on Rapsodie, its first experimental breeder reactor, which ran on plutonium. CEA engineers argued that Chinon's plutonium might go to Rapsodie, and eventually to future breeders. Since these breeders would generate electricity, it would be in EDF's financial interest to produce plutonium. Compensation would have to take this into account. Undeterred, EDF engineers calculated the financial benefit that the CEA would derive from processing the spent fuel and argued that this had to diminish whatever price EDF might eventually pay for breeder fuel.

CEA engineers tried another tack. Starting up any reactor involved adjusting the amount and distribution of fuel in the core, which in turn required removing some fuel rods before they were fully irradiated. Thus, the CEA argued, any reactor startup led to the production of at least some plutonium that would, almost incidentally, be of weapons grade (this was known as *plutonium fatal,* in the sense of "inevitable" or "fated"). Since EDF reactors would produce this plutonium on startup no matter what, the utility should not include it in its compensation calculations. But EDF responded that it could devise a startup phase that would *not* produce such plutonium. Besides, argued the Nuclear Energy Committee, it was still too early to know for certain how much military plutonium Chinon would produce, since EDF1 had not even begun operating yet. "Right now," it argued in September 1962, "it is not a question of proving anything, but of determining *very objectively* the different losses of information that could result from the presence of sub-irradiated fuel."[58]

"Very objectively" in this case meant that EDF engineers wanted to assign a financial value to the loss of information they would suffer by giving plutonium to the CEA. This information had value on two counts: it would help them design EDF5, and it would enable them to develop better estimates for the economics of future reactors. The value of knowledge thus had to be quantified. In a process similar to the technopolitics of the competitive nuclear kilowatt-hour, EDF engineers were trying to reshape the terms of the debate in order to turn the liability represented by the CEA's plutonium demands into an asset.

Indeed, it occurred to them that plutonium production could be made to help rather than hinder the competitiveness of nuclear power plants. After all, the CEA agreed that making plutonium had financial consequences for power generation. What if these could be brought into the calculation of the nuclear kilowatt-hour in a systematic fashion? Excited by this possibility, the Nuclear Energy Committee proceeded to appropriate and refashion the notion of a "plutonium credit" in order to make EDF's reactors more cost effective.[59]

The plutonium credit had both technological and economic dimensions. The amount of plutonium produced depended on how long fuel rods stayed in the reactor core and on their level of irradiation. Too long a stay or too high an irradiation level would not produce useful weapons-grade material, and these conditions also held for material destined for breeder reactors. Producing a large amount of electricity involved keeping fuel rods in the core much longer and irradiating them at a higher level. But privileging the cost of the kilowatt-hour over the quantity of electricity produced meant that EDF engineers did not necessarily want to maximize the irradiation levels or the lengths of stay. Rather, they wanted to *optimize* these quantities in order to get the lowest possible kilowatt-hour cost. Once the fuel left the reactor, it eventually went back to the CEA for treatment or processing. EDF engineers wanted to create a scale that would assign a value to this fuel according to the amount of plutonium it contained. They could then take these figures into account in calculating a reactor's fuel cycle. Thus EDF would get to run its reactors under optimal economic conditions, and the CEA would still get its plutonium.[60] Although exactly how to calculate this scale remained unclear, EDF engineers appeared confident not only that they could come up with such a scale but also that the resulting plutonium credit, together with a new fuel rod design then under development, would make EDF6 competitive.[61]

The matter became even more complicated as the definition of "competitive" began to change and as shifts in the nation's political and economic climate raised the stakes even higher. The economic competitiveness of French technologies on foreign markets was becoming an important political issue. For the nuclear program, this meant that the gas-graphite design now had to compete economically with foreign designs. EDF's nuclear engineers were increasingly confident that they would, in the not-too-distant future, make a nuclear kilowatt-hour that could compete with the conventional one. But could they make a gas-graphite kilowatt-hour that could compete with kilowatt-hours produced by light-water reactors? The greater challenge posed by this goal soon became clear. Once again, EDF engineers invoked the plutonium credit to help them in this task:

The competition that our system is likely to encounter in the near future from the boiling water system leads us to reconsider . . . pricing irradiated fuel [in light of] the prospect of breeder reactors. . . . Pricing breeders makes preparing a stock of plutonium economically interesting. This interest should translate commercially into a "plutonium credit" *on the order of magnitude of the differences in cost between the French system and the American system.*[62]

The plutonium credit, in other words, would also help EDF's gas-graphite reactors compete with American light-water reactors (which did not, after all, produce nearly the quantity or quality of plutonium that gas-graphite reactors did).

The CEA did not deny that a competitive kilowatt-hour (of either type) was a desirable goal for the French nuclear program. But its engineers and directors resisted implementing EDF's plutonium credit. Perhaps because CEA leaders had come to recognize the ever-increasing techno-political importance of economic modeling as a mode of reasoning, the CEA had started its own small division of economic studies. Economists there set to work refuting the EDF's figures. Their studies argued that the high degree of uncertainty about plutonium's technical characteristics and economic future did not justify fixing a price for it. For one thing, a liberal, supply-and-demand type of market for plutonium did not exist. Even the United States and Britain—the countries with the most experience in producing and using plutonium—constantly changed their plutonium prices. For another thing, the "use value" of plutonium depended on which kind of reactor had produced it, the operational conditions of the specific reactor, the cost of processing the plutonium, and the element's end use. In other words, these reports implied, EDF had gotten it backwards: the value of plutonium could not be determined until the metal had completed the cycle for which it was destined. It was *"neither necessary, nor desirable* and in any case difficult," one report concluded, "to assign plutonium a price and to base a development policy for nuclear energy on [that price]."[63]

The two institutions had reached an impasse that could not be resolved by the men directly involved with the research, development, and operation of the reactors in question. EDF engineers had managed to renegotiate the terms of the plutonium dispute to include broad economic criteria, but this renegotiation did not necessarily arbitrate in its favor. Indeed, the economic language just highlighted the differences between the two regimes once again: the CEA advocated a more market-oriented approach to pricing plutonium, while EDF in effect wanted to use the price of the metal as another kind of state subsidy for nuclear power generation. These competing ideas about the market undergirded the systems analysis conducted by each regime, which meant that neither could prescribe a policy agreeable to both regimes. The matter (along with several other disputed issues) traveled to the highest administrative reaches of each institution. Finally, in mid 1965, the CEA's Administrator-General and EDF's Director-General signed a series of accords governing their financial relations. In the end, the plutonium agreement favored the CEA's approach more. The two men agreed to determine the price that the CEA would pay EDF for spent fuel on a yearly basis, using the year's current international rate for plutonium as an index. There would not, in other words, be a fixed, predetermined plutonium credit. But EDF engineers did receive some compensation: the CEA guaranteed that it would provide EDF with fuel of a specified quality; if the fuel didn't yield as much energy as promised, it would reimburse EDF accordingly.[64]

Industrial Competitiveness, Exporting Reactors, and the Future of France

In each case that I have examined so far, the technopolitics pursued by the two regimes gave them a distinctive voice in a broader debate about France's industrial and political future. Some of the technopolitical entities they created—including the competitive nuclear kilowatt-hour—became common currency in this broader discourse, reshaping the parameters of debate. Others, including the plutonium credit, did not.

Though necessary for political survival, reshaping the parameters of debate could be dangerous. Once a technopolitical entity became common currency, more players could enlist it (or even change it) to support their goals. This happened to the competitive nuclear kilowatt-hour. In the mid to late 1960s, it was redefined: the reference point for the gas-graphite kilowatt-hour was moved from conventional power to non-gas-graphite forms of nuclear power. This redefined comparison point, in turn, entailed a shift in the focus of debate back to the organization of EDF's reactor projects and its contracting industrial methods. This shift proved fatal to the programmatic authority of gas-graphite engineers, because the new focus involved elements that many other people and institutions could legitimately shape. Ultimately, this shift would cause engineers to lose control over program development policy altogether.

By the mid 1960s, most political and economic leaders agreed that France had recovered from the ravages of the war. But Charles de Gaulle's goal of national grandeur through technological prowess still seemed far off. Most French technologies did not appear to offer the diplomatic possibilities of which de Gaulle had dreamed. If the nation was to steer its own course in the Cold War, then it had to move beyond merely developing an industrial infrastructure. France had, most agreed, to develop technologies that could hold their own in the world's most advanced industrial sectors.

The articulation of this problem within the state had two dimensions, one represented by de Gaulle and one by his prime minister, Georges Pompidou. De Gaulle emphasized national grandeur: for him, it seemed, the symbolic value of industrial and technological development reigned supreme. Above all, he felt, this development had to promote French national independence. Pompidou emphasized instead the economic side of the issue. He felt that, above all, French industry had to become economically "competitive" in the international market: for him, the nation had to forge a distinctive identity primarily through its economic activity. The contrast between these two approaches would become particularly important for the nuclear program at the end of the 1960s. In the middle of the decade, however, these approaches were two sides of the same coin, for both advocated the same kinds of directions for industrial development. Both favored the development of large-scale technological programs (not just in the nuclear arena, but also in aerospace, computers, and electronics).[65] Both also favored a version of the "policy of champions," in which the state encouraged the formation of national industrial "champions" through industrial concentration in key sectors such as steel, electronics, automobiles, or aeronautics.[66] By 1964, de Gaulle had even conceded that the Plan could not, by itself, direct all of France's economic and technological progress. Some impetus had to come from private industry, which in turn needed more financial incentives to make long-term investments.[67] The private and public sectors had to cooperate in order for France to achieve political and economic greatness in the industrial arena. Such ideas had the full support of Pierre Massé and his staff at the Plan, who embedded them within the goals and strategies of the fifth plan just before Massé himself returned to EDF as the utility's new president in 1966.[68]

The most immediate consequence of this emphasis on foreign competition for the nuclear program was an increasing pressure to make French reactors competitive on the international market—in other words, to make gas-graphite plants exportable. The private companies involved in building the reactors had wanted to sell their expertise abroad for a long time. They now had increased state support for such ventures, and they began pressing EDF for help.

In fact, all the main players in the nuclear program fully supported the idea of making their technology exportable. But once again they had different ideas about how best to achieve this goal. Industrial contracting became the main point of dispute. Recall that in the early years of the program, the CEA's regime had favored a "policy of champions" and had hired a private consortium as prime contractor for its Marcoule site, while EDF's regime—mistrusting private companies—had preferred to manage

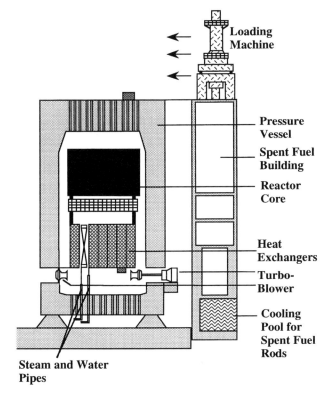

Steam and Water
Pipes

Figure 3.7
A schematic diagram of EDF4 (not to scale). Source: J. Grand and J. Hurtiger, "Aspect de radioprotection pendant les interventions de Saint-Laurent-des-Eaux," *Bulletin de l'ATEN*, no. 91 (1971). Drawing by Jay Slagle.

the bidding and construction process itself. Except for a brief flareup when EDF1's containment vessel cracked in 1959, this issue had lain dormant since the mid 1950s. It was reawakened by the loud, powerful voices clamoring for new kinds of competitiveness. The ensuing debate among engineers in the two regimes mixed personal animosity, anxious defenses of technical turf and expertise, and invocations of the national interest.

The debate started with the negotiations over EDF4. After going through several rather different possibilities and running multiple optimization studies, EDF engineers had decided to go with an "integrated" design. In each of the three Chinon reactors, the heat exchangers stood outside the pressure vessel that contained the core. In EDF4, destined for the utility's new Saint-Laurent-des-Eaux site, the heat exchangers would stand inside the pressure vessel, directly under the core (figure 3.7).

Optimization studies had shown this design to have several advantages, including lower construction costs and increased reliability.[69] Engineers expected EDF4 to have a longer life than its Chinon predecessors, so they reduced their initial payments by extending the amortization period (thereby increasing the reactor's short-term competitiveness).[70] Finally, their primary objective for EDF4's design was a nuclear kilowatt-hour that could compete with conventional power.[71] At the CEA, meanwhile, Jules Horowitz and some of the engineers in his Direction des Piles Atomiques had argued that, clever as the integrated design was, the program would do better simply to build an improved version of EDF3. The third Chinon reactor had encountered several construction problems, which CEA engineers worried had created "a very unfavorable impression of French nuclear technology." "This point," they continued, "seems essential to the CEA, which feels that it is indispensable that the next project demonstrate that a technically viable nuclear plant can be built in France in less than five years under satisfactory economic conditions."[72] The CEA thereby equated the national interest with demonstrable technical competence and argued that EDF4 should entail incremental improvements, not radical innovations. However, the EDF team, which equated the national interest with commercial viability, felt that Chinon's third plant simply could not serve as the model for a series of plants.

Though the EDF4 matter was settled in favor of the utility, engineers there continued to feel plagued by what they saw as Jules Horowitz's increasing encroachment on their territory. They felt he was engaging in "subversive action," trying to worm his way into EDF's work—"often without official instructions from the top, which ratifies [his actions] in case of success, but does not support [them] when EDF reacts violently."[73] The goal appeared to be "CEA hegemony over everything nuclear, and, in particular, the leadership of Mr. Horowitz in the domain of power reactors."[74] In one memo, one of these engineers spent four pages listing Horowitz's transgressions, accusing the CEA of withholding information and not working hard enough on the technical problems that most mattered to EDF. He gave a bitter analysis of the means by which the CEA succeeded in exerting influence (these included the scientific expertise of its personnel, its direct connection to the Prime Minister, and its role in the military program). Now, at a time when it had become extremely important to build reactors that held their own both politically and economically on the international stage, Horowitz wanted to interfere more than ever. EDF engineers hoped to use the issue of export to silence their opponent once and for all. "Where would the CEA proponents of export be today," one engineer asked, "if EDF had adopted Mr. Horowitz's point of view? EDF3 would be limited to 375 MW. . . , [and] EDF4 would be a duplicate of EDF3, a design completely surpassed by the British projects at Olbury and Wylfa."[75]

Horowitz felt equally hostile toward EDF, whose engineers, he said, opposed the CEA out of sheer stubbornness. EDF engineers did not have the proper respect for the CEA's expertise, and sought only to minimize the agency's role in power plant development as quickly as possible. They underestimated the technical significance of their early plants and were overly eager to implement innovations. This eagerness was especially foolhardy since EDF had only about 200 engineers working on the plant projects, "a high proportion of which are recent recruits or of a fairly low level (for example, from the Ecole Polytechnique Féminine)."[76] Clearly, Horowitz felt that women engineers (of which there were scarcely a handful in EDF's nuclear teams) did not have the ability to engage in so complex a task as reactor design.[77] EDF, in short, needed a humbler attitude.[78] The quarrels between the two regimes continued as teams fought over the design of cores and fuel rods for the next two reactors, and over the order in which they should be built.[79] Not everyone at the CEA shared Horowitz's willingness to quarrel over small details, but skepticism about EDF's ability to act in the national interest appeared widespread: "It would be regrettable," wrote the CEA's public relations manager, "if, instead of fighting for a more intimate role in these projects, the CEA . . . lost face in a non-existent battle. The real problem is not the relative order of the two plants, but the degree of confidence in EDF's commitment to the five year plan for developing natural uranium plants, on which hangs the future of civilian nuclear development in our country."[80]

Clearly the quarrels between the two regimes had gotten out of hand. Indeed, as the nuclear program became increasingly important in high-level government debates about France's political economy, these fights became outright embarrassments to top administrators in the two institutions. In an effort to make peace between the two groups of engineers, Robert Hirsch, the CEA's Administrator General, wrote to André Decelle, EDF's Director General, that "the French efforts to export" led the CEA to consider the development of an improved version of EDF4 urgent. Clearly, he conceded, EDF's integrated design had to form the basis for the future of the gas-graphite program over the next five years. The CEA was conscious of the increasing pressure of foreign competition and willing to undertake the research necessary to push gas-graphite design as far as possible.[81] Decelle reacted favorably to this overture, and the two leaders drew up guidelines for cooperating in the effort to make French gas-graphite reactors internationally competitive.

This truce did not have a lasting effect on the engineers in the two regimes, however. By mid 1965 they had begun quarreling again. This time, their confrontations did not revolve around what counted as good design criteria—Hirsch and Decelle had at least managed to silence them on that issue. Instead, they argued about the nuclear program's industrial contracting policy.

EDF's Direction de l'Equipement had continued to function as the prime contractor for reactor projects. For this purpose, it had created two Régions d'Equipement Nucléaire (REN1 and REN2) to supervise reactor construction. Both were run by engineers who had been around since Marcoule. They continued to believe that their nationalized utility could serve the public interest in a way no private company ever would or could. Blending in with the national political climate fostered by Pompidou and de Gaulle, REN engineers did not evoke the leftist dimensions of nationalization; instead they cast their arguments in terms of competitive pricing for foreign markets. EDF, they argued, should remain the prime contractor in order to start bidding wars between companies and keep construction costs down. Leading to cheaper reactors, this structure would serve the interest of foreign competitiveness as well as the public interest.[82]

Backed by CEA engineers, private companies argued otherwise.[83] They said that the policy did not allow them to get the experience they needed to export turnkey reactors. Because no single company had yet had the opportunity to coordinate the construction of an entire reactor, none could actually sell one to a foreign country. At best, they could put in bids for reactor parts. But the organization of nuclear programs in other countries differed substantially from that in France, and the opportunities for such bids were rare or nonexistent.

REN engineers did not accept that only private companies could export French nuclear technology, but they did not pursue this point. They even agreed that the companies should have something salable to export. To this end, they offered some concessions. Designs for three reactors were on the table: two very similar ones at Saint-Laurent, and one rather different reactor at the new Bugey site. Rather than divide the designs up into dozens of subunits and request bids for each one, engineers offered to create larger subunits. Companies could then create medium-size consortia in order to bid for the subunits. For example, in the case of EDF4 (now called Saint-Laurent 1 or SL1) this meant that only seventeen contracts would cover 80 percent of the reactor. REN engineers quite liked this solution: though it made vigilance over cost overruns more difficult, it did continue to foster competition between the consortia, and it left the overall management of the project in their hands.[84]

Private industry and the CEA wanted to push EDF further. They wanted the utility to launch bids for "nuclear boilers." In this scenario, once an EDF team had drafted a preliminary design, it would accept bids from large consortia for the reactor's central heat-generating system: the core, the loading machine, the pressure vessel, and even possibly the control systems (what counted as part of the "boiler" was not completely clear). REN engineers hated this suggestion. One wrote in an angry memo: "Increasingly one hears, especially in the high spheres closer to Politics than to Industry, that EDF is not fulfilling its expected role with respect to French Industry, and in particular that the way it divides contracts prevents the birth or impedes the growth of powerful Consortia, the only ones capable, it appears, of exporting plants abroad. This affirmation, repeated so much that it is becoming dogma, is but a vulgar untruth."[85] The CEA had let itself be influenced too much by politicians, who should not be involved in industrial policy making. And private companies certainly could not be trusted to act in the national interest. In the matter of power plant development, only EDF engineers—by virtue not just of their expertise but also of their place in French society—were trustworthy. Reducing the number of reactor subunits would give private companies the responsibility of coordination, which rightfully "belonged to the state corps . . . because we are the only ones who can do this in an efficient manner."[86] Indeed, proposing the very concept of a "nuclear boiler" revealed the ignorance and incompetence of private companies: whereas boilers represented only 25 percent of a conventional power plant and therefore could reasonably be contracted out as a whole, the equivalent in a nuclear reactor represented 70 percent of the plant and was far more complex.

Defending their contracting methods, REN engineers wove the recent technical problems with their Chinon reactors (especially EDF3) together with their arguments about the social role of state engineers and their ideas about the national interest. The constant breakdowns and delays, they argued, were the fault of private companies. "First of all, you have to have something to export. Whether we like it or not, as long as we in France cannot offer nuclear plants that function normally and give their user, in other words EDF, full satisfaction, then only political pressure or exorbitant financial advantages can lead to the export of nuclear plants."[87] Just as EDF1's cracked vessel had caused political problems for the utility, so had EDF3's heat exchanger problems. EDF engineers, however, did their best to blame the builders, whose lack of experience had caused seemingly unending delays in the reactor's startup.

Ultimately, claimed the engineers, the real issue was who could best represent France. In a passionate statement, one REN2 engineer declared:

> With two exceptions . . . all the plants exported to normally developed countries [*sic*] are American, that is to say designed by a development office and constructed by specialized Builders, most often hired after open, international requests for bids[.] I emphasize this point, because the way that GE and Westinghouse function is much closer to EDF's [methods] than to the way the members of French or English consortia "divide the pie." . . .
>
> When Parliament created EDF in 1946, it made a national Consortium composed of Companies specializing in the construction and operation of power plants, of whatever type. In one fell swoop, this brave gesture put a French Company at the same level as the largest American Companies, with means that no French Industrialist could have dreamed of before. . . .
>
> I think that the success of our Establishment, which is stunning despite all the sarcastic comments . . . just proves everything I have been saying. . . .
>
> It would thus be only natural that we could, of our own accord, sell plants abroad the day that we think it's reasonable to do so, and with the Builders that the Customer and we will have chosen. We know from experience how much confidence we inspire abroad, we who are not tied to any bank, to any factory, [or] to any Consortium—and this is not the case for the Americans. Any other solution would certainly be doomed to failure. . . .
>
> Let us thus play our role, both Abroad and in France. . . . For twenty years, EDF has forged for itself competent, devoted, dynamic, and disinterested teams. To substitute for them less dynamic, less competent, and definitely not disinterested Industrialists is not a policy but an abdication.[88]

This passage shows how many heterogeneous issues EDF's nuclear engineers marshaled in defending their contracting policy. They fully accepted the export agenda outlined by the Plan and the government. They did not, however, agree with the policy of national champions—at least, not champions who came from private industry. Comparing EDF's methods with those of American companies dissociated industrial practices from their political overtones. Thus, EDF's contracting policy appeared well tested in the capitalist world. EDF engineers invoked the utility's glorious history, but instead of stressing the social and political benefits of its nationalized status they stressed the industrial benefits: only EDF was large enough to stand up to the Americans. Likewise, only EDF was "disinterested" enough to manage the export of power plants—both because it had no ties to private financial concerns and because its state engineers were inherently interested only in the good of the nation. In the face of the threat to their expertise and authority, the engineers jettisoned the leftist dimensions of their regime. They attempted to reconfigure the political meaning of their practices to make them appear more in line with the national policies emanating from the Plan. In so doing, they tried to walk the line between Georges Pompidou's emphasis on economic competitiveness and Charles de Gaulle's emphasis on French cultural superiority. Walking this line made it all the more crucial that the EDF engineers reaffirm their identity as loyal and impartial public servants whose only thought was the best interest of the nation.[89]

EDF's nuclear technopolitical regime had taken a risk in trying to set an example for the rest of French industry. In the process of trying to enroll the state in its industrial policy, the regime ended up invoking the wrath of a government far more interested in closely directing that policy than its Fourth Republic predecessor had been. The president, the prime minister, and the Plan all pushed for "international competitiveness," and all believed that the promotion of "national champions" could best achieve this goal. For de Gaulle, they symbolized national independence and glory. Pompidou and the planners thought they would spur economic growth.[90] The Plan had quite explicitly emphasized "the need to pursue the . . . concentration of French industry to increase its competitiveness and allow it to confront the powerful foreign companies as economic opening occurs."[91] Further, nobody was particularly happy with EDF's nuclear program—especially not Charles de Gaulle. In 1966, he hoped to inaugurate EDF3 personally. Because of the technical problems experienced by the reactor, Pierre Massé (who by then had returned to EDF as its president) suggested that de Gaulle inaugurate the new tidal power plant at La Rance instead. De Gaulle reluctantly agreed, but he was angry with the utility. "I must say," he declared coldly during one ministerial meeting, "that we would not be where we are had EDF followed the wise counsel of the CEA."[92] His increasing displeasure with the utility's nuclear technopolitics echoed throughout the government.[93]

The regime faced threats from within the utility as well. A change of the guard that had been underway at EDF for some time had been accelerated by Massé's return. Massé himself strongly supported the notion of international competitiveness. He had chosen as his second in command Marcel Boiteux, whose economic expertise has already been mentioned.[94] These men and others under them had more sympathy with Pompidou's vision of France's future than with that of their own nuclear engineers. They certainly did not share a belief in the inherent superiority of state engineering or nationalized companies, and they had a markedly different domain of expertise. They did not, therefore, advocate the same technopolitics. Instead, the new economist-managers advocated a strategy closer to that laid out in the fifth plan and supported by Pompidou. Against the vociferous protests of their engineers, these managers took a first step toward changing EDF's nuclear contracting policies. For the two projected reactors at Fessenheim, EDF's newest projected nuclear site, the utility would launch two kinds of requests for bids: the first would follow the utility's traditional policy, and the other would ask consortia to bid on "nuclear boilers."[95] This way, the consortia might get the experience they needed in order to export their skills and technologies.[96]

This decision assuaged private industry, the government, and the CEA, but it angered the engineers and labor unions at EDF. Engineers continued to write furious memos defending their previous practices.[97] The political dimensions of this decision were such that even labor representatives, who had remained silent on this issue as long as the policies advocated by engineers dominated, got involved in the fray. The Confédération Général du Travail, the communist-dominated union, saw itself as the last defender of the social and political dimensions of nationalization. Claude Tourgeron, a CGT militant who sat on EDF's board of directors, argued furiously against the decision before it became final, declaring that it was bad for the nuclear program, bad for EDF, and bad for the nation. The companies bidding for the smaller sub-lots in the traditional request for bids, said Tourgeron, belonged to the very consortia that would submit bids for the "nuclear boiler." They would play with the numbers so that the consortia bids would—artificially—appear to produce cheaper reactors than the individual bids. One thing would lead to another, and this mode of bidding would prevail. Private consortia would become prime contractors, and hundreds of EDF employees would be out of work. Finally, Tourgeron argued, France would never realistically export gas-graphite reactors. "Underdeveloped" countries would not want nuclear plants, because such plants were profitable only when they were very powerful, producing more energy than such countries would ever need. And industrialized countries would want American plants because they seemed cheaper. Gas-graphite plants made sense for France only because France had its own natural uranium supply. Thus the two consortia proposed would have only one customer: EDF. This, in turn, would artificially inflate the cost of nuclear power plants. In sum, consortia bids were bad for the national economy. Other union members echoed these sentiments.[98]

A deeper issue also ran through the dispute over contracting: the changing meaning of nationalization within EDF. The CGT and the nuclear engineers shared an understanding of nationalization that rested on a leftist vision of the socioeconomic order. For them, acting in the national interest meant that a nationalized firm would supersede any and all private companies. As we have seen, though, by the mid 1960s the engineers who directed the reactor projects were not committed to this leftist interpretation. Indeed, most (particularly those who did not belong to the CGT) felt that private industry should grow strong in order for the French economy to flourish—as long as helping industry grow strong did not mean relinquishing control over plant development. The CGT, of course, did not concede these points. Roger Pauwels, another militant who sat on the board of directors, snidely remarked that some of the other directors seemed to think that EDF's mission was to facilitate the development of private industry. Indeed, this was not so far from the mark. Pierre Massé responded to this accusation primly: "That is indeed one conception of the role of the nationalized firm, and . . . we could spend a long time debating this." Such, at least, was the response registered in the first version of the minutes. The official version of the minutes amended this formulation: "The proposal [for bidding by consortia] has the benefit of the entire national economy in view, a benefit in which EDF will share."[99] We can only surmise that EDF's president wanted the economist-managers' competing claim to the best interests of the nation to go on record in a stronger way. Clearly their understanding of nationalization and public service had no leftist connotations. For them, the main mission of a nationalized firm was to foster the nation's industrial growth—and that, necessarily, meant implementing policies favorable to private companies. The fact that the Plan supported such policies served as proof that they were in the national interest.

One of the problems that engineers encountered in the battle over industrial contracting was that the hybrid nature of their two most important technopolitical tools—the competitive nuclear kilowatt-hour and optimization models—made these tools easier for others to appropriate. Unlike reactor designs, these two entities did not fall under the exclusive purview of engineers. Both were entities that had first emerged in economic spheres. Engineers had reshaped them—indeed, they had made the existence of the competitive nuclear kilowatt-hour possible— but they did not fully control them. Industrial leaders, planners, and utility economists did not agree that engineers were the most qualified people to define the relationship between industrial contracting and the competitive nuclear kilowatt-hour. During the Fourth Republic, the government was too unstable and had too many other preoccupations to address industrial policy in any depth. A decade later, however, industrial policy had become one of the Fifth Republic's top priorities. Private companies had recovered from their wartime loss of reputation, and economists had thoroughly penetrated the state and its institutions. In the end, engineers did not (or could not) devise technopolitical tools of the resiliency they had managed earlier. The forces arrayed against them were too strong, and they were too weak.

Like the engineers, EDF's leaders had been trying to escape the increasing interference of the CEA. Their solution, however, was one that engineers could never have considered seriously: abandoning gas-graphite reactors altogether, buying an American license, and pursuing the nuclear program with light-water reactors. The quarrel over this issue became known as the *guerre des filières*—the war of the systems.

*

In the Fifth Republic, as in the Fourth, engaging in technopolitics helped engineers expand their sphere of influence. Designing and building nuclear reactors posed real and difficult technical problems; it also posed difficult political problems. Often these were one and the same. EDF1's cracked containment vessel was not simply a puzzle in steel welding; it was also an embodiment of EDF's development strategy and of its relationship with the CEA. The competitive nuclear kilowatt-hour was not a transparent technical concept; it was the product of the complex technopolitics of optimization. Producing plutonium in civilian reactors was neither a purely technical nor a purely political problem; it was a process subject to negotiations among two technopolitical regimes and the state. In all these cases, engaging in technopolitics provided a way for engineers and managers not just to solve the problems at hand but also to spread their authority beyond the confines of their institution and to promote their vision of France's (technological) national identity. The government and the Plan might define "competitiveness" as a desirable goal for France, but they could not materialize that competitiveness. De Gaulle might use the nuclear program as a symbol of French power, but he could not single-handedly determine the myriad ways in which nuclear technology would become entwined with France's future. Engineers produced material manifestations of these visions for France's future and symbols of its identity. Technopolitical practices and artifacts thus provided means not only to exert influence over other institutions but also to stake out a place in larger debates about industrial policy, international competitiveness, and the future of the French nation.

At no point were engineers reluctant to expose the political dimensions of their work. On the contrary, they were explicitly addressing their work to broader political issues. Sometimes, indeed, they were using those broader issues to legitimate their practices: in the debates over industrial contracting, for example, engineers recast the political meanings of their contracting policy to place it in the context of Fifth Republic priorities. In that case, politics was not just a goal but also a resource for engineers.

That example, together with the introduction of optimization techniques and the invention of the competitive nuclear kilowatt-hour, also reveals the risks of technopolitics for engineers. Admitting heterogeneous practices and criteria into their work opened the door to other forms of authority and expertise. EDF engineers adopted optimization studies in part to strengthen their regime in its programmatic battle with the CEA. But those very techniques made their authority in their own regime vulnerable by privileging economic modes of reasoning, and thus privileging the authority of economic experts. EDF's economist-managers used those same kinds of practices to discredit the gas-graphite design altogether.

4

Technological Unions

To what extent did social groups other than technologists, engineers, and scientists incorporate technological prowess into their visions of French national identity? How did they do so, and to what ends? All too often, historical accounts of technological change in the twentieth century confine themselves to the designers of artifacts and systems. We know a tremendous amount about the creation and spread of technological systems, but relatively little about how the people who work in those systems think about technological change and its role in their lives. Yet their point of view is crucial, not only for its own sake but also for the sake of understanding the multiple dimensions of technological change.

Labor union discourse provides a good entry point into these questions. Unions have a powerful voice in French society. From an American perspective, they function almost like political parties: each of the three major unions has a distinct ideological platform, which they rehearse in their public statements and wield in their strikes. The unions have a complex structure: they are organized into national confederations, each including numerous trade federations, which in turn are divided into local sections. For example, the Confédération Général du Travail groups together trade federations for many sectors: electricity and gas, aircraft, chemical, metallurgy, banking, etc. The other two unions also have their own trade federations for each sector. Unions express general ideological positions through their confederations, leaving specific sectoral demands to the federations.[1] It is at the confederation level, therefore, that unions contribute to national political discourse. And it was at the confederation level that unions participated in the national conversation about technology, politics, and national identity in the 1950s and the 1960s.

In chapter 1, I examined how technologists as a group presented a vision of French national identity that revolved around technological prowess. In the present chapter I look at the visions offered by France's three major labor unions. Just as technologists sought to conflate technology and politics in order to legitimate themselves as creators of the nation's future, so too labor militants conceptualized relationships between technology and politics that would give workers and their unions agency in shaping the nation's future. At times, these conceptualizations criticized (implicitly or explicitly) the visions and actions of technologists. But unions also saw political opportunity within technological development. For them, artifacts and systems had the potential to become vehicles for social change. Their official discourse spelled out how this might occur.

In chapters 2 and 3 I showed that different technologists had different visions of how to conflate technology and politics, which they enacted by shaping distinct technopolitical regimes. In this chapter, I will show that militants in each union also had different ideas about how technology and politics should interact, which in some cases arose from efforts to distinguish their confederation from the other two. In order to understand the positions they articulated, and hence how the range of ideological options available in French society shaped the meanings of technological change, we must grapple with these differences. But while unions were more directly involved in industrial development than other critics of state technologists (such as the social scientists we encountered in chapter 1), they were not in a position to *create* technopolitical regimes. Nor were they on a mission to do so. Instead, militants and other workers had to function within technopolitical regimes. As will be discussed in chapter 5, this necessity tended to erase the differences among the unions as workers struggled to find their place within the nuclear program of the 1960s. Not until the 1970s would the distinctions in how each union conceptualized the relationship between technology and politics reassert themselves.

In the meantime, though, the first step toward understanding labor visions of a technological France lies in examining the ideology of each union and the links that each imagined among technology, politics, national identity. In general, the three unions shared a faith in the promise of technological progress and a belief that the future of France depended on such progress. They did not always agree, however, on what that future should be, or on how technological development might enter into it. For example, all three unions supported the basic concept of national planning. But planning by whom, and under what conditions? For the communist Confédération Générale du Travail, proper planning could not occur in a capitalist society; only a socialist revolution could produce structures that would ensure that planned industrial development would benefit the working class. The other two unions adopted a more reformist approach, arguing that participation in existing planning structures, however imperfect, was still better than exclusion. For the CFTC (later renamed CFDT), planning industrial development could provide the means toward a more egalitarian political system. For Force Ouvrière, which claimed to be apolitical, planning could provide an objective means of shaping the future. The three unions also enunciated distinct conceptions of national technological prowess, military nuclear development, and European atomic collaboration. In each case, the ideas expressed by unions both emerged from and further articulated their broader ideological agendas.

The bulk of this chapter explores the articulation of these ideas. Union discourse does not make sense, however, without an understanding of the political layout of postwar French labor. First, therefore, let me sketch out a political map.

The Politics of Unionism

The Vichy government outlawed two bastions of working-class politics: the Communist Party and the Confédération Générale du Travail labor union. Militants were forced underground, and many joined the Resistance. After the liberation, these militants—like other Resistance fighters—became national heroes. In 1944 the reinstated CGT and the Communist Party launched the so-called battle of production, intended as the working class' patriotic contribution to ending the war and beginning national reconstruction. Its goal was to raise production levels in order to defeat Nazism, then ensure postwar national independence through industrial self-sufficiency. Militants asserted that class and national interests had converged during this difficult period, and that, for the sake of both, workers should avoid strikes and stoically pour all their energies into rebuilding the nation.[2] The CGT thus emerged from the war with impeccable nationalist credentials.

The dominance of left-wing parties in the postwar government initially gave worker organizations high hopes for the future of French social relations. Nationalization seemed to bear out these hopes. In most cases, nationalization entailed a tripartite directorial structure, in which management, workers, and consumers were all represented on the board of directors. Of course, the meaning of nationalization varied significantly for different groups and changed over time, as we have seen in the case of EDF.[3] But in the immediate postwar period, nationalized industries appeared to be concrete manifestations of a new social contract between the working class and the state, and evidence that workers' roles in (re)building the nation would be suitably recognized and compensated.

By 1947 this initial optimism had begun to fade, all the more so as Cold War politics permeated France. Frustration grew over poor living conditions and the lack of basic supplies. Defying the slogans of the battle of production, severe strikes—initially championed by non-communist unions—erupted all over the nation. Communist ministers were dismissed from the coalition government. The Communist Party and the CGT, now firmly in the opposition, denounced the Marshall Plan as an illusion that deceived the French into an unequal alliance with the imperialist United States. Evoking their wartime heroism, communist organizations declared themselves the true and unique defenders of the French national interest.

Although its leadership was dominated by communists, a substantial non-communist minority existed in the CGT before 1947. These two factions had often disagreed. The onset of the Cold War prompted the communist majority to adopt a range of pro-Soviet, anti-American positions, which deeply disturbed the minority. Despite their own sympathies with the Socialist Party, minority leaders strongly advocated the separation of labor unions from party politics. They rejected communist assertions that the only true path to better conditions for the working class passed through socialist revolution and the overthrow of capitalism. Minority leaders maintained that unions had to defend the working class's interests— such as salaries, job security, and benefits—regardless of the political system in place. Their role was not to overthrow that system. Minority leaders had hesitated over supporting the battle of production precisely because it sacrificed traditional union practices (such as strikes) in favor of national political goals.[4] By the end of 1947, the two groups clashed too profoundly to continue functioning within the same organization. The minority faction split off to form the CGT-Force Ouvrière, an independent confederation.[5]

From then on, Communist Party members dominated the CGT's leadership, and most of the union's positions echoed those articulated by the party. The nation provided an enduring theme for both the party and the union as they strove to articulate a nationalism distinct from de Gaulle's.[6] Throughout the 1950s and the 1960s, the CGT fiercely and consistently defended the notion of French national independence, by which it meant autonomy from other capitalist nations (especially the United States) and industrial strength relative to Germany. This stance prompted the denunciation of emerging plans for European cooperation on the grounds that cooperation would threaten French autonomy. CGT militants argued that the Common Market was an instrument of capitalist hegemony that would continue to exploit workers. They insisted that the nation's current industrial structures pauperized the working class. This too they linked to national independence: an autonomous France required a strong working class.

Force Ouvrière leaders defined their union in opposition to the CGT. Rather than link union doctrine to party ideology, Force Ouvrière would be an "independent" union. Force Ouvrière should aim to defend the working class, militants insisted, not to overthrow capitalism. Leaders strongly emphasized their political autonomy. They gave this autonomy roots in the prewar history of French unionism and claimed that it made Force Ouvrière a more legitimate labor union than the CGT. Did autonomous really mean apolitical, though? The answer depends on the meaning of "political." In the narrow sense of party and revolutionary politics, militants repeatedly professed that union actions were fundamentally apolitical.[7] They could not deny their sympathies with the Socialist Party, but they contended that these stemmed from nothing more than "converging opinions."[8] Politics in the broad sense (especially anti-communist rhetoric) seemed acceptable, however, and many militants did articulate positions on several national and international issues, most notably European cooperation. Force Ouvrière viewed a united Europe as the only realistic alternative to the spread of Soviet communism (a point on which the union and the Socialist Party "converged"). It supported proposals to institutionalize European cooperation and called for the participation of European labor unions in managing these institutions.

The other major non-communist union, the Confédération Française des Travailleurs Chrétiens (CFTC), had roots in late-nineteenth-century Christian trade unionism. Many of its leaders sympathized with the left-wing Christian Democratic party. Again, there were no official ties between the two organizations; in fact, the union's rank-and-file tended to vote more to the right. In the 1940s, the CFTC contained two major subgroups: a majority who wanted to retain the union's Christian orientation and a strong minority, led by the Reconstruction group, who wanted to abandon references to Christianity.[9]

Despite its minority status, Reconstruction was extremely active in shaping the union's policies—for example, by successfully promoting the notion of *planification démocratique*, an approach to national planning that would give the working class a greater role. The issue of whether to remain explicitly Christian dominated internal union debates in the early 1960s, as increasing numbers of militants expressed interest in secularizing. In 1964, after extensive debates, reports, and questionnaires, members voted to change the union to the Confédération Française Démocratique du Travail (CFDT). Around 60,000 members refused to endorse the change; they split off and retained the union's original name. Most members stayed, though, and after 1964 the union faced the difficult problem of defining its identity and establishing its legitimacy. (I will not discuss the post-1964 CFTC. In order to make my argument easier to follow, therefore, I will henceforth refer to this union as the CFTC/CFDT.)

Like Force Ouvrière, the CFTC/CFDT rejected the pro-Soviet communism of the CGT. But unlike Force Ouvrière, it did seek an explicitly political, ideological anchor. CFTC/CFDT militants did not find Force Ouvrière's pragmatism congenial. Many retained a strong sense that moral, humanist values had to underlie any CFTC/CFDT position. Thus, for example, their fascination with modernity was often accompanied by denunciations of crass consumerism. At first, the notion of democratic planning provided an ideological anchor. In the late 1960s, particularly during and after the 1968 strikes, the notion of *autogestion*—self-management—began to supplant democratic planning. The fundamental goals of both were the same: the leveling of social class and the participation of workers in managing not only businesses but also the nation. The degree to which individual militants believed that such a shift should entail a fully socialist system varied considerably.

The politics of French labor unionism were thus fractious and complex. Each of the three major unions identified with a different flavor of left-wing politics. For historical as well as ideological reasons, the unions had difficulty cooperating. Both Force Ouvrière and the CFTC/CFDT perceived the CGT as a mammoth organization whose dominance had to be actively resisted. Force Ouvrière not only opposed the CGT's ideological stance but also suspected that the communist union's offers of cooperation were imperialistic attempts to recapture Force Ouvrière members. After secularization, the CFTC/CFDT became more amenable to cooperating with the CGT. In the mid 1960s these two unions—while retaining fundamental ideological and strategic differences—developed a common platform. In the late 1950s, meanwhile, some members of Force Ouvrière had begun to consider a rapprochement with the CFTC's Reconstruction group. After 1964, however, Force Ouvrière pulled back, viewing the newly secularized CFTC/CFDT as a direct competitor for the attention of non-communist workers. Such inter-union politics often helped to shape the positions adopted by militants.

Quantifying the constituency of these unions in the 1950s and the 1960s is notoriously difficult. Many social scientists do not trust the figures provided by the unions, but there appear to be no reliable surveys for this period. Furthermore, membership figures do not tell the whole story. Most workplaces had at least one committee devoted to personnel issues and composed of elected representatives. Most representatives were union militants, but those who voted for them were often simply union sympathizers.[10] Despite these problems, we can still offer a rough distribution. The CGT remained the largest labor union in France, with a membership of around 2 million workers in the 1960s. The CFTC/CFDT was the second largest union in France, with a membership of roughly 600,000. (Its numbers dropped in 1964, but climbed again during the rest of the decade.) Force Ouvrière did not lag far behind; its membership stayed around half a million throughout the 1960s.

The two institutions of the nuclear program had somewhat different numbers. EDF proportions roughly followed national figures. The CGT dominated EDF, garnering about 60 percent of the vote in personnel elections throughout the 1960s. After the war the CFTC/CFDT attracted only 10 percent of the vote in EDF, but by the mid 1960s this number had risen to around 20 percent. Force Ouvrière's popularity remained around 15 percent throughout this time period. CEA proportions differed significantly from the national figures. Despite the communist purges, the CGT won 30 percent or more of the vote there throughout the 1950s (though the CEA's military sites, including Marcoule, did not allow the CGT to establish local sections until the 1960s). By the mid 1960s, however, the CGT vote had dropped to 18 percent. The CFTC/CFDT benefited from this change: its popularity grew from 20 percent in the late 1950s to nearly 40 percent by the mid 1960s. Force Ouvrière hovered around 15 percent throughout both decades. About 30 percent of CEA employees voted for an independent union of nuclear workers that had no nationwide confederation.[11]

Conceptualizing National Technological Progress

None of the three unions fundamentally challenged the concept of a technologically radiant France. In some cases, they even used the same rhetorical archetypes as state technologists, such as France's backwardness, or the Malthusianism of private industrialists. But sharing a fundamental

Table 4.1
Unions' political affiliations and positions on selected issues.

	Political sympathies/ nationalist outlook	Position on French nuclear strike force	Position on Euratom	Conception of relationship between technology and politics
CGT	Communist/ nationalist	Strongly against	Strongly against	Straightforward: Technology is a political tool. Its development must be directed by nationalized companies in order to prevent capitalist exploitation and engender socialist revolution.
Force Ouvrière	Socialist/ internationalist	Mostly silent	Strongly in favor	Tension: Technology is inherently neutral and apolitical. At the same time, it can lead to an internationalist future.
CTFC/ CFDT	Christian Democrat/ blend of nationalist and internationalist	Against	In favor	Complex and ambivalent: For some, technology is uncontrollable and oppressive. Others argue that its development can and must be shaped by humane values.

belief in the political, economic, and cultural importance of technological progress did not mean sharing state technologists' visions of France's future. Instead, each union conceptualized national technological progress in terms of its political framework.

In the 1950s and the 1960s, most union discourse about technology did not address the relationship between technological change and workplace experience directly. Except for a few scattered discussions of automation,[12] most discussions of technology centered on national industrial development and its relationship to the sociopolitical order. In this regard, the CGT and the CFTC/CFDT envisioned a nation different from what most state technologists had in mind. For the CGT, technological development could pave the road to a socialist society. For the CFTC/CFDT, it opened the way to greater worker participation in running the nation. (To some this meant a socialist system; to others it did not.) True to form, Force Ouvrière did not link technological change to France's internal socio-political order. But its militants did argue that technological development could help situate France within an international cooperative framework that would strengthen the nation. These three visions of technology were entwined with different pictures of how technological development occurred, the extent to which it could be controlled, where its potential lay, and how it mediated or represented the place of workers in the French nation. In these pictures, technology appeared sometimes as a thoroughly (and desirably) political entity, at other times as an unstoppable deterministic force, and sometimes even as both.

For the most part, I have limited my discussion to discourse published in official union newspapers, magazines, and journals. This discourse represents the unions' official platforms, and thus their formal contributions to national debates. It suffices for my purpose here, which is simply to outline the alternative visions of technological France offered by labor. A word of caution, however, to those interested in internal union affairs: these sources tend to mask differences within each union. This is particularly true for the CGT and Force Ouvrière, whose confederation publications created the illusion of internal unanimity; it is less true for the CFTC/CFDT. For those not familiar with French labor politics, table 4.1 offers a rough guide to each union's position on the questions examined here.

The CGT: National Independence through Technology
The CGT's representation of technological progress—like its politics more generally—was the least ambiguous. Technology, for CGT militants, was crucial for the future and independence of the nation. But not just any form of technology would do. The right people had to control technological development. In the hands of capitalists, it served as merely another tool of exploitation. In the hands of workers or of nationalized companies, it could lead to the economic growth and political prestige that France craved.

Like state technologists, CGT militants appropriated the postwar trope of "Malthusianism" to make this point. Militants did not give this notion a very specific meaning. "Malthusianism" functioned as a shorthand for evil French capitalists who shunned modernization in order to preserve their economic power. In a speech to technologists and political scientists at the 1958 "Politique et Technique" conference, CGT Secretary Pierre

Le Brun denounced the continued existence of an industrial class that was "fundamentally anti-progressive and Malthusian and for whom technology is essentially a way to push the exploitation and domination of others even further."[13] Eight years later, the CGT militant Henri Beaumont accused "Malthusian" electronics firms of curtailing research that would enable France to remain independent of the American electronics industry.[14] In the first case, "Malthusian" behavior consisted of using technology to exploit workers. In the second, it consisted of impeding the realization of national independence through technological development. In both cases, "Malthusianism" gestured toward inappropriate forms of technological development. Good technology, for these CGT militants, could only come from good politics. "Labor syndicalism," Le Brun wrote, "can only wish for a fruitful interpenetration of politics and technology, [and] for political deliberations and decisions to be enlightened and enriched by all that technology can bring to them."[15] The right blend of technology and politics would lead to improved working conditions, higher pay, and better scientific and technical education. Beaumont focused on national politics. Private firms, he argued, could not be entrusted to do research in the French national interest. A nationalized electronics industry was "an imperious necessity for a nation that wants to be independent and not depend on foreign [nations], in this case the USA."[16]

The CGT's analysis of technological change proceeded on this dual front of social and national politics, often conflating the two. Articles in *Le Peuple* and *La Vie Ouvrière* showered praise on French technological achievements.[17] The only major difference between the CGT's praise and that of state technologists was that the union usually included some mention of workers. Witness this 1957 ode to a new suspension bridge: "Magnificent work of art in a grandiose natural setting, the bridge of Tancarville bears witness to the worth of French technology. It will bring honor to the engineers, technicians, and all the workers who will have worked on it."[18] Most worthy of praise were the achievements of the nationalized companies. SNCF and EDF workers received frequent acknowledgment in the pages of the CGT's publications, as did the technological systems that they created. Such articles also provided an occasion to indict the capitalist system. In 1967, EDF's Rance power plant was described as "a prowess that owes nothing to men of money: the tidal power station of the Rance [is] the first in the world to have domesticated tidal energy."[19] Similarly, the success of the Caravelle commercial airplane—which even an American airline pilot, capitalist though he was, admitted was "the best [he'd] seen in 27 years of flying"—was due to the nationalized status of its manufacturer, Sud-Aviation.[20] Renault, meanwhile, had pushed "the avant-garde of technology" since its nationalization in 1945.[21] Its postwar achievements demonstrated the "vitality and ingenuity of French technology and science, despite the terrible handicap of four years of occupation."[22]

CGT militants thus privileged national independence every bit as much as Gaullists, albeit from the other end of the political spectrum. Both representations of the nation shared referents (the Occupation), and both groups derived legitimacy from similar credentials (the Resistance). But CGT militants constructed a somewhat different association between technological prowess and national identity. They juxtaposed praise for French technology with indictments of capitalism, support for nationalized companies, and commendations for workers. Technological radiance thus derived from the political superiority of nationalized industry and the labor of French workers. Even when they accused state technologists of perverting the original social mission of nationalization, CGT militants continued to fantasize about the revolutionary possibilities of nationalized structures. The technologies produced therein were inherently better, and more national, than those produced by private companies. When the confederation's press did evoke the technological achievements of private companies, it did so as a way of indicting capitalism and praising worker contributions to French grandeur. For example, when the "the atomic millipede" (the convoy that brought heat exchangers to the Marcoule reactors) caught the imagination of mainstream journalists, one CGT militant wrote caustically: "Fine. Let us also applaud this beautiful achievement of French science and technology." But, he continued, who applauded the men who had built these huge machines with their hands? "It's an old habit of the 'free, objective,' etc., etc., mainstream press to ignore systematically those . . . whose work is at the base of national wealth."[23] For the CGT, French grandeur thus rested on nationalized companies and the labor of workers.

The CGT thus did not challenge either the concept of grandeur or its link to technological prowess. But it did challenge the manifestation of this link in de Gaulle's *force de frappe*. Following the lead of the Communist Party, CGT militants vociferously opposed the construction and testing of atomic weapons and supported the "peaceful use" of atomic energy. Militants rested their arguments on the same concept that nuclear leaders and de Gaulle used to promote both the military and the civilian nuclear program: the radiance of France. For the CGT, however, only peaceful uses of atomic energy would enhance French grandeur: "The prestige of France would be considerable among peoples the world over if it took a solemn decision to devote itself uniquely to peaceful applications of atomic energy."[24] Furthermore, military uses bolstered the capitalist system: "It is obviously impossible to place useful industrial applications, like the production of electricity from a new source of energy, on a par with the manufacture of the most murderous war machines ever. The development of modern technology opens remarkable possibilities for man, but the use to which the capitalist world puts [this technology] has nothing to do with improving the living conditions of the people."[25]

In making such arguments, CGT militants frequently invoked Frédéric Joliot-Curie, a communist and a member of their union. Recall that American pressure had led to Joliot-Curie's dismissal from the CEA in 1951, after he publicly refused to build a French atomic bomb. For the CGT, this dismissal elevated the scientist from a Resistance hero to a martyr.

Joliot-Curie's name legitimated CGT writers by enabling them to appropriate French nuclear history. The confederation press portrayed Joliot-Curie, his wife Irène, and his in-laws Marie and Pierre Curie as heroes of French science. A series of articles in 1956 told the story of this great scientific dynasty, emphasizing the hardships imposed on the family by the French state.[26] "On three black canvas notebooks, Pierre and Marie Curie recorded the phases of a discovery that would completely change humanity. They had the right to the gratitude of the Nation. But the State did not give them the decent working conditions for which they had hoped. In order to obtain them, they had to struggle."[27] In the next generation, the struggle became overtly political: Irène was part of Léon Blum's leftist government before the war, and "Fred" (the nickname indicated camaraderie) joined the clandestine Communist Party during the war. The whole French nuclear power program was a "great idea advocated as of 1945 by Frédéric Joliot-Curie."[28] Rehearsing this triumphant history of French nuclear efforts made that history into a communist morality play and asserted the CGT's right to pronounce on the future of the program. History averred CGT workers as the moral and material guardians of the French nation.

This history also legitimated the CGT's accounts of the devastating effects of nuclear explosions. Militants gave horrifying descriptions of victims of Hiroshima and Nagasaki and accounts of US testing in the Pacific. Bringing the matter closer to home, one writer exclaimed: "AN EXPLOSION LIKE THE ONE IN THE PACIFIC WOULD DESTROY ALL OF

Figure 4.1
Charles de Gaulle and his nuclear armor. This cartoon appeared in the 8 January 1964 issue of CGT's weekly, *La Vie Ouvrière*. Courtesy of Henri Sinnot, Institut d'Histoire Sociale, CGT.

PARIS [and its] GREATER SUBURBS ALL THE WAY TO MANTES, MELUN, FONTAINEBLEAU: SIX MILLION DEAD."[29] None mentioned the Soviet weapons program. To those who wondered whether the union should expend so much effort on a political (rather than a specifically working-class) question, one militant responded that the defense of peace concerned union members on all possible fronts: as humans refusing to condone suffering, as proletarians refusing fratricidal struggles with workers in other nations, as producers who did not want the fruits of their labor hijacked for nefarious purposes, as consumers who would find the money better spent on higher salaries and improved housing, and as citizens for whom war was a basic violation of democracy. In short: "There are no longer any questions that are purely 'syndical' or 'economic.' Everything is imbricated."[30]

Indeed, militants argued, de Gaulle's aspirations to a *force de frappe* were immoral not just because of the inherent immorality of atomic weapons but also because money spent on them came from housing, food, schools, hospitals, or more beneficent industrial development. France's world technological standing had suffered: "A century ago, one out of every two engineers in the world was French. Today, only one out of fifty is French."[31] De Gaulle's military aspirations were anti-modern: "To be part of our times, we must disarm, not arm."[32] A cartoon showing de Gaulle in medieval armor (figure 4.1) illustrated this message. Gaullist military ambitions hurt even the civilian nuclear power program. Making the same argument as many EDF engineers, one CGT writer noted that using Chinon's reactors to make weapons-grade plutonium prevented those reactors from producing the power, and engineers from gaining the technical experience, that France needed so badly. "The grandeur of a country," he asserted, "is measured by its economic power and its intellectual radiance; Gaullist chit-chat on the grandeur of France through the *force de frappe* is terribly weak."[33]

The independent pursuit of technological prowess also meant that France should not join forces with other European nations in industrial matters. Specifically, it should not sign the Euratom treaty for cooperative nuclear development: "Euratom is merely the fashionable word to camouflage the small Europe of six, the Europe of the Atlantic pact and German hegemony."[34] Euratom would only help German rearmament by assisting German industry. Ultimately, Euratom represented a kind of treason, both because it would lead to German dominance and because it would upstage France's own nuclear program[35]:

> [Euratom's] family council is dominated by Aunt Germania. . . . France, the largest producer of uranium in Europe, would thus deliver raw and refined minerals, information and technologists, and its overall technical experience to the other nations, but essentially to Germany. This would be a new Kollaboration, still in one direction only. . . .
> It is the partisans of Euratom who have hindered the spread of large hydroelectric dams, who have closed the mines, who have delayed oil prospecting in France, who removed Joliot-Curie from Atomic Research and who denied substantial funding for the peaceful production of atomic energy.[36]

Thus, for the CGT, technology was thoroughly political. It was not a neutral feature of the social order. Capitalists used technology as an instrument of exploitation. Military technological development would drain resources from social programs (and threaten the USSR, though union militants tactfully left this point to their colleagues in the Communist Party). But, properly developed by truly nationalized companies (i.e., companies that did not secretly serve capitalist interests), technology could provide the route to national independence. And, championed by peace-loving scientists, it could contribute to the radiance of France.

In the end, the abstract dimension of the CGT's conceptualization of the relationship between technology and politics strongly resembled that of many state technologists. For both, conflating technology with politics provided a means to extend their political purview and legitimacy. But the difference in their political program mattered. For CGT militants, conflating the two provided a way to stake out a place for the union and for the working class in the modernizing nation. From this place, true credit for French technological prowess went to workers, especially those who worked for nationalized companies. And from this place, the representatives of the working class could lay out their program for national technological development.

Force Ouvrière: International Cooperation through Apolitical Technology

For the CGT, good politics produced good technology, and vice-versa. In contrast, at the foundation of Force Ouvrière's supposed lack of ideology lay the claim that it did not pursue politics. Under these circumstances, the union could hardly propose technology as a political instrument. Instead, Force Ouvrière followed the pattern it had established for so many of its positions and did the opposite of the CGT: it attempted to portray technological development as fundamentally neutral. Technological progress itself was "collective, universal, and irreversible."[37] Only its social effects were subject to control. "If we know how to handle it, technological progress can be the cause of a happier life, of a new stage of social progress, of human progress."[38] Technological change was the "fundamental fact of our era. As the result of intelligence and know-how, in this century it has attained an unparalleled . . . rhythm of development. We often hear that it contains the potential for the best and the worst. . . . Social progress will march to the same beat as technological evolution, or there will be no social progress."[39] For Force Ouvrière, "social progress" was an apolitical concept, in the sense that its achievement was not tied to a specific political system or party: capitalism could lead to social progress as easily as socialism. In this scheme, technology, disembodied and neutral, set the beat. Militants had to concentrate on ensuring that workers benefited from technological change; they should not waste time fighting the system which produced that change.

Force Ouvrière's representations of technological prowess often deemphasized national origins in favor of an internationalist vision—again, in reaction against the CGT. Militants did portray technology as central to France's future, but they situated both technological change and the nation's future in an international context. The union's press managed this by focusing on the technical and scientific details of industrial systems. Concentrating on these supposedly neutral details made technology appear apolitical, universal, and international. Through artifacts and systems, France could become a player in new types of international collaborations. So while national achievements mattered in Force Ouvrière's discourse, their significance derived primarily from their ability to make France part of a trans-national, post-political, non-communist system.[40] Force Ouvrière's representations thus contained a fundamental tension: on the one hand, militants claimed technology was apolitical; on the other, they saw it as the conduit to an internationalist future. This tension came from a parallel tension in their ideology: while Force Ouvrière supposedly did not engage in politics, its anti-communism and its internationalism constituted an undeniably political stance.

Force Ouvrière's internationalism shaped its rendition of nuclear history. Its version of this history contrasted sharply with the CGT's. Articles that discussed specifically French nuclear research ignored Frédéric Joliot-Curie—even when describing the experimental reactor that he masterminded.[41] But the union's enthusiasm for nuclear development certainly matched that of its communist nemesis. Exuberant predictions about the social, economic, and medical benefits of nuclear technology filled the pages of FO's weekly newspaper, especially in the mid to late 1950s.[42] What mattered for Force Ouvrière was the effect of technology, not its sociopolitical origins.

In view of this consciously trans-national approach to technological development, it will come as no surprise that Force Ouvrière ardently supported the Euratom treaty. Euratom was "a vital and urgent necessity for France and for Europe. Our syndicalist movement has committed itself to this without reticence."[43] In the rhetoric of Force Ouvrière militants, Euratom transcended politics because it transcended the nation. Atomic energy spurred industrial development: nuclear plants provided the energy required for development in other sectors and offered a new market for existing sectors (such as the chemical or metallurgical industries). A single nation the size of France did not have the resources to build a truly

Figure 4.2
A cartoon from the weekly *Force Ouvrière*. One man says to the other: "Do you think that with the atomic bomb we will no longer be a 'diminished great power'?" Courtesy of Jean-Pierre Alliot, Force Ouvrière.

competitive nuclear program; it could never compete with the super-powers. To succeed in this domain, European countries had to work together. Euratom would leave petty nationalisms behind and would result in improved economic development for all. By getting involved, Force Ouvrière and other European labor unions could ensure that workers' health and safety concerns were incorporated into the fabric of Euratom's institutions. They could also join the inevitable discussions of labor markets that accompanied all efforts to institutionalize European cooperation.[44] Euratom thus represented an ideal example of neutral technological development that promoted international collaboration, not just among industries, but also among labor unions.

With so much attention to Euratom, Force Ouvrière had little energy left for France's national nuclear program. Occasional articles mentioned French power reactor sites, but usually tangentially. Force Ouvrière had quietly supported de Gaulle's return to power in 1958, largely because it viewed Gaullism as the only viable alternative to fascism and communism.

Figure 4.3
Another cartoon from *Force Ouvrière*. The poster proclaims: "Come to Tahiti, Dream Island." The atom bomb responds cheerfully: "Here I come!" Courtesy of Jean-Pierre Alliot, Force Ouvrière.

When de Gaulle announced the *force de frappe*, Force Ouvrière remained silent—perhaps because of its tacit support for de Gaulle, or perhaps to differentiate itself even further from the CGT. The only acknowledgment of the atomic arsenal appeared in a couple of gently mocking cartoons in the weekly *Force Ouvrière* (figures 4.2, 4.3). According to one editor, such oblique references provided the only manner in which editorial policy could differ from confederation policy.[45] These images aside, the weekly presented a benign view of the military atom. Fallout from nuclear testing had not been too noxious. Good protection against radioactive exposure existed. "This new industry, in constant evolution, does not neglect to elaborate its own safety regulations."[46]

Force Ouvrière's cultivated abstinence from French politics supposedly freed militants to focus on workers' issues. Indeed, in the mid to late 1950s *Force Ouvrière* writers paid considerably more attention to workplace conditions in the nuclear industry that did the CGT. Again, though, Force

Ouvrière did not focus on French nuclear workers per se. Instead, the union presented reports on studies conducted by the international labor organizations to which it belonged.[47] It expressed optimism about the atomic industry's working conditions. Though the industry could certainly benefit from "a good syndicalist education," it was, writers claimed, also one of the least dangerous, because of a heightened awareness of its potential risks.[48]

Force Ouvrière's press did sometimes praise specifically French technological prowess. But on these relatively rare occasions, it usually either dissociated technology from politics or situated the value of France's achievements in an international framework. (Sometimes it did both.) One article in the weekly *Force Ouvrière* praised the dams, bridges, and airports built by French engineers in Africa and South America. It explained that "sympathy for underdeveloped countries seeking to establish their economic progress on solid, independent foundations" constituted a "salient trait of the export of our engineering knowledge." France could help other nations precisely because its technology was apolitical. "France can guarantee independence to the countries which appeal to its engineers [for aid], *because our technologists do not double as politicians or propagandists.*"[49] Force Ouvrière even located the success of France's *nationalized* industries in an international context. One article praised EDF for the high-voltage transmission lines it had built to Germany, Switzerland, Italy, and Spain. What was good for Europe was good for France. "National egotism would be ridiculous and disastrous. Whether we like it or not, the people of Western Europe are linked by these interconnections. The flag of Europe flies above the high-tension networks. And Electricité de France steadfastly pursues its work in the service of our economy, the French economy of tomorrow."[50] Even these mild patriotic outbursts thus situated French national identity firmly within a united Europe. Technology would literally bind France to the rest of the continent.

This did not mean that the sole virtue of nationalized industries stemmed from their role in European cooperation. Nationalized industries mattered for France too. Their significance, however, derived not from their revolutionary potential but from pragmatic considerations. The public sector furnished the most economically rational means of developing industry, which meant that it provided the best means of promoting the material interests of the working class. EDF and the SNCF garnered the most praise in these domains, and Force Ouvrière argued that the oil industry and other sectors should follow their example. In one respect, of course, such arguments in favor of spreading nationalization

resembled those of the CGT. But there was a crucial, if subtle, difference. The CGT located the importance of nationalized industry in the
domain of political ideology. Its interpretation of nationalized achievements portrayed technology and politics in a mutually constitutive relationship. Force Ouvrière located the importance of nationalizations in
the domain of economic and social rationality, a notion that essentially
referred to the standard of living. Its depiction of nationalized achievements located the motor of economic and social progress in technological development.

In a sense, maintaining a careful separation between technology and
politics provided the only doctrinally justifiable means for Force
Ouvrière militants to pronounce on technological matters. The confederation's identity rested on the official rejection of national party politics. Conceiving of technological development as political in any deep
sense would therefore remove technology from the union's self-defined
purview. Witness Force Ouvrière's official silence on the *force de frappe*, a
technological system no one could legitimately label apolitical. European
industrial cooperation—which for the CGT carried a heavy political
charged—appeared neutral in Force Ouvrière's press because it was
trans-national. Only by portraying technological development as neutral
could Force Ouvrière militants justify discussing it.

Force Ouvrière's discourse implicitly rejected the notion of French
grandeur. For Force Ouvrière, however, French national identity was still
bound up with technological development. France's future rested with its
ability to develop industrially in an international context. Technology
would bind the nation to the rest of Europe, and the working class would
help that process. Once again, the portrayal of the relationship between
technology and politics served to delineate a place for the union in shaping the nation's future.

The CFTC/CFDT: Complex Relationships between Technology and Politics
Both Force Ouvrière and the CGT offered fairly unambiguous portrayals
of technological change in their publications. Analytic differences aside,
militants for both confederations expressed little uncertainty or ambivalence about technology's value and its place in French society. In contrast,
CFTC/CFDT representations of technological change appear more fragmented. The main reason is that this union made more space for conflicting voices within its ranks; indeed, it consciously differentiated itself
from the other two unions by proudly incorporating dissent and dialogue
in its official discourse. One manifestation of this breadth was the larger

number of union-sponsored publications, in which militants could express a wide range of opinion.[51]

The pages of CFTC/CFDT publications offered a correspondingly broad range of representations of technological change. Some writers portrayed technology as ineluctable and uncontrollable. "Self-nourishing," said one militant, technological progress "contributes to its own acceleration and expansion."[52] For this author, the outcome of technological change remained uncertain. A similar uncertainty transpired in a special issue of the weekly *Syndicalisme* entitled "Industrial modernization: menace or hope for the world of work?" Would technological progress lead to unemployment or reskilling? Oppression or liberation? The answer was often mixed. Ambivalence also accompanied descriptions of technological prowess. *Syndicalisme* covered the same prowesses as did the CGT press: electronic calculators, the Tancarville Bridge, and the SNCF's rail system all received attention. And even more than the CGT, CFTC/CFDT militants argued that success in these domains rested on the workers.[53] But focusing on the "human dimension" of technological change, they raised anxious questions about the benefits of modernization. Even the most sanguine writers—such as Pierre Papon, who believed that science and technology were "factors of the *national independence* of a country"[54]— evoked the dangers of pursuing progress unreflectively, and spoke of the need to elaborate socially responsible development policies.

The CFTC/CFDT's approach to the atom exemplified the ambivalence of its militants toward technological change. Atomic achievements appeared as the paragon of modernity in CFTC/CFDT publications, much as they did in the other confederation periodicals. Nuclear matter carried the potential of tremendous social and economic change. "A kilo of Uranium—smaller than a pack of Gauloises—would suffice to drive an atomic train around the world five times," affirmed one article. Nuclear technology could improve food preservation and thereby offer creative solutions to world hunger. "So let us hope that the face of the world will be changed and that every man, every woman, and every child will know new living conditions."[55] The CFTC/CFDT also expressed pride in specifically French atomic achievements: "With G1, France has attained industrial atomic achievement, the essential mark of the modern era."[56]

But unsavory incidents within the nuclear program tempered such hope. In early 1956 *Syndicalisme* reported on a labor relations incident on the Marcoule construction site involving a private contractor. It was well known by this time that workers who applied for jobs at Marcoule had to undergo a security check designed to keep communist sympathizers off the site. The contractor had apparently tried to recruit local farmers, hoping that such men would have little interest in party politics of any kind. Once workers had signed on, the contractor went further and tried to prevent them from unionizing. Complaints to the Ministry of Labor did little to improve labor relations on the site. Management countered a 25-day strike by hiring scabs and bringing in the police. Despite the marvelous technologies under construction (in 1956 the CFTC/CFDT still labored under the illusion that these reactors had civilian destinies), Marcoule had become "the site of fear."[57]

Nearly twenty years would pass before the union would make nuclear safety a national cause célèbre. But militants began discussing workplace issues in the late 1950s. Some called attention to the same international studies that Force Ouvrière cited.[58] Others focused on French efforts to understand the nuclear workplace. In 1958, the CFTC/CFDT militant Alfred Williame presented a report to the state's Conseil Economique on technical, financial, and regulatory aspects of radiation protection. Formulated in general terms, the report described the risks of radiation exposure and outlined the measures required to deal with these risks. These included training nuclear personnel in safety practices; defining hours, leave times, shift rotations, and a retirement age appropriate for atomic workers; and measuring radiation levels in the workplace and around nuclear sites.[59] The report won the unanimous approval of the Conseil Economique et Social, which forwarded its recommendations to the government. In 1960, another militant recalled Williame's report and described the operation of health and safety commissions on EDF's first nuclear site at Chinon. I shall examine the CFTC/CFDT's involvement with specific nuclear workplace issues in greater depth in chapter 5. For now, the point is that the CFTC/CFDT, however sporadically, paid greater attention to these matters than did the other two confederations— perhaps because it deliberately encouraged militants to exhibit greater creativity and independence of thought. In any case, the French nuclear program appeared neither as a disembodied symbol of glory or perdition nor as a weak precursor to more ambitious international efforts. Instead, it was represented as a complex industry, rich in positive symbolic value yet facing real difficulties.

The CFTC/CFDT also straddled the other two unions on matters of nuclear policy, coming out both in favor of Euratom and against the *force de frappe*. Though somewhat less enthusiastic than Force Ouvrière, it made similar arguments in favor of Euratom. Supporting the international endeavor did not, however, lead its writers to dismiss the national program.[60] Initially, the CFTC/CFDT situated its opposition to the weapons program in the context of Christian peace doctrines.[61] As the union moved toward secularization, writers increasingly associated this opposition with that of the moral scientist. Despite their clear desire to maintain a safe distance from the CGT, CFTC/CFDT writers made many of the same arguments as their communist counterparts—albeit in a more detailed and technically sophisticated manner.[62] Rather than merely claiming that money spent on the *force de frappe* robbed worthier causes, for example, unionized scientists and technicians from the CEA's Saclay research center presented careful calculations to demonstrate the precise economic effects of pursuing the military atom.[63] CFTC/CFDT militants also argued that French prestige was better served by peaceful economic development within a European context. They thereby conflated arguments for European cooperation with those against the *force de frappe*:

... these days independence is in fact more tied to healthy economic structures, high scientific and technical potential, and a certain cultural radiance. In other words, the independence of France and of Europe rest much more on their ability to oppose American economic penetration than on the installation of an autonomous defense system that is ineffective and ruinous.[64]

While the CGT wanted a France autonomous from all other Western nations, the CFTC/CFDT situated the nation in a third political space that would resist both superpowers. France would derive radiance from an alliance with the rest of Europe that focused on peaceful technological development.

Technology thus appeared as a double-edged sword. It was certainly not apolitical, even though the CFTC/CFDT (like Force Ouvrière) claimed to be above party politics. For the CFTC/CFDT, rising above party politics did not mean eschewing politics altogether. Its militants did not shrink from viewing Euratom, national sovereignty, or nuclear policy as broadly political matters that legitimately demanded their involvement.[65] At the same time, though, the politics of technological change was more nebulous for the CFTC/CFDT than for the CGT. CFTC/CFDT militants did not label technologies as good or bad based on institutional or political provenance. Rather, technological change appeared as a messy process over which unions and workers had uncertain and uneven control.

Nonetheless, the CFTC/CFDT clearly considered technological development to be central to modern society, and to require deep examination. This attitude became particularly apparent in the internal debate preceding the union's secularization. A commission established in 1960 to discuss the union's ideological future polled leading militants on four basic questions. Here is the first of these:

In order to better accomplish its syndicalist mission, what are the essential problems that the CFTC needs to confront in the areas of:

a) French realities

b) the intersection of the social, the economic, and the political

c) the increasingly international aspect of all issues

d) technology

e) consumer culture[66]

A report based on the answers went out to every CFTC/CFDT local along with another, similar questionnaire. Every union member thus had the opportunity to express an opinion on these matters.

The responses to these questionnaires are preserved in the CFDT's archives. At first, they appear to provide a unique source, an expression of what these unionized workers "really thought" about technological problems in the abstract. Yet, while local militants did provide their own opinions, many of their answers echoed—and sometimes copied exactly—the ideas articulated in the report. As the historian Frank Georgi argues,[67] the questionnaire answers are better read as the means by which union members participated in the building of a collective identity rather than as a pure reflection of their raw opinions. The responses, therefore, cannot be understood separately from the report.

The report painted a bleak picture. Technology presented a tremendous threat of alienation. It could crush workers and rob them of their individuality. Radically new technologies changed class structure by requiring more skilled workers and technicians and fewer manual operatives. Such changes could lead to layoffs, the "depersonalization" of work, and adaptation problems for older workers. Thankfully, remedies did exist in better and more democratic education and in continued fights for the universal right to unionize.[68]

Some militants cited in the report worried about the encroaching dangers of a depoliticized, technocratic society. Their language here resembled that of the social scientists I examined in chapter 1. "In a 'technological' society," one of them said, "politics tend to be devalued and relegated to the realm of technology. Technological evolution is necessary for economic expansion, but it should not be the sole determinant [of such expansion]." He affirmed the primacy of politics, "which includes a conception of man and of social life."[69] Technocracy might lead to the loss of democracy and the concentration of power in a few expert hands. The report's authors commented in another section: "Citizens and workers are at a loss in front of the complexity of the problems faced by the State and by business; to a certain extent, leaders themselves are obliged to have confidence in their technologists. These all-powerful technocrats present serious dangers, and the question arises of how to balance their power."[70] In order to confront such problems, the authors suggested, the union might need to enlist engineers and managers.

In the questionnaire that accompanied the report, union members were asked the following:

1. Based on your regional and professional experiences, can you try to explain the consequence of technological evolution on the mentality and on the very structure of the working class?

2. How can syndicalism adapt to this?[71]

The union local of one Paris bank answered: "We are seeing the replacement of 'politics' by the cult of technology."[72] Other responses expressed a similar frustration: "In the relationships between *technology* and *politics* it seems to us that 'technologists' complicate the parameters of problems at will, in such a manner as to make them unintelligible to the masses and thereby remove these from making political choices."[73] Many respondents also supported the conclusion that technological development was changing class structure. Widening, skill-based differences in jobs and salaries created divisions within the working class, worried one respondent. Individualism might supplant working-class solidarity: "Though this situation abolishes Marxist theses about class struggle and the pauperization of the proletariat, it is also in danger of substituting . . . individual well-being at the expense of collective progress."[74] This comment dug at both the CGT (which promoted the pauperization thesis) and Force Ouvrière (whose pragmatism seemed too materialistic to many CFTC/CFDT militants).

Still, not all members agreed with the report's representations of technological change. Some answers—particularly those from sections with a large proportion of technicians and highly skilled workers—said the report presented "too negative an attitude" toward new technologies.[75] These respondents emphasized the need to recruit highly trained employees:

. . . syndicalism cannot be the enemy of technology. Progress marches on! Syndicalism must adapt and try to benefit as much as possible. Labor organization must include all the personnel of a company: employees, technologists, skilled workers, managers, manual operatives. Together, they can develop solutions that will correct for technological dangers. . . . It is therefore up to us to train technologists as militants.[76]

One group of telecommunications technicians went so far as to issue a warning: "We can already see a syndical fissure. Technologists do not like the old-fashioned methods of unions. The technologist is a man who is above all realistic. 'Patter' does not affect him much. . . . He wants actions, not promises. If working-class syndicalism does not adapt to modern methods, technologists will abandon it."[77] Such threats were rare, though. Most responses which argued in favor of a more positive attitude toward technological change and a kinder view of technical experts also evoked the dangerous temptations facing technicians and others. The CEA's Saclay local—unusual in that it included primarily technicians and scientists—did so with the most eloquence:

Technology is the motor of today's civilization; it is a stranger to all moral ends. . . . The technologist is generally scrupulous and honest in his conclusions, but he often wears blinders. Yet irrational values should not inspire decisions after the technical examination of a problem is completed by infirming the conclusions of this examination. On the contrary, [they] should be present from the start, in as explicit a manner as possible, in order to give meaning to the effort of [developing] a technical solution. Then the so-called contradictions between the 'ideal' solution and the 'technical' solution would disappear by themselves.[78]

Unions could thus help scientists and engineers build better, more responsible technologies. Saclay's militants demanded that "irrational" social and political values shape development from the outset. This conception of politically malleable technology strongly resembled that of the engineers, and no wonder: some of the CFTC/CFDT militants from Saclay shared a professional background and culture with those engineers (including a few Polytechnique graduates), and most had probably worked with them on some aspect of gas-graphite research and development. As militants, they wanted to inspire their non-unionized colleagues to design technologies based on more "human," non-capitalist values and to make them aware of the power and responsibility associated with their work. They also wanted to show their confederation that technological development was malleable and controllable, and to demonstrate the benefits to the working class of recruiting technical experts.

Meanwhile, the questionnaire responses make clear the tremendous range of views within the CFTC/CFDT. Whether or not they reflected on this subject independently, most union members agreed that technological change was pivotal to the future of French society and should therefore be of paramount concern to their confederation. Their responses depended in part on their level of involvement with creating technological change and in part on how they interpreted the report, how they positioned themselves within the union, and a host of other factors. Some saw technology as a dangerous and divisive force, others as a useful and controllable entity. The manner and extent to which militants thought technology could come under political control varied, as did their understanding of what political control might mean. But all agreed that technological development, both at the national level and at the workplace level, was a legitimate concern for labor union politics.

The three labor confederations thus concentrated primarily on general issues of national (or international) technology policy. They were mostly interested in large-scale systems: electric power, energy, railways, aviation, electronics. Their concern with these systems revolved around matters of sociopolitical power. Who would direct the design of these systems? How could unions shape their deployment? How should the social effects of technology be managed? In addressing these questions, each union offered a distinctive representation of the relationship between politics and technology. These representations reflected broader union platforms and became part of their efforts to establish a distinctive doctrine. In this fashion, union wove notions about technological change into their self-conceptions and into their ideas about how to shape the future of France.

The CFTC/CFDT report and the responses to the questionnaire raised one of the most important issues unions had to contend with, intermittently throughout the 1950s and the 1960s, and increasingly in the 1970s: how to recruit technicians, managers, and engineers to the labor movement. Taken up by all three unions, this issue went to the very heart of what it meant to be a labor union in an advanced technological society. It deserves a brief closer examination, not only for this reason but also because the rank-and-file engineers of the CEA and EDF were part of the elite which they sought to recruit.

Recruiting Technical Elites

As numerous scholars have observed, the figure of the *cadre*—the midlevel manager—became increasingly important in the decades after World War II.[79] In the words of one militant: "The *cadre* is he who defends both the interests of the factory and the interests of the work force. . . .

The *cadre* is thus boss, worker, and technician at once."[80] For unions, the emergence of the *cadre*—an undeniably masculine archetype—signaled fundamental changes in the skill-based structure of the working class. At the same time, unions joined the national chorus proclaiming the need for even more technically trained personnel to ensure that France could participate fully in the modern world.[81] How should they handle such changes? The three unions agreed on this issue more than most, even if pride and principle forbade them from acknowledging their commonalities. They focused their discussions on the need to make elite education more broadly accessible and to enlist experts in the cause of the labor movement.

All three confederations bemoaned the lack of French engineers, technicians, and scientists. *Force Ouvrière* writers estimated that France needed to produce between 10,000 and 17,000 new engineers every year in order to match the "Anglo-Saxon nations."[82] The CGT agreed, adding that everyone should have access to the appropriate training and demanding fellowships for students whose families could not afford to support them through extensive education.[83] All three unions called for better *promotion ouvrière* programs, which would give workers the opportunity to retrain for more technically sophisticated jobs within the same company.[84] Force Ouvrière militants even seemed willing to break their usual abstinence on attacking "the system" in order to critique the structure of French education: the problem, said one writer, was that "the prestige of the Grandes Ecoles exerts a real dictatorship which victimizes [individuals with] real intellectual abilities."[85] Instead of emulating the examination system of the *grandes écoles*, other engineering schools should admit any student with an appropriate high school diploma. Furthermore, scientists and engineers did not receive sufficient compensation. According to one of the CFTC/CFDT's science policy specialists, low salaries and unsatisfactory labor contracts discouraged young people from pursuing research careers.[86] These criticisms did not attack elite state technologists per se; rather, unions demanded more recognition for rank-and-file experts and better access to the institutions that produced them.

At the same time, unions debated how to approach the existing and growing population of technical experts. Clearly these men could not be left to the mercies of management, or of the CGC (Confédération Générale des Cadres, a union for mid-level management and staff). Engineers, technicians, and *cadres* had to be recruited to the cause of the working class.

This mission appeared to hold particular importance for the CGT, which had sought—with very modest success—to attract *cadres* since early in the twentieth century. After the war it sponsored several *cadre* syndicates, loosely grouped into a Union Générale des Ingénieurs et Cadres. Not until the 1960s, however, did the UGIC gain the support required to have a strong voice within the CGT.[87]

In rallying union-wide support, militants presented the UGIC as a necessary weapon in a class war over the loyalty of engineers and other *cadres*. In particular, it could help combat the influence of the CGC, which the CGT saw as a pawn with which upper-level management inculcated *cadres* with capitalist ideology. The UGIC could make engineers and technicians understand that they shared more interests with workers than with management:

The evolution of technology—leading to the employment of a steadily increasing number of technicians, *cadres*, and engineers under conditions that are often similar, from the perspective of work discipline and intensity, to those imposed on blue- and white-collar workers—makes the maneuvers of employers more and more difficult and helps [these employees] become conscious of the solidarity of their interests with those of workers as a whole.[88]

A 1960 survey indicated that 14 percent of private-sector employees fell into the category of technicians, *cadres*, and engineers; this proportion reached 20 percent in the public sector.[89] The CGC duped *cadres* into thinking their interests lay with the ruling class. Only the CGT could demonstrate the importance of an alliance with the working class.

Belonging to the CGT would help engineers keep their priorities straight even when they rose to positions of power, thereby averting the ever looming danger of technocracy. Engineers, wrote one militant, had "a very healthy feeling of being creators, which can nevertheless stray towards technocracy if we don't show them who really profits from the . . . technology that they develop."[90] In particular, engineers needed to understand that blindly proposing methods to increase productivity did not serve anyone's interest. An alliance between *cadres* and workers would strengthen the struggle against capitalism.[91] Predictably, the most successful example of the benefits of such an alliance came from nationalized industry—especially EDF. One article in *La Vie Ouvrière* asserted: "Electricity and gas employees, whose right to strike . . . Pompidou contested recently, have for a very long time offered the example of a seamless solidarity between workers and *cadres*. From the manual operative to the engineer, they present a united front."[92]

Force Ouvrière militants made similar arguments in favor of recruiting *cadres.* Force Ouvrière had had a large federation for *cadres* and white-collar employees since its creation.[93] The only way, it claimed, for "relations between labor and 'technocracy' to be more profitable [for labor] than those with 'capitalism'" was to recruit the rank-and-file of "technocracy" to the cause of labor.[94] But recruiters had to keep in mind both the similarities and the differences between *cadres* and workers. On the one hand, an engineer, like a worker, would always be someone else's subordinate. On the other hand, engineers underwent a longer apprenticeship, spent more time keeping up with technical developments, and generally had more people under their command than even the most highly trained foremen. Successfully integrating them into the union required taking all these factors into account.[95]

The CFTC/CFDT did not form a separate organization for engineers and *cadres* until 1967, but engineers, scientists, and technicians had been an important part of its constituency throughout the postwar period. As we saw earlier, the subject made for lively discussion within the union. Indeed, despite the relatively strong presence of this constituency, CFTC/CFDT militants expressed more reservations about technical experts as a group than did their counterparts in other unions. For them, the danger of technocracy resided not only in the structure of the state but also in the mentality of engineers and *cadres*. "*Cadres*," wrote one militant in 1967, "have a very developed sense of order. They put this order in place, and they benefit from it. Hence, they do not want to saw off the branch on which they are seated. Our problem as syndicalists is to react against this state of mind."[96] Engineers often did not find strikes congenial modes of action, because strikes challenged order and efficiency. Militants therefore had to proceed gently. They needed to teach engineers how to actively incorporate the proper values in their work: "There are *cadres* who call themselves leftists yet who are rotten technocrats; for them, mathematics win out over human values."[97] Such attitudes had also appeared in the questionnaire responses—for example: "The technologist is necessary, but [let's] take the levers of command away from him, because the technologist is often blind. Leave the control to MAN."[98] Though some technicians resented pejorative comments that represented them as inhuman and unmanly,[99] others readily agreed that technical specialists, while meaning well, often ignored the social consequences of their actions.[100] Joining the CFTC/CFDT would make engineers and technicians better men who would produce better technology.

For the three confederations, therefore, the changing sociotechnical world demanded transcending traditional class boundaries. They saw technological knowledge as a locus of power, and they sought access to that power. For all three, this meant broadening their constituencies and their appeal, while at the same time blurring skill-based class distinctions.

*

Labor unions offered distinctive visions of France's future technological identity. The CGT glorified French technology but argued that its true apotheosis could only come after a socialist revolution. Force Ouvrière lauded technological progress as a trans-national phenomenon that would situate a new France in a non-communist international community. The CFTC/CFDT expressed ambivalence about technological change, placing it at the center of French modernization and seeing its development as a complex process that required careful control.

Each scenario incorporated a distinct relationship between technology and politics, which in turn articulated how each union envisaged its role in shaping France's future. The CGT gave a fairly straightforward rendition of this relationship: The right political system yielded the right technological system. In order to truly shape France's future (technological and otherwise), militants should advocate revolution. The CFTC/CFDT took a more nuanced approach: Technological decisions were political because they were about power and social order. Values necessarily guided technological choices. Only by explicitly acknowledging this could unions and workers acquire a voice in shaping the nation's future. For both unions, tight links between technology and politics thus entailed ways to shape the nation. The obverse held true for Force Ouvrière: Only by separating technology and politics (conceived in the narrow sense) could the union pronounce on technological change.

Such distinctions aside, labor unions' scenarios for a French technological future rested on a different vision of the sociopolitical order than that imagined by state technologists—a more inclusive vision, one that gave workers a central role. In this sense, their scenarios worked as alternatives to those of state technologists. In another sense, however, the mere fact that unions imagined the scenarios they did strengthened the general proposition that France should have a *technological* future, a *technological* identity. In other words, the implicit agreement that France should define itself in technological terms was as significant as the disagreement over what those terms should be.

This significance transpires most clearly in the unions' discussions of how to recruit the technical elite, and in their basic agreement on this issue. Not even the CGT sought to wage class warfare with engineers;

instead, it, like the other unions, strove to include the technical elite in its struggles. Having embraced rapid technological development as part of France's future, militant leaders felt that the best way for their unions to participate in that future was by recruiting those in charge of designing that development. How to effect this recruitment posed a puzzle that unions could not fully resolve in the 1960s.

Their fundamental agreement on, and puzzlement about, the issue of technical elites may help to explain the relatively minor role played by the unions' ideological differences on nuclear sites during that decade. The daily operation of nuclear reactors did not require large numbers of manual workers. Most of the men who operated reactors were, on the contrary, skilled workers who underwent further training for their jobs. Those near the top of the workplace hierarchy received the designation of "technician." Most nuclear workers thus belonged to the elite of the working class; the engineers who supervised them, meanwhile, belonged to the technical elite which unions sought to recruit. Furthermore, the training and conditions of work in nuclear reactors raised new issues for workers, technicians, and their unions. Under such circumstances, the question of what problems unions should address (and how) apparently took precedence over articulating ideological differences. The youth of these early nuclear workers (most were in their twenties) probably contributed to the relative lack of union disagreement on reactors sites: even workers who had unionized before arriving at a site did not necessarily have the investment in rehearsing doctrinal differences that older militants might have had. Once unions had gained some experience with the nuclear workplace, their ideological differences resurfaced.

5

Regimes of Work

In first three chapters I discussed the development of two technopolitical regimes, one centered in the Commissariat à l'Energie Atomique and the other in Electricité de France. "Technopolitical regimes" are linked, inter-defined, mutually constitutive constellations of engineering practices, technological artifacts, political programs, and visions of the sociopolitical order. The artifacts of these regimes provide the basis, and sometimes even the mechanisms, for their political power. At the same time, political agendas both drive and are constituted during the process of designing technological systems. The narrative that CEA scientists, engineers, and administrators developed about their institution as the guardian of French scientific and military autonomy was not merely rhetoric; that narrative was cultivated and acted out in the reactors they designed, in the long-term development plans they advocated, and in their efforts to shape EDF's reactors. The same was true for EDF technologists and their image of the nationalized utility as the nation's foremost public service institution. Each institution's cultural self-image was expressed, reshaped, and solidified by the material practices of its members. Similarly, the strength of each institution's political program rested on its technological practices and artifacts. The notion of "technopolitical regime," then, captures not only the fundamentally hybrid nature of the goals and activities of these institutions but also technologists' efforts to use these hybrids as instruments of power, models for state politics, and expressions of French national identity.

To varying degrees, France's three major labor unions all constructed scenarios in which workers could participate in this redefinition of French national identity centered on technological prowess. In theory, at least, workers could see themselves playing an active part in the construction of a new technological France. How did workers conjugate the scenarios offered by their unions with the technopolitical regimes in which they worked? In the present chapter I explore this question by examining hierarchy, authority, and work practices in the CEA and EDF. How was authority distributed within these technopolitical regimes? How did the programs and visions of engineers and administrators shape the spheres of nuclear workers (construed broadly to include manual operatives, skilled workers, and technicians)? How did these regimes regiment work practices? What space existed for dissent or difference? I argue that, in order to understand how nuclear workers made sense of their socio-technical roles, we must look at the manifestations and meanings of these technopolitical regimes in the nuclear workplace and at local unions' responses to these regimes.

Let me begin with a preliminary sketch of the ideological place of workers in EDF and the CEA. As the labor unions themselves so often repeated, EDF was the joint creation of management and labor. The communist labor leader Marcel Paul was one of the most prominent heroes of the utility's earliest years. With varying degrees of intensity and insistence, both management and labor agreed that Paul played a large role in the success of the company's nationalized structure. From the beginning, the utility cultivated a strong sense of its social mission: to redefine the relationship between the French worker and the French state.[1] EDF's labor contract—which guaranteed paid leave, made pay scales public, and incorporated union representatives into the utility's managerial structure—was portrayed as the key element in reifying this new relationship.[2] Even after the onset of the Cold War, the communist-dominated Confédération Générale du Travail retained a strong union presence in the utility, consistently capturing over half the votes in elections for personnel delegates.[3] And, as we saw in chapter 4, EDF continued to figure as the model for a new industrial and social order for labor unions through the end of the 1960s.[4]

While workers dominated EDF numerically, they made up a relatively small proportion of CEA employees. Indeed, the CEA sought to carve out a rather different place for itself on the landscape of French social politics. Its original heroes were scientists and Charles de Gaulle. True, one of its most prominent founders—Frédéric Joliot-Curie—had been a communist; but he was ousted by the government in 1951, together with many other communist scientists and technicians. Unlike EDF, the CEA did not welcome labor unions; indeed, they were denied entry to some sites until the early 1960s. A sizable minority of research scientists objected to the CEA's military aspirations, but few if any challenged its image as an institution dominated—and appropriately so—by scientific and technical experts. In short, workers simply did not figure in the CEA's portrayal of its mission and identity.

These are the bare outlines of how workers were supposed to figure in these two technopolitical regimes. How did these outlines map onto everyday practice? How did the technologies and organization of work enact or test these symbolic roles? Did workers abide by the prescriptions set for them by their regimes?

To address these questions, I will examine two nuclear sites: the CEA's Marcoule site and EDF's Chinon site. Each site housed three gas-graphite reactors. In addition, Marcoule had a plutonium extraction plant that prepared weapons-grade material for the French atomic weapons program, as well as numerous laboratories and testing facilities. Chinon too had a laboratory, though one smaller than Marcoule's. I will confine myself here to reactor work. More specifically, I will examine the heart of reactor operations: the shifts which rotated around the clock to maintain and control the fission reaction. I focus on the physical layout of work spaces, the structure of authority, the production of operational guidelines, and the degree of initiative allowed workers at each site.

Managing radiation-related risks was an essential part of nuclear work. Industrial-scale nuclear plants were new technologies in the 1950s. Just as there were no predetermined ways of structuring the reactor workplace, definitions of nuclear risk were highly contested terrain.[5] The Cold War made even Western nations suspicious of one another when it came to nuclear matters. Despite the Atoms for Peace conferences in the mid and late 1950s, the institutions of the French nuclear program, like their counterparts elsewhere, essentially developed their own workplace practices.[6] Thus, in each section of this chapter, after surveying work organization and practices, I turn to the structure and practices of risk management.

By now it has become banal to conclude that risk is culturally constructed. Numerous authors have argued that an individual's responses to danger depend on that individual's place in one or more systems of social organization[7]; others have argued instead that differing understandings of risk stem from conflicting social or political ideologies.[8] Less common are attempts to use the cultural construction of risk, not as a conclusion, but as tool with which to analyze broader themes. I argue that responses to risk did not reflect static, preexisting identities, but rather helped workers to develop, shape, and enact their identities in counterpoint to the technopolitical regimes in which they functioned.

Finally, I examine union activity at Marcoule and Chinon. Here, the rhetorical structure of my argument follows the structure of union-management relations at the sites themselves. Unions were not welcome at Marcoule, where they functioned largely outside the institutionally sanctioned limits of work. Accordingly, I grant them a separate subsection. Chinon, however, integrated unions into many of the ordinary workings of the site. I do the same in my discussion, adding a subsection on union opposition at the end.

Marcoule

The Site

The CEA's Marcoule site lay some 30 kilometers from the medieval papal city of Avignon at the base of a steep, tooth-shaped hill known locally as the Dent de Marcoule. The Rhône flowed along the eastern frontier of the site, providing a handy depository for waste effluents. Anxious to avoid alarm or confrontation, CEA administrators had deliberately chosen a sparsely populated region to accommodate the nation's first gas-graphite reactors. They also hoped that the Mistral, an exceptionally strong wind characteristic of the region, would whisk away any noxious elements that might accidentally escape from the plant.[9]

The CEA began buying land in 1953 while the initial blueprints for its first gas-graphite reactor, G1, were being drawn up. Construction proceeded at lightning speed. By 1958, an on-site plutonium factory was converting the spent fuel from G1 into weapons-grade material. The second reactor—G2, the one I focus on here—began functioning in 1959, and G3 was ready the following year. The year 1958 witnessed one of Marcoule's biggest hiring efforts, as site managers searched for men[10] to operate and maintain the second two reactors and the plutonium factory. By the end of that year, 1200 employees—engineers, technicians, and workers—worked at Marcoule. Around 200 were assigned to the operation and maintenance of G2.[11]

Many prospective workers and technicians arrived at Marcoule in response to ads published by the CEA in local newspapers. French public education in the 1950s did not offer training programs for reactor technicians or operators—indeed, many applicants barely knew what a reactor was. Engineers expected that they would have to train workers, and they primarily sought men who had followed some kind of technical curriculum in school. They did not, however, expect the content of this curriculum to have much relevance to reactor work, which they conceived

Figure 5.1
The western entrance to the guardhouse at the Marcoule site in 1959. Employees
and visitors had to pass through the guardhouse and show identification before
being allowed onto the site. Source: CEA/MAH/Jahan.

as something completely new under the sun. Upon arrival, prospective
employees took a written test and submitted to a background security
check, intended in part to weed out communist sympathizers.[12] Those
who passed were assigned to a specific location on the site—the reactor,
the plutonium factory, or the laboratory—and put through an intensive
training course.[13]

Most of the individuals hired at Marcoule in 1958 were not explicitly told
that they would be helping to produce plutonium destined for an atomic
bomb, even after the French government formally declared its intention to
pursue the military atom. Nevertheless, plenty of clues pointed in that
direction, and most workers quickly guessed the site's purpose. Perhaps the
most obvious of these clues was the presence of numerous former military
officers among the engineers and high-level technicians. In part, this was
because in the 1950s few experiences other than military service provided
knowledge of remote control and automation technologies like those used
at Marcoule. Furthermore, the Cold War atmosphere made former military
officers appear less of a security risk than civilians. Other clues to

Marcoule's military mission lay in the site's security measures: no one suspected of communist affiliations was to enter the grounds, and employees were forbidden to discuss their work when off the site. One engineer recalled a particularly ironic example of these measures:

I worked with an engineer who had designed all the command [and] control mechanisms for the Marcoule pile [G1]. But his father—or his uncle—was a communist *conseiller général* from Corsica. He was never able to enter Marcoule. He had designed it, but he couldn't get into it. That didn't matter. . . . We held meetings in the guard house. . . . Yup, he couldn't enter because they discovered that he had a communist father. . . . That was just totally military.[14]

Another engineer, this one of Eastern European origin, recalled an investigation following a minor accident that had put G1 out of commission for several weeks:

After the accident, they asked themselves if I hadn't perhaps sabotaged [the reactor], if perhaps I wasn't a communist. I wasn't, but there was a little inquiry commission. . . . It wasn't very pleasant. The military goal wasn't officially known, but of course I knew about it.[15]

Incidents like these made Marcoule's military mission easy to guess.

But Marcoule was not a military site per se. The former military officers who ran it also had to follow expectations set by the CEA's culture of elite scientific expertise. Thus, the structure of Marcoule's hierarchies was inspired by military experience, but the basis for authority within those hierarchies rested on scientific and technical expertise. Workers were expected to obey orders unquestioningly.

All in all, then, Marcoule was the quintessential expression of the CEA's technopolitical regime: a fortress impenetrable by communists, a civilian site thoroughly imbued with military goals and methods, a domain reigned by scientific and technical expertise. How exactly did this regime manifest itself in the organization of work?

Working in the G2 Reactor

Work organization at G2 followed a strict hierarchy of expertise: engineers executed the conceptual work, technicians manipulated the complex instruments central to the reactor's operation, and workers handled simpler instruments scattered throughout the reactor. A closer look at the jobs performed by these employees will help us understand how this organization worked and how these employees interacted with each other. This, in turn, will shed light on the social meanings of authority and expertise at the site.

G2 had six engineers supervising its operations. The *chef de centrale* (a former naval officer) and his assistant coordinated the work of the rest of the staff and maintained links with the CEA's research division. The other four engineers each headed one team, and the four teams rotated eight-hour shifts over a four-week period. In addition, each engineer had a field of specialization and remained on call in case a problem developed that required his expertise.[16] The engineer on duty tried to solve any unexpected problems on his own, but if something became too complex for him he could call in the relevant expert for help.[17]

The engineer on duty had an office just off the "command room."[18] Although he ultimately had responsibility for the smooth functioning of his shift, he did not supervise the routine work of his staff. That task fell to his immediate subordinate, the *chef de quart* (shift foreman).[19] Instead, the engineer worked on more general operational tasks: writing or modifying operational guidelines, redesigning problematic components, keeping track of plutonium production in the core, and so on. The *chef de quart*, a skilled technician, was based in the command room. He supervised the workers and technicians on his shift and served as liaison between them and the engineer on duty. Workers and technicians reported malfunctions or other problems to the *chef de quart*, who in turn reported them to the engineer.

The command room lay on the top floor of the "command building," right next to the reactor building. A spacious room with a high ceiling, it was dominated by the control board, the design and operation of which reflected the CEA's tendency to subdivide tasks into discrete areas of technical expertise. The control of Chinon's plants integrated different aspects of reactor operation as much as possible. G2's control board, however, contained three separate sections, each manned by a different technician. The *conducteur de pile*, who was in charge of "piloting" the reactor core, sat on the left side of the board.[20] In the center sat the *conducteur CO_2*, whose task involved monitoring and regulating the flow of cooling gas in the reactor. On the right side sat the *conducteur IRE* (energy recuperation installation pilot) who controlled the functioning of the turbines and heat generators annexed to the reactor.

After the engineer on duty, the *conducteur de pile* enjoyed the highest status of all those involved in operating the reactor. His job had its own slot in Marcoule's official personnel categorization: neither an "engineer" nor a "technician," he was, quite simply, a *conducteur de pile*—a minor expert in his own right.[21] Significantly in an institution that ascribed the highest value to anything it labeled "nuclear,"[22] the instruments at his

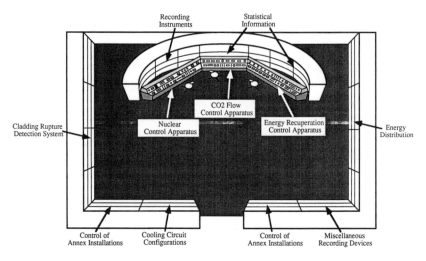

Figure 5.2
The floor plan of G2's control room. Drawing by Carlos Martín.

Figure 5.3
G2's control room in 1959. Source: CEA/MAH/Jahan.

disposal fell collectively under the rubric "nuclear control." These included scales and switches that permitted him to read and regulate the power of the fission reaction, a drum to lower or raise the control rods, and an array of recording devices that informed him about the condition of the reactor's crucial components. He thus monitored and regulated the fission reaction at the very core of plutonium production.[23]

The two other men stationed at the control board did not benefit from the special status of the *conducteur de pile.* The *conducteur CO₂,* who was responsible for more conventional technology, was a "technician"; the *conducteur IRE,* as an EDF employee, did not fall into any of Marcoule's personnel categories. But even though their work was less valued in part because it was, in some ineffable way, less nuclear, the instruments they confronted daily at their stations were equally sophisticated. The *conducteur CO₂,* responsible for continuously verifying and regulating the flow of cooling gas in the cooling system, had instruments that told him the temperatures of hundreds of reactor components and gave him measurements of G2's thermodynamic power. Such information enabled him to control the rotation speed of the turbo-blowers that blew carbon dioxide through the pipes, thereby controlling the flow of coolant in the core. Because the rate at which uranium transformed itself into plutonium depended heavily on temperature, this job was as crucial as that of the *conducteur de pile* for producing weapons-grade plutonium in G2's core.[24]

The main responsibility of the *conducteur IRE* lay in ensuring that the energy recuperation installation evacuated the heat generated by the fission reaction in the core without interfering with plutonium production. Although this installation was theoretically designed not to impinge on plutonium production, it required operation and monitoring. The EDF pilot had to ensure that his installation would not interfere with the core's stability.[25] Otherwise, he could run that installation as he saw fit; that, however, did not leave him much leeway, for the amount of energy he could generate in his installation depended on the behavior of the core, which did not lie within his control.

In addition to the engineer and the *conducteurs,* a shift included seven men listed as workers in Marcoule's personnel categories. These were stationed throughout the reactor building to run an array of devices associated with G2's cooling system, the electricity distribution network, and the loading machine.[26] A description of the job performed by the loading machine operator provides an illuminating contrast with those carried out by the *conducteurs* in the control room.

Most of G2's regular shift workers moved in and out of the reactor building, taking measurements, performing maintenance work, and so on. The operator of the loading machine, however, spent all his time perched atop the machine, right next to the reactor core. His control board enabled him to retrieve new fuel slugs for loading, move the loading machine into position, and load the new slugs in while pushing the old ones out. Yet this intimate contact with the core, with all that was most "nuclear" in G2, did not confer special status upon this worker, for, unlike the *conducteur de pile,* he did not have to make decisions about how to interpret complex instrument readings. His devices were extremely simple, consisting of buttons arranged on colored bands. Each band corresponded to a task, each button to a discrete operation within that task. Lights and arrows on the button told the operator when to push the button, how long to leave it pushed in, and when to proceed to the next specified task.[27] The job left little room for initiative. Together with that of the worker who managed the spent fuel in the cooling pool, this job commanded the lowest status in the reactor: by all accounts, loading machine operators were the *ouvriers*—the laborers—of G2.[28] Clearly, intimate knowledge of the core brought status, but intimate contact did not.

Thus the high value placed by the CEA's technopolitical regime on scientific—especially nuclear—expertise shaped the definition of tasks and jobs. How did authority flow within Marcoule's version of that regime? To what extent could employees at different hierarchical levels shape their jobs? Let me address these questions by examining how the creation and use of written operational guidelines mediated both the technical work of employees and the social relations between them.

Throughout the 1960s, formal operational manuals for G2 did not exist. The novelty of the technology made the creation of manuals before reactors went on line impossible. Nor did France's fledgling regulatory system require such manuals. Nonetheless, the scientists and engineers in charge of G2's initial startup had drafted preliminary operations guidelines with the help of the builders. The task of formulating more complete instructions fell to G2's four shift engineers, who wrote guidelines in conjunction with their two supervisors. These instructions fell into two categories: *règles* (rules) and *consignes* (guidelines). Rules never changed; once formulated, they remained constant for the life of the reactor. Guidelines, however, could change with operational experience and design modifications.[29]

G2's shift engineers implemented numerous design modifications for the reactor during the 1960s, and could therefore easily write adjustments to the guidelines. During routine operation, however, technicians and workers came into much closer contact with the instruments and devices that made up G2 than the shift engineers. These men regularly identified flaws in existing guidelines. But they could not implement modifications independently. One engineer I interviewed explained this by saying that a worker might notice a problem but might not have sufficient knowledge of the entire reactor to modify guidelines safely. Safety, however, was not the only reason for restricting who could write guidelines. Inspired by their military antecedents, the behavioral practices that reigned in Marcoule's hierarchy prevented everyone but the *chef de quart* from even suggesting modifications directly. When they identified a potential problem, non-engineers either told the *chef de quart* or recorded it in the *cahier de bord*,[30] the log book where the personnel kept track of the operations carried out during a shift. The *chef de quart* then investigated the problem and reported his findings to the shift engineer. Depending on the problem's magnitude, the engineer either devised a solution independently or consulted his three peers.[31] They reified any modifications by writing a new guideline, which then journeyed back down the hierarchical ladder: the *chefs de quart* signed the document to prove that they had read it, and then the relevant worker or technician received the guideline, which he was expected to follow to the letter until a subsequent guideline replaced it.[32]

For the engineers who ran the reactors, the opportunity to rewrite guidelines embodied a fundamental professional distinction. G2's engineers called themselves *exploitants*, meaning production engineers. In contrast with Chinon, at Marcoule only engineers identified themselves thus.[33] Design engineers thought that *exploitants* had the least interesting work in the profession.[34] But the latter scorned the design engineers, who did not appear to understand the complexities of devising appropriate operational procedures and who seemed to think that designing a plant was enough to make it work. Marcoule's *exploitants* took particular pride in their special expertise, thinking that their work with nuclear reactors placed them in the elite of production engineering.[35]

The same process that gave Marcoule's engineers their sense of worth, however, undermined that of workers. In contrast with engineering positions, their jobs did not live up to romantic ideas about working for a pioneering industry. The rigidity of the hierarchy, combined with G2's single-minded purpose (producing weapons-grade plutonium), meant that, at least officially, their work consisted predominantly of following written orders. Whether or not they stuck to those guidelines we do not know, but the fact that they were not expected to provide creative input into reactor operation contrasted sharply with the expectations that EDF had of its reactor workers. The room for initiative or discretion that did exist for Marcoule's workers and technicians—the opportunity to bring problems to the attention of the *chef de quart*—was robbed of its satisfaction by the hierarchical journey that any suggestion had to take. Suggestions that eventually changed guidelines returned to the workers as orders. Many workers came to view their job as a *boulot de con*—roughly translatable as "ass's work."[36]

Marcoule's workers resented the military-style chain of command that governed their workplace and the constraints placed on any enterprising impulses they might have. Its engineers, along with the rest of the CEA, took the need for such hierarchies and constraints for granted. This pattern of formalizing expert power and curtailing non-expert initiatives intensified with the organization of radiation safety practices.

Risk and Radiation Protection

Radiation was nothing new to the CEA. But in France, as elsewhere, most of the work done in potentially radioactive environments had been done in labs or other research settings. The dangers of nuclear work had not yet begun to cause widespread concern, and formal guidelines for working in such environments had yet to be developed. EDF and the CEA took different approaches to this problem. Not surprisingly, the solution devised at Marcoule relied on creating experts empowered to organize and monitor radiation protection. I will now turn to the role of these experts, and more generally to the cultural meanings of risk at Marcoule.

The initial design of the Marcoule reactors incorporated some protection against radiation: thick concrete walls surrounded radioactive sources, and detectors placed near these sources measured radiation levels. But before early 1959, the men who ran the reactors had neither individual detectors, nor protective clothing, nor training in the basics of radiation protection.[37] The arrival of hundreds of new employees in 1958 prompted Marcoule's administration to create an on-site service to handle radiation protection: the Service de Protection contre les Radiations (SPR). The administration requested a few scientists from CEA headquarters to help train SPR men, most of whom had some form of university-level science education. This training process conferred upon SPR men the status and title of experts: they became *ingénieurs de radioprotection*, even though few had formal engineering education. These *ingénieurs*, in turn, trained a staff of technicians, who themselves came to qualify as experts, albeit of lesser status; they became *agents de radioprotection.*

The SPR began functioning in 1958, during G2's year-long startup phase. Initially, SPR agents could do little other than observe reactor work: the urgency that the CEA placed on plutonium production prevented its agents from implementing protective measures that might slow down the startup.[38] Once G2 began routine operation, though, the SPR had more power. Marcoule's administration did require that protective measures not seriously interfere with plutonium production, but within that limit the SPR had complete authority over radiation protection guidelines and practices.

The first measures taken by the SPR involved developing protective clothing and reconceptualizing the work space. Its *ingénieurs* felt that the choice of protective clothing had to complement the layout of the work space in the reactors and the plutonium factory, especially since the risk of radioactive exposure varied greatly throughout the installations. In G2, for example, the operator of the loading machine ran a greater risk of exposure than did the *conducteur de pile,* and most workers traveled between higher-risk and lower-risk areas. After considering various options, the SPR decided to provide two types of protective outfits: the "universal outfit," which consisted of a heavy shirt and pants, and the "work outfit for zones at permanent risk of contamination," which included, in addition to the "universal" version, an extra jacket, a cap, and overshoes.[39] It also subdivided each facility into two work zones—"regulated" and "unregulated"—and a third "forbidden" zone. "Regulated" zones fell into three categories. The green zone included areas where the level of ambient radiation always fell within the accepted norms[40] and a risk of contamination existed only in the event of an accident. G2's control room fell into this category. The yellow zone covered areas where radiation levels remained within the norms but a perpetual risk of contamination existed. G2's loading machine fell into this category. In the red zone, radiation levels were high enough that all human work had to be done quickly and under constant surveillance. Employees typically entered red zones only under unusual circumstances (such as an accident). Finally, the SPR provided every Marcoule employee with an individual dosimeter and a personal film badge to measure his daily and monthly exposures to radiation.[41] These measures were in place by late 1960.

Apparently, Marcoule's employees did not take well to the new rules. SPR agents protested the "negligence" of the personnel in respecting regulations and wrote about the difficulties in striking the proper balance

Figure 5.4
Men donning the "universal outfit" before entering the work zone. Source:
Photothèque.

between making employees understand risk and alarming them unduly.[42]
To address these problems, the SPR instituted a policing system, which
involved regular employee surveillance by SPR agents. An *agent de radioprotection* assigned to each shift took responsibility for ensuring that workers and technicians complied with protective guidelines. The SPR insisted
on building a separate cubicle for the radiation detection instruments
formerly housed in the command rooms of G2 and G3. When he was not
following personnel around, the SPR sentinel sat in this cubicle, keeping
one eye on the instruments and the other on men entering and exiting
the reactor.[43] The SPR also launched a series of efforts to teach the rest
of the personnel about radiation protection. It sponsored lectures about
the dangers of handling radioactive materials, and it put up warning
posters featuring cartoon characters and clever slogans. The remarks of
three SPR *ingénieurs de radioprotection* on the importance of designing
amusing posters reveals the how these experts thought about the workers
they sought to protect: "Most often the worker reacts ironically to a serious poster that brutally shows the risk being taken. . . . Furthermore, the

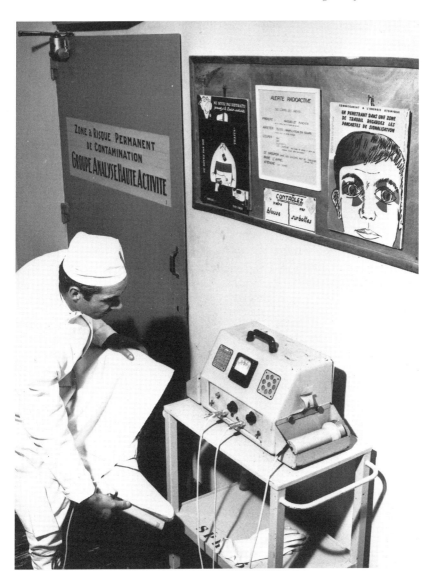

Figure 5.5
Scanning for radioactivity after leaving a yellow zone. The two cartoon posters warn employees to pay attention to radiation in their work zones. Source: Photothèque.

worker, during his work, will more willingly look at a poster that for him is a sort of recreation."[44] As far as the SPR agents were concerned, workers could not be trusted to take safety issues seriously.

Interviews with former G2 workers suggest that the SPR's complaints about workers' disregard for safety measures were well founded. One worker put it succinctly: SPR agents were "a pain in the ass."[45] But that comment alone suggests that workers did not circumvent regulations out of ignorance, carelessness, or stupidity—as the SPR seemed to imply—but out of a kind of defiance.

A partial explanation for how employees reacted to safety guidelines lies in the cultural meanings of radioactivity in the CEA in general and at Marcoule in particular.[46] The CEA's "basic" researchers, it appears, treated radioactivity with disdain. One researcher recalled that he and his colleagues had deliberately dragged a radioactive source around the halls and outside the building, laughing when the lab technicians next door worriedly followed the radioactive trail a few hours later with their Geiger counter. Nostalgic tales of such pranks abounded in the interviews. Often, these tales ended with sighs about the "good old days" and hasty assurances that such practices were no longer tolerated. Marcoule engineers told similar stories about their exploits in the 1950s and the 1960s. The dominant theme of these tales, however, was not mischief but heroism. One engineer recalled an incident in which a piece of cloth had gotten stuck in one of G1's channels. Rather than build a mechanical device to remove the cloth while keeping people at a "safe" distance from the core, this engineer decided to fetch the cloth himself. It only took a few minutes, he assured me, and although he absorbed more than the accepted norm of radioactivity, he offered his present good health as proof that he had not done anything overly dangerous. At the same time, he affirmed that he would not have sent "one of his men" to do the job: he would never have asked someone to take a risk he himself was not prepared to assume. His narrative could easily have been that of a military officer describing a battle scene.[47]

Courting danger in the form of radioactive exposure was thus heroic and manly, but only in exceptional cases. An engineer who would approach a reactor core to dislodge a wayward piece of cloth for the sake of France's nuclear program was a noble man, worthy of praise and admiration. Of course, Marcoule's *exploitants* could afford to engage in such heroics: they spent most of their work time in unregulated zones. Furthermore, the nature of Marcoule's hierarchy ensured that engineers had the most opportunities to behave heroically. When an engineer flaunted radiation safety guidelines he took an exceptional risk and became a hero, but similar actions carried a very different meaning if performed by workers during the ordinary course of their work day. A worker who removed his film badge to conceal the dose he received under routine conditions took—in the view of the SPR—an unnecessary risk. If caught, he would not get the admiration reserved for the heroic engineer; instead, he might be subjected to a good scolding from an SPR agent, and his case would provide supporting evidence in an article complaining that workers did not respect safety rules.[48]

Exceptional risk taking contributed to the social recognition granted to an individual, but ordinary risk taking did not. G2's loading machine operators and with the men who worked in the disactivation pool under the reactor (where spent fuel was put to cool) risked radiation exposure more regularly than their colleagues. Yet they also had the lowest status at G2. The regard accorded them by their peers was in no way enhanced by the fact that their jobs were more dangerous—indeed, quite the opposite. It was not solely because the jobs required little training that they carried little prestige; it was also because their work made them dirty rather than heroic. Unlike Marcoule's heroes, these men accumulated radiation doses slowly over time. They were thus, in a sense, permanently unclean.[49] A man who happened to accumulate more than his allotted dose in a given year would be temporarily transferred to another post, and the transfer itself would be experienced as a kind of humiliation—which could well explain why some preferred to remove their badges to conceal their exposure. Taking risks brought glory only when danger loomed suddenly. When the danger consisted merely of commonplace radiation, the work required to face it became banal.[50]

Marcoule workers did not remove their badges out of ignorance or love of danger. Somewhat ironically, badge removal was an act of (social, if not physical) self-preservation. It was also an act of rebellion against a system that accorded them little respect and no autonomy. An examination of labor union activity at Marcoule reinforces these points.

Unions
Because of the CEA's continuing Cold War suspicions of anything that smacked of Communism, labor unions had considerable difficulty establishing themselves at Marcoule. The two non-communist unions—Force Ouvrière and the CFTC/CFDT—succeeded in creating locals around 1960; the communist CGT followed, with a notably smaller constituency, about a year later. Only the CFTC/CFDT appears to have kept docu-

mentation from this period, in the form of its bi-monthly newsletter *Rayonnement*.[51] It seems clear, however, that union solidarity at Marcoule—and indeed within the CEA as a whole—was much stronger than it was at the national level, at least through the end of the 1960s. The Marcoule locals apparently even considered joining together to form a single trade syndicate, though in the end they did not feel it wise to make such a move when union relations at the national level remained problematic.[52]

The unions might have succeeded in establishing locals at the site, but Marcoule's administration remained actively hostile to anything that remotely resembled revolutionary activity. Not that union actions at the site—or indeed anywhere else in the CEA—were particularly revolutionary in the 1960s. Unions focused primarily on fighting for higher salaries and better benefits. The CFTC/CFDT in particular portrayed itself as a modest, reasonable organization that was anxious to avoid needless confrontation. "Today's technological world," it proclaimed, "demands a constructive [approach to] syndicalism."[53] After a CEA-wide strike for higher salaries in May 1961, one Marcoule militant wrote: "We must remove anything that [smacks of] revolt from this strike; but we must remember the rise in consciousness and the affirmation that [we are] a collectivity. . . . We will retain neither fever nor passion . . . but Realism and Determination."[54] Despite such moderate rhetoric (which may well have been designed to reassure and recruit non-unionized workers), Marcoule's directors wanted to curb all strike activity. They denied production bonuses to anyone who had participated in the May strike.[55] In subsequent years they sought to keep an official record of who went on strike.

Such moves angered the unions, which accused the site's director of trying to intimidate workers. By recording who participated in strikes, the administration was undermining the workers' fundamental right to strike.[56] Marcoule's director sought to play with workers' sense of pride and masculinity, telling the unions: "If agents are not capable of taking responsibility for their decision [to strike] by signing a declaration, then these are children that you represent."[57] To which the CFTC/CFDT replied bitterly: "If any of us thought that in our Modern Society (and doesn't the CEA owe itself to be at the forefront of tomorrow's society?) class struggle was an idea to be overcome, here is a severe reality check."[58]

The tension between directors and employees was much worse at Marcoule and at the CEA's other "production" sites than at its research centers, in part because the employees in question were workers rather than trained researchers. "The hierarchy in the production centers is much more marked than in the research centers," remarked one CFTC/CFDT militant. "There is little contact between technologists and workers, which leads to a different climate, one which makes the work of personnel representatives difficult."[59] Visitors from the research centers also noticed the sharp contrast in atmosphere. One Saclay researcher recalled his surprise at discovering that Marcoule engineers used a different cafeteria than workers, while top-level administrators entertained visitors at a small nearby château that the CEA had purchased for this purpose.[60] The CEA's technopolitical regime thus found its most extreme expression at Marcoule, where military hierarchical habits reinforced the elitism of experts to create a strict separation between workers and engineers.[61]

It is this social tension, far more than carelessness or indifference, that explains workers' attitudes toward radiation protection guidelines. Indeed, left to their own devices, unionized workers appeared to be actively concerned with workplace safety and complained that the administration did not take this issue seriously enough. Like any other industrial site, Marcoule was legally required to maintain a Commission of Health and Safety (CHS) which included personnel as well as management representatives. Marcoule's CHS met only twice a year, the minimum prescribed by law; in contrast, Saclay's CHS met monthly. This laxness, argued the CFTC/CFDT, reflected an indifference to health and safety issues which stemmed from the site's military "vocation."[62] Marcoule's directors, said the unions, felt that the production of fissile material for the nation's weapons program exempted them from adhering to normal health and safety practices. "How does Marcoule's CHS work?" asked one militant. "Badly! Very badly! The Administration, proud of its double electric fence, isolates itself. Ostrich-like, it doesn't want to see anything or hear anything; perhaps without even realizing it, it takes the law into its own hands."[63] Apparently even the state's work inspectors were cowed by the aura of weapons-grade plutonium and stayed well away from the site.

The CFTC/CFDT found management's attitude toward workmen's compensation plans equally egregious. As long as they did not receive radiation doses in excess of internationally established norms, Marcoule's workers were not entitled to compensation. The union, however, argued that lack of conclusive proof that radiation exposure caused certain illnesses did not exempt the CEA from responsibility.[64] The notion of maximum admissible dose mattered for work organization and for daily practice, said union militants, but did not bear on compensation issues. They added bitterly: "At Marcoule, some agents are removed from an active zone and assigned to a non-active [zone] just before they reach their irradiation threshold. Is this a new assignment, or a disguise?"[65]

Once again, such behavior showed that Marcoule considered itself a "closed establishment, on the fringes of legislation."[66]

Evidently, unionized workers—and probably non-unionized workers too—were concerned with their health and safety. They did not blindly trust the CEA's experts to look after their best interests. At least some of these workers were the same ones who defied radiation protection guidelines. Clearly, then, this defiance represented not a blanket disregard for health and safety but a rebellion against a regime that gave them little freedom and even less credit. When even strikes could lead to penalties, the only source of autonomy left was to break the rules.

A comparison between Marcoule and EDF's nuclear site at Chinon will yield a better appreciation of how the CEA's technopolitical regime produced distinct work spaces and practices and a better understanding of how responses to risk served as a means of enacting identity in counterpoint to that regime. There was more than one way to organize work in a gas-graphite reactor, structure authority on a nuclear site, and face and interpret the dangers of radiation exposure.

Chinon

The Site

The Chinon site lay on the banks of the Loire, within the territory of the village of Avoine, about 10 kilometers from the château of Chinon. EDF chose this location because of the region's paucity in local energy resources. Flat and distant from major seaports, the Loire Valley could not develop hydroelectric energy and had difficult access to coal or oil imports. A major river was needed to supply cooling water for the reactor's heat exchangers. And, like the CEA, EDF wanted to place its first nuclear site in the countryside, away from a metropolitan area.[67]

Reactor designers at EDF liked to say that a nuclear power plant was just a conventional power plant with a different heat source. By downplaying the nuclear dimension, they differentiated themselves from their CEA counterparts and asserted their own particular expertise in building power plants. This attitude prevailed even more among the men who ran the three reactors at Chinon (EDF1, EDF2, and EDF3).[68] Many had previously worked at other EDF power plants and had developed a strong sense of profession: that of electricity producer. The reactors offered them a new and more exciting means of developing their skills and continuing their vocation. The spirit of electricity production hence shaped the work at Chinon, much as the military spirit had done at Marcoule.

EDF did much to encourage this sense of profession throughout the utility—not just among engineers but also among workers. As we have seen, at the foundations of its technopolitical regime lay an ideology of public service and worker participation. The utility carefully maintained and cultivated this ideology, most obviously in its training programs and in its monthly in-house magazine *Contacts électriques.* EDF trained and professionalized young men in its own schools. One description of the school in Gurcy-le-Châtel frankly acknowledged:

. . . it is not just a matter of inculcating . . . young people with the theoretical and practical knowledge needed to train a good worker. That is not even the essential part. What we want above all, is to make men of them in every sense of the word: develop their moral value, give them a sense of their responsibilities and a sense of "public service," and make them understand that regardless of the specific position that they will be called to occupy, their work will contribute to the well-being and development of their country.[69]

Work was a "calling," and masculinity, morality, and patriotism were intertwined in definitions of the good EDF worker. Physical exercise, team sports, and a system in which students elected representatives from within their own ranks to maintain discipline within the school all contributed to developing such values. Equally important to the ideal male worker was an inclination to heroism. Heroic deeds performed by EDF agents were regularly reported in *Contacts*, and workers were often portrayed in heroic postures—for example, perched atop a pylon, with a triumphant arm in the air.[70] In 1961 the utility's production and transmission division made a documentary film, entitled "Workers of Light," that covered the work life of a group of EDF agents in a remote rural area. The film described "the hard work of winter, the snowstorms, the downed lines which had to be repaired in record time so that the population would not be deprived of current."[71] This stirring portrayal of selfless, masculine heroism was shown to plant and line workers throughout the country.

Workers thus arrived at the Chinon site with a clear sense of who they were expected to be and how they were expected to fit in. Had any of them ever been to Marcoule, they would immediately have seen how such expectations fitted into EDF's technopolitical regime. Rather than emphasizing distinct areas of specialization, Chinon's work structure—although still very much a hierarchy—demonstrated a more technically and socially integrated approach to reactor control than did Marcoule's.

Work organization changed regularly as engineers and workers learned more about how to operate reactors (indeed, workers had input into such changes). The same principles, however, held throughout the

1960s. Whereas at Marcoule each shift was headed by engineers, at Chinon four engineers were responsible for three reactors: one oversaw operations as a whole, and one was in charge of each reactor.[72] Skilled technicians and workers took care of the daily operation of the reactors. Five teams worked in eight-hour shifts. Each shift thus had one more week off per rotation than its counterpart at Marcoule, and at the end of the 1960s a sixth shift was introduced. Shift organization also differed significantly between Marcoule and Chinon. At Chinon each shift was headed by a *chef de quart*, who was in charge of supervising the operation and safety of the reactor. A *chef de bloc* sat at the control board and operated the reactor. He was seconded by a *technicien de radioprotection* (TRP), who monitored the radiation detectors and offered safety and protection advice to personnel working in regulated zones. But in contrast to Marcoule, where the equivalent of a TRP would do nothing else, at Chinon the *chef de bloc* position and the TRP position were interchangeable: men took turns each week filling these two posts. Another technician, labeled an *opérateur*, ran the fuel loading machine. Finally, the *chef de bloc* had help from several manual operatives known as *assistants conducteurs* and *rondiers*.[73]

The Chinon employees also differed from their Marcoule peers in background and in training. The CEA, as we saw, did not expect the previous knowledge of its workers to be relevant. In contrast, Chinon sought men with power plant experience, and it put them through seemingly more rigorous training programs than Marcoule workers underwent.[74] The upper echelons of the hierarchy—from the engineers down to the *chefs de bloc*—arrived at the site when the reactor they would operate was still under construction, helped foremen supervise the construction, and participated in the initial startup, thereby gaining intimate knowledge of the plant.[75] Such experience made it reasonable for them to think that they might eventually qualify for substantial promotions—a hope that was further encouraged by EDF's "labor advancement" program (a product of labor union demands), which provided the requisite formal training for such promotions.[76] Though I did not find out how many Chinon employees participated in this program, it was apparent from my interviews that the program gave employees a sense that their institution afforded them serious opportunities for advancement—a sense not shared by Marcoule workers.[77]

Working in the EDF2 Reactor

To better illustrate the differences between Marcoule and Chinon, let me offer a snapshot of work organization and practice at EDF2 (Chinon's second reactor) in 1965. As will become evident, the differences between work there and work at G2 cannot be explained by design differences, particularly since EDF2 was technically far more complex than G2.[78]

Most of the early *chefs de bloc* of EDF2 had become acquainted with reactor work at EDF1 and were officially assigned to EDF2 in 1963. In part because they had witnessed much of its construction, they developed a deep loyalty to EDF2. One of these *chefs de bloc* nostalgically recalled the first few years of operation as a time of great excitement, discovery, and creativity—a time when the *chefs de bloc*, in tandem with the reactor's designers and the *chef de quart*, developed their own methods and guidelines for operating the reactor.[79]

Indeed, in stark contrast to G2, where only engineers wrote operational guidelines, at EDF2 the *chef de quart*, aided by his *chef de bloc*, was specifically in charge of developing such guidelines. Codified, standardized operation manuals still did not exist. Each reactor employee, regardless of his level in the hierarchy, carried a little book in which he noted peculiarities of various machines or instruments and wrote down the operational sequences he had agreed upon with the *chef de quart*. To a significant extent, these handwritten notebooks constituted the operational guidelines. Their personalized nature greatly promoted a feeling of responsibility and pride in the men who ran EDF2.[80]

Although EDF kept the CEA's military nomenclature, the control room where the *chef de bloc* sat to operate the reactor had a very different layout from G2's command room.[81] EDF2's control room was a direct expression of the utility's technopolitical regime, dialectically relating the technological blending of control functions to the social blending of responsibility. The control board reflected the utility's conception of its reactors as fulfilling a single purpose: electricity generation. Recall that G2's tripartite control board reflected the CEA's conception of a reactor as comprising three distinct parts: a nuclear core, a cooling system, and an energy recuperation system. In EDF2, the three sections of the control board—"principal," "secondary," and "automatic"—corresponded not to different reactor parts but to different phases of operation, and the *chef de bloc* and his assistants each worked with all three. The principal section contained all the instruments and switches needed to run the reactor "manually" and under routine conditions.[82] From it the *chef de bloc* obtained information about the reactor core, the carbon dioxide cooling gas, and both heat exchangers; operated the control rods and regulated the flow and temperature of the cooling gas; and started up or shut down the reactor. In the secondary section were all the instruments not

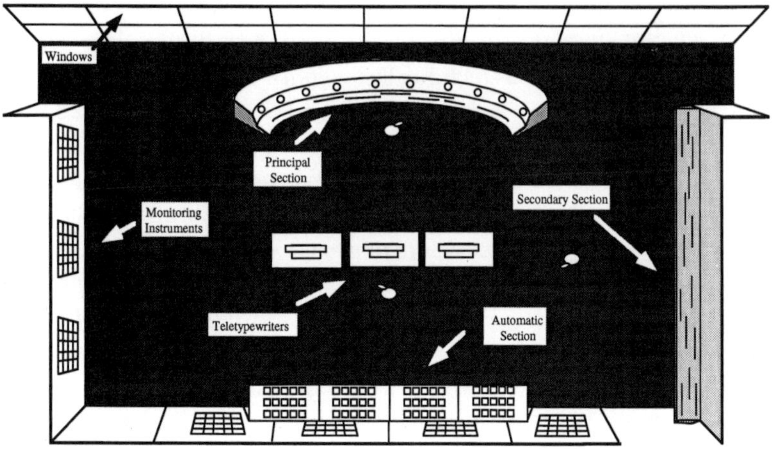

Figure 5.6
The floor plan of EDF2's control room. Drawing by Carlos Martín.

Figure 5.7
The principal control board of EDF2 in 1970. Photograph by Michel Brigaud.
Source: EDF Photothèque.

Figure 5.8
Teletypwriters in EDF2's control room, 1970. Photograph by Michel Brigaud.
Source: EDF Photothèque.

directly needed for the routine operation of the reactor.[83] The automatic
section received and displayed all the signals needed to verify the reactor's "automatic" functioning mode. Instruments in that section displayed
the core's reactivity, the temperature and flow of the cooling gas, and the
amount of electricity generated. The automatic section also included an
emergency shutdown button.[84] Thus, whereas at G2 each pilot had
responsibility for a different system, at EDF2 an employee in the command room might control any or all of the reactor's systems.[85]

EDF2's system of loading and unloading fuel also differed from G2's in
name and design. In both cases the machine itself was right next to the
core; however, G2's was operated from the machine itself, whereas EDF2's
was operated via a control room located a considerable distance away. This
difference, and the difference in the types of encoded instructions
received by the loading machine operators at each site, probably reflected
little more than changes in technologies of command and control that
had occurred in the five or so years that separated the design of the two
control systems. Whereas the Marcoule operator's instructions were
encoded in the blinking lights on his control panel, Chinon's operator
received instructions in the form of punched paper tape.[86] He positioned
the loading machine above the appropriate channel openings by means
of push-button switches located on his control board. He then fed the
appropriate tape into his auxiliary "programmer" and started the programmer up. The auxiliary loading machine automatically undid the

opening's seals and positioned the unloading arm. The operator next positioned the main loading machine (which contained the new fuel slugs) above the loading arm, started up the main programmer, and fed it the appropriate tape. He then sat back and watched his instruments while the loading arm automatically unloaded and reloaded the fuel slugs according to the instructions encoded on the tapes. Once the loading sequence was completed, the operator guided the loading machine full of used slugs to a chute through which they fell into the disactivation pool.[87]

Both Chinon's *opérateur DPM* and his Marcoule counterpart followed coded instructions and guided loading machines. But the Marcoule operator sat right next to the reactor core, while the Chinon operator had a separate control room. Chinon's loading machine operators enjoyed much higher status than their Marcoule counterparts. Marcoule operators were known as "laborers." But Chinon operators belonged to a kind of elite. Though subservient to the *chefs de quarts* and the *chefs de bloc,* they outranked the *rondiers* in Chinon's official hierarchy; at Marcoule, the functional equivalent of the *rondiers* outranked the loading machine operators. Other Chinon employees even suggested that the *opérateurs DPM* formed a snobbish lot.

What accounts for this difference in status? During the 1960s, remote control technologies, which became increasingly widespread in industrial activities, carried the prestige that new technologies so often do; perhaps the fact that the *opérateur DPM* at Chinon exercised, quite literally, a more "remote" control over his machines than his Marcoule equivalent lent the former greater prestige. The increased distance of his workplace from the reactor core also meant that he did not run as constant a risk of radiation exposure. In the Chinon work scheme, the men in closest and most consistent proximity to radioactive sources were the *rondiers,* who often entered the reactor building to check instruments or arrange spent fuel slugs in the deactivation pool. In one of the few parallels between Marcoule and Chinon, this proximity appeared closely related to the fact the *rondiers,* not the *opérateurs DPM,* held the lowest rank (both officially and socially) at EDF2.

One result of labor-management negotiations over the shape of EDF after the war was the creation of Comités Mixtes de Production (CMPs). These "mixed production committees" existed at various levels of the firm and brought personnel and management representatives together to discuss workplace issues. Such committees were not unique to EDF—they existed in other nationalized industries as well—but they did not exist at the CEA or in most private firms. In theory at least, these committees acted as the means through which labor participated in the management of the utility. In practice, their effectiveness and power varied enormously. By the late 1960s, unions complained that management had not taken the advice of the CMPs for years, and that these were little more than shams to preserve the illusion of co-management.[88] The plant-level subcommittees (Sous-Comités Mixtes de Production, abbreviated SCMP) could work quite effectively, however, and this was the case at Chinon.

Chinon's SCMP met several times a year, on an ad hoc basis, to discuss work organization, operational and safety guidelines, job classifications, and any unusual problems that might arise in the workplace. Presided over by the site director, it included engineers, technicians, and workers. Some were appointed to the subcommittee; others were as union representatives elected by the personnel. The site director or one of the top-level engineers typically presented proposals for changing or reshaping some aspect of work practice or training, which the committee then discussed as a whole.

More than any other documentary evidence to which I gained access in my research, the minutes of Chinon's SCMP show the craft nature of nuclear work in the 1960s. Work organization underwent constant evolution as committee members discussed which technicians were required for particular maintenance functions, how many *rondiers* each reactor shifts needed, or what were the appropriate skill-based qualifications for various positions.[89] Specific procedures were elaborated as the need arose, and committee members openly discussed the ad hoc creation and modification of operational guidelines. Evidently, the inclusion of workers in the guideline-writing process, combined with the urgent sense of mission that pervaded the site, led some to take alarming liberties with existing guidelines. The site director issued a stern warning at one meeting:

> The condition of the equipment has not always allowed [us] to respect guidelines. Because of this, engineers [*cadres*] have been forced to ignore certain guidelines. Operators have sometimes thought themselves authorized to do the same thing on their own recognizance. . . . Only the Operations Engineer or his hierarchical superiors are entitled to make such decisions if they themselves established the guidelines in question. . . .[90]

Further, said the director, if an engineer specifically asked a *chef de quart* to ignore a guideline, the *chef de quart* had to demand a written, signed release before complying. Clearly, then, there were limits to the freedom granted shift workers at Chinon. But equally clearly, these limits lay far beyond those at Marcoule. Management did not always accept the suggestions made by the SCMP, but it did approve many. All in all, worker participation—illusory in some other parts of EDF—was real at Chinon.

Risk and Radiation Protection

Because the CEA was so obviously the national institutional expert on radioactivity, we might expect EDF to have deferred to it in matters of radiation protection. But EDF engineers and managers, skeptical of over-specialization, saw no reason why they too couldn't learn about radiation safety. Reluctant to cede control of work practice to the CEA, they preferred to develop their own approach to handling risk. The utility therefore created its own radiation protection division, the Service Général de Radioprotection (SGR), which fell under the general purview of plant operations.[91]

The SGR would eventually supervise radiation protection at all EDF's nuclear sites, but in the early 1960s its only responsibility was Chinon. Initially, the SGR viewed its main task as the demystification of nuclear risk. Dr. Delpla of the SGR told the utility-wide health and safety commission: "Industrial medicine applied to the problems of a nuclear power plant is not fundamentally different from conventional industrial medicine." Though a new type of risk was involved, its parameters had been known for some time. "Nuclear energy," said Delpla, "is reputed to be mysterious, extraordinary, and extremely dangerous. One of the medical doctor's roles, therefore, is to correctly define nuclear danger and to specify its nature and extent." Delpla's "personal conviction," which he admitted others might not share, was that "one could work with absolutely no risk in a radioactive environment as long as one kept rigorous track of the doses absorbed by the organism and kept these within reasonable limits."[92] These arguments did not fully persuade CGT militant leaders, however. Jean Thomas, a unionized engineer who sat on the same committee as Delpla, worried that radiation protection fell under the organizational purview of production rather than industrial medicine. At the plant level, said Thomas in a 1964 meeting, this meant that the radiation protection service would not enjoy "the independence of the site's medical doctor; as a result, any proposals it might make to improve safety guidelines and working conditions would be linked to the opinion of the hierarchy."[93] In the end, though, management ignored Thomas's objections. The development of radiation protection guidelines remained subsumed under the general province of plant operations. The issue did not arise again at the utility-wide level for the rest of the 1960s.[94]

Although SGR men first went to the CEA for radiation protection training, they soon developed a distinctive approach to worker protection. Instead of following the procedures established at Marcoule, the SGR men decided to base their protective measures on EDF's general worker safety policy, developed at conventional power plants. In keeping with other manifestations of the utility's technopolitical regime, this approach was based on a sense of shared responsibility; in the mid 1970s, it would acquire the label "autoprotection." Essentially, this meant that no single, separate group of safety experts supervised employees.[95] Instead, each worker, technician, and engineer was in charge of his own protection in his specific domain of activity. A *chef de quart*, for example, had overall responsibility for what happened on his shift. At the same time, however, each individual employee under the *chef de quart* was in charge of his own radiation safety and that of his immediate surroundings.[96] All EDF employees received training in radiation protection, though the level of training varied with hierarchical position. Engineers learned radiation protection theory, as well as how to define the three basic work zones of a reactor (red, yellow, and green—EDF kept Marcoule's zoning methods). A *chef de quart* or a *chef de bloc* learned some theoretical aspects of radiation protection, as well as more practical points such as how to seal off a radioactive area, or where and how to place warning signs.[97] According to those who taught the safety classes, the notions of radiation protection were "stripped of the mathematical formalism that surrounded them like a halo, giving place to notions . . . closer to our everyday concerns and better understood by the agents."[98] Each shift designated a radiation protection technician, but that job was interchangeable with that of the *chef de bloc*, and two men in any given team took turns taking these jobs. The role of the *technicien de radioprotection* was more advisory than supervisory, at least with respect to his EDF colleagues (radiation protection practices differed for non-EDF contract workers).[99] This also gave him the opportunity to identify any need for new or additional protection guidelines, which he would then elaborate in cooperation with the Service Général de Radioprotection.[100] Because he belonged to the shift and not to a separate division, he always worked with the same men as teammates. This greatly increased the trust and respect between the man in charge of radiation protection and those he aimed to protect. The same man could be called upon to take risks or to manage them. This too contributed to a greater sense of shared responsibility among all the reactor employees.[101]

Chinon's employees felt proud of the control they had over work and risk practices. Unlike their Marcoule counterparts, they had a strong sense of professional identity that was partly created and heavily reinforced by EDF's technopolitical regime. Although EDF had its share of engineers and research labs, it was mostly composed of workers and production facilities. Everyone employed in the daily production of electricity, regardless of educational background, identified as an *exploitant*. (Recall that at Marcoule only engineers identified as *exploitants*.) Chinon workers therefore had a professional identity recognized throughout the institution that employed them. In addition, the nuclear employees formed an elite among EDF's *exploitants* because they worked with the most avant-garde of all utility technologies. They were not just *exploitants*; they were "men of the atom."[102] They nourished the belief that only the best employees of conventional power plants could qualify for nuclear work. This belief was bolstered by the image presented by *Contacts électriques* to the rest of the utility. Chinon's first reactor "demanded a tough apprenticeship from the men of the atom: everything was new, everything posed a problem that required resolution."[103] One *chef de quart* wrote: "You're always under some tension here. . . . Keeping watch on the temperature [of the core] will obsess you. You pilot the reactor with the control rods and the turbo-fans; so you move from one [control] panel to the other. And then you have to keep the temperature under control. I tell you, just keeping the power stable at the same level for two days running is an entire job. Sometimes, you won't have touched anything and the power changes, whimsically. And then there are breakdowns, tests, so you have to start all over again."[104] "Routine?" asked another worker. "I don't know what that is."[105] In the interviews I conducted, workers told stories of men who, unable to handle the novelty and uncertainty of nuclear technologies, had quickly returned to work at conventional plants. Proudly, they noted that "in those early days" EDF employees outside the nuclear arena viewed those inside as slightly crazy.[106]

By giving them a reputation for mild insanity, facing the risks of radiation exposure also earned Chinon workers social recognition among EDF's *exploitants*. Radiation was mysterious. Men who dressed in special work suits looked like "strange ghosts"; one could "hear their breathing through their masks."[107] The work of the radiation protection technician was exciting, adventurous: "[He] hangs over the concave chasm of the sphere [the EDF1 reactor], crosses the succession of shining corridors, climbs iron staircases or takes the quiet elevator that leads to mysterious rooms where the fuel cladding rupture detectors think, where the giant fans snore."[108] Work with radiation was romanticized within EDF in a way that it wasn't within the CEA. Radiation protection technicians figured as the explorers and tamers of the giant slumbering, rumbling creature that was the reactor.

At Chinon as at Marcoule, those in closest and most constant proximity to radioactive sources (the *rondiers*) had the lowest-skilled jobs and carried the lowest status. Routine risk taking did not enhance social status at Chinon any more than it did at Marcoule. But the *meaning* of routine was relative. In the eyes of their non-nuclear peers, the work performed by Chinon's *rondiers* was anything but routine, whereas their counterparts at Marcoule had no such peers within the CEA. Thus, even the lowest-status technical workers at Chinon—men who routinely worked in the "yellow" zone and whose risks were not especially valued in their workplace— could get recognition for their bravery from the larger institution that employed them.

Furthermore, at Chinon reactor workers at all levels could earn greater social recognition by behaving heroically. Indeed, Chinon, like Marcoule, had its heroes: men who performed exceptional tasks and ran risks in order to keep the reactors running. But at Chinon, opportunities for exceptional risk taking were distributed in the same way as opportunities for shaping operational guidelines and responsibilities for everyday risks. In other words, anyone who worked in a reactor had the potential to be a hero.

"The concept of safety," recalled one *chef de bloc*, "did not exist. It was production. . . . The priorities were different [then]."[109] When things got busy, non-urgent projects—like safety training sessions—would be postponed.[110] Some former Chinon employees explained unorthodox work practices by referring to the "pioneer spirit" that pervaded the site. One engineer, who had first come to Chinon to help develop radiation protection guidelines, fondly recalled an incident at EDF2:

Sometimes . . . the [fuel] rod would get unhooked from the grip [of the loading machine arm], and when it fell, its graphite case would break, and we would have to stop the reactor to clean the channel. One day we were fishing around inside this channel, and we had already taken all the fuel slugs out. But I had counted wrong, we thought there was still one in there. And we re-introduced the arm. . . . When we gripped something and took it out, we discovered it was the waste bin [that lay at the bottom of each channel to collect debris][111] . . . . When we saw that waste bin! We couldn't reintroduce it into the channel, and we wondered what we were going to do with the thing. We put our arms around it, unhooked it, and passed it to each other [taking turns] so that we could go put it in a lead coffin. . . . One could never do such a thing today. . . . We did it because—well, we probably could have done something else, but we had been scratching around in those channels for a while. I got one rem, that wasn't bad. I wouldn't do that now. . . , but, well, there were two of us who did that maneuver.[112]

He smiled mischievously and went on to talk about how people viewed radiation exposure at Chinon:

We were very attentive to the doses caught by the personnel, and sometimes we had to quarrel to get people to wear [their detectors]. They wore them when there were no problems, but for delicate operations—they'd put them in a corner. They felt that they had a job to do, and in order to do it within a reasonable period of time, one had to take a certain number of—one had to act in such a way that led to doses, and that couldn't be seen. . . . It was the pioneer spirit, you know. I don't know that the end result was as bad as all that, because if you perform a delicate operation in a very short time period, taking risks—the doses you get are not very much higher than those you would get doing things more correctly but taking a longer amount of time. These were always very short operations, and we always knew the risks we were taking. It didn't happen that often, but I did see it happen.[113]

At Marcoule, workers who put their detectors "in a corner" rebelled against a restrictive workplace and, if caught, were subject to a scolding by radiation protection experts. At Chinon, where responsibility for radiation protection rotated among foremen and where workers counted as *exploitants*, a man who shed his detector was heralded as a pioneer. Casting off his detector was an affirmation (perhaps even a celebration?) of his participatory role in EDF's technopolitical regime, not an act of rebellion. If the maneuver was sufficiently dangerous, he might even become a hero.

Still, heroic behavior and production imperatives notwithstanding, unorthodox practices sometimes worried Chinon's Commission of Health and Safety. In contrast with Marcoule, at Chinon CHS meetings were, more often than not, occasions for agreement between management and unions. Regular meetings helped to ensure a climate of cooperation. All manner of things got discussed in these meetings: accidents, radiation protection training, and safety guidelines. The fact that radiation protection guidelines were under constant evolution did not appear to alarm the unions, whose representatives simply seemed glad to have input into identifying and correcting problems. This equanimity may have also stemmed from the fact that the majority of workplace accidents did not involve radiation. They tended instead to have more mundane origins. For several years, for example, the CHS tried to persuade workers to wear gloves on the job, having calculated that more than three-fourths of minor mishaps could be prevented with this simple measure.[114]

Still, not all accidents were minor. There were eye injuries, broken bones, and even one death (a construction worker fell from a high platform). As more and more workers came to Chinon, the site's safety record deteriorated. In 1964, Chinon ranked thirteenth in overall workplace safety among EDF's power plants. By 1967, it had dropped to twenty-fourth. The CHS tried to encourage safer work practices by sponsoring a series of events for an "accident-free month"—including a contest—every year, but worker response was slow at first. Only 15 percent of Chinon employees participated in the 1967 safety contest, which led the CHS to unanimously "deplore agents' lack of enthusiasm for safety problems."[115]

Perhaps part of the problem stemmed from the fact that Chinon personnel got mixed signals about how they should behave. On one side, the "pioneer spirit" pervaded the climate at work. Peer and management approval clearly encouraged employees at all hierarchical levels to do nearly anything necessary to keep the reactors functioning smoothly. On the other side, the CHS issued mild admonishments for bending safety rules. In 1969, for example, the CHS began to notice that some agents failed to turn in their film detectors every month as they should have. The committee decided to keep a list of the offenders, and one member proposed sanctions for anyone who failed to turn in his badge more than twice in twelve consecutive months. The rest of the committee decided that actual "sanctions" would go too far, however, and voted instead simply to call offenders into a CHS meeting for an explanation—hardly a strong deterrent.[116] Even when the safety record improved (as it had by 1970, when EDF2 reached a new record of 822 consecutive accident-free days), the CHS suspected that one reason lay in underreporting. One member ruefully pointed to a "climate of competition that has developed among the [reactor] units and the services which may encourage certain accident victims not to stop working." "One agent," he continued, "seems to have preferred to take two days of compensatory rest instead of stopping work."[117]

I could not gain access to medical records or accident statistics for Marcoule and Chinon, so I cannot evaluate which approach to workplace organization led to "safer" conditions for workers. Nonetheless, my impression is that EDF's participatory approach to safety, though clearly more congenial to workers, does not appear to have produced a less dangerous work environment. This becomes particularly clear when non-utility employees are included in the analysis.

Indeed, workers employed by private companies regularly passed through Chinon. Most of these men worked for the firms contracted to build the three reactors and auxiliary facilities at the site. Some were subcontracted by EDF for odd maintenance or repair jobs. Few traces of these men remain, either in EDF or union documents or in the memories of the men I interviewed, so I cannot say much about them. All the evidence that does exist, however, indicates that Chinon employees considered contract workers inferior to EDF agents.

Contract workers appeared in Chinon site records only when they caused problems. The most severe problems involved radioactive contamination. Sixteen cases of contamination had occurred at Chinon by December 1964. The causes and severities of these incidents were not recorded in the minutes of the CHS meetings, but apparently "most" of them involved private contract workers.[118] One case—a contract worker who received 50 rems, or ten times the allowable yearly dose—was severe enough to come to the attention of the region's parliamentary delegate.[119] Both EDF and the unions explained such accidents by asserting that the victims were not utility employees. Private companies, said CGT and CFTC/CFDT militants, did not take the same care as EDF in teaching their employees safety practices.[120] These men were careless, even dirty: their work outfits were often so filthy that they could not be laundered to the proper pristine whiteness (prompting the CHS to decide that all permanently stained work suits would be issued exclusively to contract workers).[121] Contamination occurred because the victims had ignored warning signs or failed to follow proper procedures. Some companies apparently made their employees pay for lost dosimeters, which only encouraged contract workers not to wear the detectors at all.[122] Blame for contamination, therefore, lay not with EDF's safety practices and guidelines, but outside EDF, with the inadequacies of non-nationalized companies. The contrasts between EDF agents and others—the clean and the dirty, the pure and the polluted—only reinforced the belief that management and workers alike had in the justice of their technopolitical regime.

Unions

As we have seen, a substantial amount of what union militants did at Chinon consisted of participating in commissions that evaluated work and safety organization and practices. Throughout most of the 1960s, most other syndical activities involved simply participating in strikes or other actions led by the national federations. Most demands centered around salary and benefits issues, and were no different for Chinon than for any other power plant site.[123] Very occasionally, the national unions raised other kinds of issues.[124] In 1962–63, for example, the CFTC/CFDT and the CGT lodged formal protests, at the national level, against the partial use of Chinon's reactors to produce weapons-grade plutonium for the CEA.[125] But the national federations apparently did not expect their locals to get involved in this protest, and local militants did nothing besides reprint the text of the protests. Thirty years later, the workers I interviewed did not even remember that plutonium production had disturbed anyone.

This basic agreement between unions and management on everything but salaries and benefits would persist until the 1970s, when the CFDT began sustained investigations into working conditions throughout the nuclear industry, which eventually led to serious protests. But the atmosphere began to change palpably in 1968. Joining workers and students across the nation, Chinon agents staged a two-week sit-in strike in May of that year. Like others, they called for the lowering of class barriers and for more democratic decision making. The strikes left visible traces in the records of SCMP and CHS meetings. After May 1968, worker representatives on these commissions began to demand much more information. They wanted to participate in discussions about the budget, future investments, outside contractors, and overall plant goals. They demanded better sources of technical information for the personnel, starting with a library.[126] "1968" (as the May movement is usually called) clearly did mobilize Chinon's workers. In the years that followed, their demands grew more insistent, eventually becoming, in the case of the CFDT, serious challenges to EDF's technopolitical regime.[127]

Throughout most of the 1960s, however, all three union locals at Chinon continued to see themselves as active partners in EDF's technopolitical regime. Even the strikes that did occur were viewed—by unions and management—as normal parts of work life. Unlike Marcoule's directors, Chinon's management did not try to stop workers from striking. Instead, all parties followed the basic understanding reached at the national level about which issues were legitimate causes for strikes (salaries and benefits) and which should be discussed within existing co-management commissions (work organization and practices). Unions and their workers, therefore, were active—if subordinate—participants in creating and maintaining EDF's technopolitical regime.

*

The different technopolitical regimes of the CEA and EDF manifested themselves deep in the structures, technologies, and practices of each institution. Marcoule's work hierarchy formalized the authority of highly trained experts and the military "vocation" of the site. Knowledge was compartmentalized, and it corresponded closely to the social hierarchies in place. Depth of knowledge had more value than breadth. Each expert had his domain of specialization, over which he exerted complete authority. This system thus left no recognized space for non-experts to take initiative. Although Chinon's reactor work was also hierarchically organized, the flow of technical and social authority was far less rigid there than at Marcoule. Individuals at most levels of the hierarchy knew something about several domains of reactor operation, which meant that social hierarchies had more to do with experience and responsibility than with specific knowledge. Breadth of knowledge had more value than depth. Combined with institutionalized commissions that gave workers a formal place to discuss workplace issues with management, this system left more room for men to take initiative. The differences between the two sites rested on distinct visions—of the state, of the social role of technical experts, and of the place that workers should have in postwar France—and on the technological embodiments of those visions. For the CEA, workers were mere cogs in a regime run by experts for the greater glory of France. For EDF, workers were powerful motors in a regime steered by nationalized institutions for the betterment of France.

The contrasts in the command rooms of the G2 and EDF2 reactors and in the organization of radiation protection at the two sites exemplify these differences. G2's control board was organized according to three types of functions and required people with three distinct types of knowledge to run it. EDF2's control boards integrated these functions and split them into two levels of operation, requiring people with the same kinds of knowledge to run them both. Marcoule had specialized radiation protection experts to make and enforce rules. These men, who did not get involved with the actual operation of the reactors, had responsibility for the safety of all others. At Chinon, those who supervised radiation protection and created rules were interchangeable with those who ran the reactors. Each worker, technician, or engineer was ultimately responsible for his own safety. Thus, technologies and forms of technological knowledge not only reflected but also constituted forms of social organization.[128] They were both the instruments and the outcomes of social politics in these two technopolitical regimes. Imbued with cultural meanings, they directly linked the social and political ideologies of Cold War

France and the daily activities of the workplace. Through these instruments, the organization of work and risk at Marcoule and Chinon was not just about how best to run a reactor but also about how best to run a nation and organize a society.

Workers at Marcoule and Chinon fully understood the technologies and practices of the workplace as the instruments and outcomes of the social politics of technopolitical regimes. Workers at Marcoule resented their lack of importance in the CEA's ideological scheme. They received this message every time they wanted to take initiative and couldn't do so, and every time they saw the radiation protection agent take note of their movements. Their response was rebellion: if they couldn't be part of the new technopolitical order, they could at least defy it. Thus not conforming to the rules became an act of rebellion, a way in which they could play an active part in defining their workplace identities. In contrast, Chinon's workers understood and took pride in their own importance in EDF's regime. Their workplace responsibilities and practices showed them to be the vanguard of a new society. They broke the rules with the full complicity of their superiors; these were acts which constituted and confirmed their identities as pioneers.

Thus, responses to risk were as much responses to larger ideological constructs as they were responses to danger. For workers, meaning-laden practices and technologies constituted the technopolitical regimes in which they functioned. These practices and technologies were, in a sense, the material manifestations of the new technological France. Workers' interactions with these manifestations became a means of defining their place in this society. Unions played an important role in these definitions. At Marcoule, unionized workers explicitly related their dissatisfactions to the site's military "vocation" (a word that evoked both technical goals and social organization) and to the CEA's conviction that its expertise and mission exempted it from all accountability. Participating in union action, along with disobeying rules, was a way to defy this broader social model that left them little official agency. Workers at Chinon, meanwhile, embraced EDF's regime, which gave them and their unions a formal place. They agreed, for example, that cases of exposure to radiation did not reflect flaws in EDF's regime, but faults in how institutions outside that regime trained and treated their workers. At both sites, responses to risk provided a means by which workers could shape and enact their own roles in the technological Frances that had been presented to them.

6

Technological Spectacles

In the France imagined and enacted by the engineers and workers I have discussed, technically trained men would—by planning, building, and operating large-scale technological systems—play a leading role in shaping the nation's future. What place did this conception leave for people not involved in creating and running those systems? How did ordinary citizens see the new technological France, and how did they fit into it?

These questions address the two overlapping domains of representation and experience.[1] We need to understand the terms in which technological change was presented to ordinary people. To a certain extent, these terms shaped how citizens viewed the making of technological France and how they situated themselves in the new nation. But people who lived near the sites of large-scale technology also had experiences that did not fit into the representational frameworks they were offered.[2] We must also, therefore, examine those experiences and their interpretations of them. In this chapter, granting that such divisions are inherently artificial, I will focus primarily on representation; in the next chapter I will concentrate mostly on experience.

Popular representations made technological change into a spectacle. This spectacle could take either of two related forms. In one form, the technological spectacle was a *drama* that played out (and intertwined) the themes of salvation, redemption, and liberation. In this drama, technology would save France from economic and cultural disaster and redeem it after the humiliation of the Occupation. Through technological achievements, the French would perform a second liberation, not thanks to American soldiers but thanks to their own knowledge and resources. In its second form, the technological spectacle appeared as a *display*. Journalists compared reactors to cathedrals or other historical monuments and described beholding them as a kind of transcendental experience. Reactors were configured as tourist sites—displays that could be visited, like monuments or museums. Both kinds of spectacle made large-scale technological change into something the public could consume—by reading, gazing, and touring.[3]

The two forms of technological spectacle operated at both the national and the regional level. The spectacle—jointly produced by journalists, technologists, and politicians—was narrated or performed in the popular media. Many articles on French technological achievements in mainstream newspapers originated with press releases issued by the CEA or by EDF. Journalists adopted and adapted metaphors used in these press releases, spreading them well beyond the confines of the original institutions. The distinction between journalism and promotion was further blurred when an industry sponsored an entire page or section of a newspaper. Typically, such pages combined straightforward ads with "news" articles about the latest technological developments.[4] The production of the technological spectacle was thus neither entirely choreographed nor completely random, but somewhere in between.

National-level representations appeared in newspapers with nationwide circulations and on radio, television, and film. We must keep in mind, however, that journalists in these media did not write from some kind of abstract national perspective: they worked in and wrote from Paris. Parisians also formed their main audience. According to a 1955 poll, only 15 percent of French citizens read these publications and no others. In contrast, half of the population exclusively read its regional press. And 27 percent of all men and 17 percent of all women claimed to read both.[5] Clearly "national" newspapers held limited interest for the majority of French citizens who lived outside the capital.

Analyzing representations at the national level is therefore necessary but insufficient, particularly when the ultimate goal is to understand the social life of these representations and the experience of ordinary people. I therefore examine representations operating at the local level as well. In keeping with the rest of this book, I focus here on the area around the CEA's Marcoule site and the area around EDF's Chinon site. Indeed, most of the discussion concerns the immediate vicinities of these two sites. For the sake of rhetorical convenience and with apologies to the rest of each region, however, I refer to these areas by the names of the larger regions within which they are located. In the case of Marcoule, this is the department of the Gard. In the case of Chinon, this is the region known as the Touraine, which includes the department of the Indre-et-Loire.[6]

The local representations co-produced by site administrators, journalists, and elected officials had a more obvious strategic dimension than their national counterparts. The departmental and municipal officials who expressed the greatest enthusiasm for nuclear sites collaborated with site administrators in efforts to persuade residents of the benefits of modernization. They expressed these benefits in terms of local social and economic interests, and their explanations appealed to a sense of regional culture and history. Local officials thus operated in a space between the region and the nation: they represented their region when they negotiated with site administrators, and they spoke for the nation when presenting modernization proposals to their constituents.

As a result, residents of the Touraine (*Tourangeaux*) and those of the Gard (*Gardois*) saw the nuclear sites as both a local and a national phenomenon. Most obviously, they were local on a physical level, and therefore had a significant social and economic impact on the two regions. Additionally, leaders and some residents in both regions sought to appropriate the sites as sources of regional pride and natural extensions of local history. This did not mean that residents framed the sites uniquely in local terms. In part, the cultural appropriation of the site on a local level worked through a dialectic between the nation and the region. Thus, for example, the Tourangeaux framed the Chinon nuclear site as a "château of the twentieth century," thereby situating the site in a historical line with the Loire Valley châteaux of the Renaissance. This gave the site a local meaning, but one that derived part of its significance from the relationship it evoked between the Touraine and France. Loire Valley châteaux were important because they had housed the kings and queens of France; the nuclear site would, by analogy, place the region in an equivalent position of leadership. The nuclear sites brought the nation into the region, unbidden.[7] In the process of making sense of the new arrivals and seeking ways to turn them to best economic and cultural advantage, local residents used the sites to resituate their region within the nation.[8]

In the bulk of this chapter, both in my national and my regional discussions, I examine the spectacle of French technological radiance produced by journalists, officials, and technologists. In French, *spectacle* refers to theater productions as well as to less structured displays. Just as critics might review a theatrical production, a range of critics assessed France's technological spectacle. These critics came from a broad cross-section of French society, including communists, Poujadistes, satirists, science journalists, and religious leaders. This wide range meant that the critics were in no sense organized. Many of them lambasted the pageantry of France's nuclear strike force, but they did so from many different perspectives and for different purposes. Despite the best efforts of some of them, these critics did not orchestrate effective opposition to nuclear technology—they failed, for example, to persuade large numbers of people to protest the French bomb. They did succeed partially, however, by offering audiences an alternative lens through which to view technological development. In the last section of this chapter I consider how the spectacle and its critics blended in popular imagination by examining a 1957 play produced by residents of the Gard about the arrival of Marcoule in their region. This play can be understood as a counter-spectacle to the dominant display of French technological radiance—one in which the actors constructed a narrative that combined dominant representations, the insights of critics, and their own local interpretations of technological change.

Salvation, Redemption, and Liberation

The Allied nations saw the US bombing of Hiroshima and Nagasaki as a spectacular end to World War II. Even as they expressed horror at the victims of radiation poisoning, Western journalists described the atomic explosions in awestruck terms. Everyone agreed that humanity had unleashed an enormous new force. Anxious to find a role for their nation in this development, French journalists immediately began to write France into the narrative of the atom bomb.[9]

They found a ready-made hero in Frédéric Joliot-Curie, the son-in-law of Pierre and Marie Curie, a Nobel Prize winner, a member of the Communist Party, and the CEA's first scientific head. Press accounts varied as to Joliot's role in the development of the atomic bomb (as did their accuracy). Some claimed he had provided the crucial link by discovering fission and chain reactions; others referred more vaguely and modestly to "important discoveries."[10] Soon, though, a more entrancing aspect of Joliot's activities came to light that made for even better drama: his role in the Allies' procurement of heavy water, a potential moderator for fission reactions. In 1947 this story was dramatized in the film *La bataille de l'eau lourde*.[11]

In the opening credits, the film claimed to "retrace faithfully the adventure of the men who participated in the battle waged by the Allies against Germany for the possession of a rare product of capital importance to the conquest of Atomic Energy: Heavy Water." The director had enlisted the collaboration of the very men who had participated in this battle: various members of the Norwegian and French Resistances, Raoul Dautry, and three scientists (Frédéric Joliot-Curie, Hans Halban, and Lew Kowarski). These men, said the opening credits, "relive on the screen, in scrupulous exactitude, the episode of the secret war that History is already calling—THE BATTLE OF HEAVY WATER." Sure enough, Joliot, Dautry, and the others played themselves in this reenactment of wartime events.

The film opens with Joliot speaking to his colleagues and students at the Collège de France about the fantastic contributions that nuclear energy could make to human progress. He then appears in his laboratory, discussing the need for heavy water with Halban and Kowarski. The only manufacturer of heavy water is in Norway. Thanks to Dautry, the scientists persuade the Norwegians both to sell them heavy water and to withhold the substance from the Germans. Dautry sends a team to Norway, and they successfully accomplish the dangerous mission of exporting the heavy water. Other missions ensue throughout the war as the Allies try to blow up the heavy water factory in order to prevent the Germans from obtaining the substance.

This movie fit into a genre of Resistance movies made after the Liberation (and indeed derives its title from the most celebrated of these, René Clément's *La bataille du rail*). It message could not be clearer. French scientists and Resistance fighters played a crucial and heroic role in helping the Allies with their atomic bomb research. By preventing the Germans from getting the bomb, they helped to win the war. The presence of real historical figures asserts the veracity of the story and underscores the heroism of the men involved. In this drama, French scientists redeem their nation by playing a part in the Allied victory.

Joliot did not remain a nationally acclaimed hero for long. As the Cold War intensified, his speeches increasingly intertwined communist rhetoric with calls for the peaceful application of scientific research. The hero became more and more controversial. *Le Parisien Libéré*, *Le Figaro*, and other right-leaning papers accused him of making the CEA a communist stronghold.[12] The communist paper *l'Humanité*, meanwhile, lionized Joliot, linking his fate with that of the nation in headlines such as "Joliot-Curie and French Science—Ramparts of National Independence."[13]

Joliot himself may have lost national favor, but the drama of his wartime heroism had many sequels. In the most widespread of these, scientists and engineers appeared as the potential or actual saviors of a declining nation.[14] Ministers, technologists, labor unions, and editorialists issued repeated calls for greater numbers of scientists, engineers, and technicians to secure the future of the nation. These appeals resembled calls to arms, evoking wartime urgency and appealing to the patriotism of their audiences: "France . . . is counting on your willingness to change your own destiny by orienting yourselves toward mathematical and scientific studies.

Our country needs engineers and technicians."[15] The Communist Party was particularly adamant on this score, but the patriotic call transcended party politics and appeared in both national and regional papers.[16] Advertising for technical training programs also linked the future of individuals to the future of the nation. Witness, for example, this 1961 ad for the Institut Technique Professionel: "Engineer.—Our country, rich in uranium, has nothing to fear from the future if it can make its youth aware of this new path. At the time when the atomic plant at Avoine (Indre-et-Loire) is being built, we can better understand the prospects offered by this new science which needs a great number of engineers right away."[17] A photograph of a young man in a white coat sitting at a control board illustrated how prospective students could simultaneously serve their nation and secure their own future.

How, precisely, could engineers serve and save the nation? General audiences often read the answer in the form of dramatic adventure tales that paralleled the dynamic of the nation's wartime experience of victory snatched from the jaws of defeat. Consider the stories told about the Caravelle jet airliner. *France-soir* ran a week-long series that told the story of the birth of this "prodigal child of French aviation." The series appeared on the same page as a serialized novel and was dramatized in much the same way. French aviation, which once had led the world, had sunk so low after the war that Air France had refused to purchase French planes and had bought American ones instead. "For the average Frenchman," said one of the articles in *France-soir*, "this was another small humiliation, another lost illusion."[18] (The initial "humiliation," of course, had been the 1940 defeat.) In 1957, however, young French aviation engineers had saved the day by developing the Caravelle's "revolutionary" engines: "No point in asking . . . which one found the 'trick.' . . . Satre [the chief engineer] peremptorily declares: 'The Good Lord was with us.'"[19] The rest of the series recounted the "epic struggles" that eventually resulted in the successful manufacturing of the plane that would serve as "France's ambassador to the world." Coverage of the Caravelle by other papers across the political spectrum followed this pattern.[20] Over the course of the 1950s and the 1960s, the mass media offered similar promises or affirmations of salvation through many other technologies.[21] In 1957, *Le Figaro*, particularly anxious that France retain its status as a colonial power, published numerous articles arguing that French technology would also provide salvation to the territories of the empire. A typical article argued that French technical personnel were essential to Algerian development and proclaimed: "We will have definitively saved

Algeria on the day when it becomes an envied model, a 'pilot-country' for the whole Arab world."[22]

In dramatic accounts of the Caravelle and of other technologies, religious metaphors of salvation and redemption were tied to reenactments of the Liberation (or promises of new liberation), as though France could atone for the unforgivable defeat of 1940 through technological development. Throughout the 1950s and the 1960s, ministers and presidents performed similar pageants in their ritual visits to inaugurate sites of French technological radiance. The live audiences for such events consisted of plant workers and local residents, but thanks to the press, radio, and television the rituals also had a virtual national audience. Toward the end of the Fourth Republic, for example, President René Coty traveled to Colmar, in Alsace, to inaugurate EDF's latest hydroelectric plant. *Le Figaro* explicitly linked the event to postwar euphoria. The president received an "indescribable welcome: the atmosphere is like that of the joyful hours of the Liberation. All along his route, [crowds] cheer deliriously for Mr. René Coty standing up in his car."[23] After the inauguration, Coty toured Colmar, attended a banquet, and watched a display of regional dances and costumes. If the inauguration served as proof of the "incessant progress of French technology," surely the other festivities reaffirmed the Frenchness of the much-disputed Alsace region. The inauguration thus simultaneously asserted French technological radiance and confirmed national borders.[24] Charles de Gaulle, a big fan of such events, offered a similar performance two years later on his visit to Sud-Aviation, manufacturer of the Caravelle. According to *France-soir*, he too was greeted ecstatically by workers, who crashed through the barriers separating them from their leader, eager to shake his hand. Presumably, de Gaulle's speech did not disappoint them:

I am profoundly impressed by what I see at Sud-Aviation. . . . The splendor, the immensity . . . and it's all of you, gathered here, who impress me. You are lucky, because with your problems, even your pain, and all your worries, you are part of a great work [*grande oeuvre*]. . . . From here, in increasing numbers, emerges the fast, the only, the sweet 'Caravelle' that takes off into the sky toward all the nations of the world, to represent France and show what she is capable of when she wishes. . . . France too is lucky, despite her difficulties, despite the obstacles which arise inside and out. Your achievement proves that she is worthy, and that she is France![25]

De Gaulle thus cast these workers as actors in "the splendor, the immensity" of the technological drama. At least for a brief moment, they were the spectacle too. Such performances were repeated throughout the

1960s.[26] Official visits and inaugurations served both to commemorate French technological radiance and assert its marvelous qualities.

The most awe-inspiring show of all, for de Gaulle and for much of the mainstream media, was the display of France's first atomic bomb exploding over the Sahara on February 13, 1960. The next day, a photograph of the mushroom cloud covered the front page of *Le Journal du Dimanche* (the Sunday edition of *France-soir*) under the headline "Here are the first photos and the first story of the French atomic explosion."[27] The paper devoted four full pages to the event. It profiled the men responsible for developing the bomb, listing their war records and their colonial experience.[28] One journalist also provided what he called a "film of the explosion":

1 a.m.: Last meeting of the military chiefs around General Ailleret. Everything is ready.

2 a.m.: The chief of staff arrives in the large room of headquarters, 13 kilometers from point zero. In the middle of the room, a table with seven buttons.

H hour – 35 min.: The automatic "program" is set off. Headquarters supervises the final stages on two television screens.

H hour – 14 min.: At Reggane, the army's radio reporter leaves his microphone on. You can hear the final sounds of the trumpets calling in the troops.

H hour – 2 min.: At headquarters, the "automatic program" continues its operations. At Reggane, the radio reporter announces, "the men have taken their safety positions, seated on the ground, with their backs to the tower, their heads between their knees. Two rockets, white and orange, have just left."

H hour – 1 min.: At Reggane, the radio reporter announces, "three orange rockets have just left. Only 50 more seconds." Watch out! a red rocket has just taken off.

7:04. H hour: A formidable roaring resounds. A ball of fire rises in the sky. But the shock wave takes 1 minute and 15 seconds to be felt at Reggane.

H hour + 1 min.: The radio reporter describes the scene. An immense atomic mushroom now rises in sky, above the barracks. The base is mauve. The mushroom is getting bigger, its base is growing. The light is blinding. "Holy cow! the moon pales in comparison," we hear.

The radio reporter continues: "the top is now taking the shape of an immense spherical ball, a kind of comet whose tail is made of smoke and whose head is made of snow.

The mushroom continues to grow. The glare is still strong. Just now, despite my folded arms, despite my special glasses, I had the impression that the light penetrated my arms, my glasses, my eyelids."[29]

In this and other descriptions, the French bomb was an amazing show, a carefully orchestrated, awe-inspiring display of strength and (literal and metaphorical) radiance. The bomb had celestial dimensions: it made the moon pale in comparison, it resembled a giant, bizarre comet. The light it emitted blinded and penetrated the human body. Articles in papers that approved of the bomb quoted de Gaulle and his cabinet waxing rhapsodic about the fabulous success of the test, and published sidebars assuring their readers that the fallout had completely bypassed inhabited regions of the desert.[30] Even the publications that denounced the French bomb and its fallout granted—and criticized—its theatrical quality.

The bomb was thus a part cosmic, part transcendental spectacle, replayed in the media for all to experience. In the weeks that followed the explosion, publications sympathetic to the Gaullist *force de frappe* reaffirmed the bomb's redemptive and salvational powers. The nation could once again defend itself. The humiliation of 1940 would never happen again. France had retrieved its status as a great nation, a world power. The Reggane test had reenacted the final wartime victory of the United States, this time with France playing the leading role.

Technological development was thus a tremendous spectacle, a drama propelled by scientists and engineers, and a display of national radiance. The exalted language used to describe technologies transformed them into redemptive acts—atonements for 1940, or reenactments of the Liberation. They thus bound French nationhood to technological achievement. Some of the enactments of this show—such as appeals for more scientists and engineers—specifically sought to enroll the audience in the drama. For the most part, however, the producers of the pageant seemed to expect their audience simply to applaud, cheer, and be awed.

How did this spectacle manifest itself at the regional level? In the Gard, near the CEA's Marcoule site, the spectacle of technological radiance presented itself, above all, as a drama of regional salvation through the reconciliation of modernity and tradition. In the Touraine, near EDF's Chinon site, the spectacle was primarily a display akin to the region's châteaux, a monument for locals and tourists to gawk at and tour.

Reconciling Modernity and Tradition

The central drama describing Marcoule's arrival was co-produced by CEA administrators, local leaders, journalists, and scholars. It blended national symbolic and ideological resources with regional ones. Its multiple variations generally featured the same basic plot: The economy and social fabric of the Gard had begun to decline in the early 1950s. The local coal mine had shut down, the purchasing power of consumers had decreased, wine and other local products were selling at a loss, and agricultural

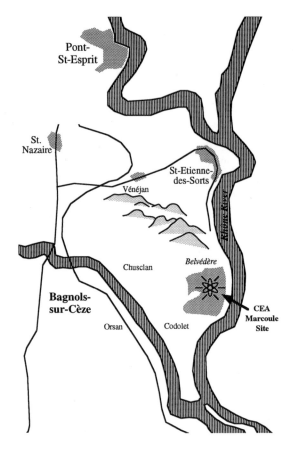

Figure 6.1
The region around the CEA's Marcoule nuclear site (not to scale). Drawing by
Carlos Martín.

workers were fleeing the land. In this dismal climate, the CEA announced
that it would build France's first industrial-scale nuclear site in the villages
of Chusclan and Codolet, on the banks of the Rhône. Marcoule brought
new people, virtually unlimited employment, and regional modernization, all of which blended harmoniously with traditional lifestyles. A
unique emblem of French technological prowess, it brought glory to the
region because it brought glory to France.

A typical production of this drama occurred when a newly created
Urbanism Prize was awarded to the town of Bagnols-sur-Cèze by the
Ministry of Construction. With populations between 4000 and 5000,
Bagnols and Pont-Saint-Esprit were the nearest sizable towns to Marcoule

(figure 6.1). Shortly after arriving in the region in 1953, the CEA discussed housing development plans with both towns. The mayor of
Bagnols, the dynamic and ambitious Pierre Boulot, saw new construction
as the perfect means by which to revitalize his town's economy. He eagerly
seized the CEA's offer, pledging to persuade his townspeople of the won-
drous possibilities of urbanization. The Ministry of Construction took
charge of the project, designating an urban planner and an architect as
project directors. Thus began more than ten years of urban planning and
development in Bagnols.

In 1960 the Ministry of Construction decided to reward these urbanization efforts. "Is it not astonishing," it asked rhetorically, "that until now,
nothing in our Country—renowned for its taste, its sense of measure, and
its methodical spirit—has attracted the attention of public opinion to the
originality of French solutions to urban problems?"[31] Creating an
Urbanism Prize would remedy this shocking state of affairs and "show the
personality of French Urban Planning to its best advantage." Bagnols-
sur-Cèze won the prize, which consisted of a plaque, a ceremony, and an
illustrated commemorative booklet that recounted the drama of its
urbanization, featuring Marcoule as the saving agent of modernization.
The project directors, the CEA, and the mayor of Bagnols produced the
ceremony and the booklet.

The head urban planner, one R. Coquerel, focused his story on the
multiple manifestations of the harmony between tradition and modernity exemplified by the new Bagnols. All the participants in the project
had cooperated in its planning and in its implementation: "No one shut
himself in his own preoccupations, in his own specialty: technical, financial, human and political problems were all studied together."[32] This col-
laborative spirit resulted in another kind of harmony: the equitable
distribution of modern amenities. Established residents of the town
would have access to the new housing. "Providing the old neighborhoods with the same equipment as the new ones is more important than
over equipping the latter," Coquerel wrote. "The community is a whole:
maintaining its unity is a sacred thing." Most important to him was the
esthetic harmony expressed by the newly urbanized town. To illustrate
his respect for Bagnols's history, Coquerel represented the "historical
evolution of the town" with four diagrams that depicted the town in
Roman times, in medieval times, before the arrival of Marcoule, and
after the start of urban planning. The modern high-rises, "symbols
and identity of the new city," expressed the "continuity of the gothic
spires of the churches and the existing medieval towers, symbols and

identity of the old town," thereby providing "organic and spiritual continuity between the two towns."

The CEA's contribution to the prize prose also told of harmony—claiming, for example, that the CEA had asked the villages of Codolet and Chusclan for hospitality before beginning to build the site (a claim that those villages hotly disputed). It placed even greater stress, however, on the role of the state in bringing the entire project—both Marcoule and the new Bagnols—to fruition. It narrated the importance of plutonium production for the nation, of the role that the CEA played in helping the government implement its policy of industrial decentralization, and of the enthusiastic participation of all branches of the state. Most of all, though, the CEA represented itself as the agent of modernity. Its personnel, representative of the most modern of all technologies, required the most modern of all surroundings. The CEA had therefore rescued Bagnols from the depths of backwardness: "A little town once in decline, Bagnols-sur-Cèze is now a modern . . . center where every month the [CEA] delivers . . . 1,500,000 new francs worth of salary to its employees."[33]

The last word in the prize document was reserved for Mayor Pierre Boulot. His enthusiasm for the transformation of his town produced a paean to modernity. He began with an exultant list of accomplishments, including not only the construction of new housing but also the installation of sewage and running water, the dramatic increase in Bagnols' population and birthrate, and the opening of new commerce. A list followed of future facilities, which included a stadium, a pool, a cultural center, and a much expanded hospital. Boulot then proceeded to an almost obsequious expression of gratitude to the agents of modernity (the urban planners, five different ministries, and of course the CEA). "Thus Bagnols," he concluded, "an atomic City, a mushrooming Town, a Town integrated in the past, . . . proud of that past and of its progress, turned toward the future, salutes the promoters of an astonishing, human, and peaceful endeavor."[34] Since Marcoule produced plutonium to fuel atomic bombs, Boulot's choice of "peaceful" is particularly ironic.

State technologists and local politicians thus *together* produced a dramatic narrative in which Marcoule descended upon the region and saved it from underdevelopment and underpopulation. This story conflated technology and the state into a single agent of modernity that would complement the traditions that had sustained Bagnols over the centuries. Bagnols's past served to situate and legitimate modern developments in a continuous, progressive regional history. The new town was but the logical and harmonious extension of the old one.

The joining of modern and traditional had been a theme of local representations from Marcoule's earliest days. In 1957 it had been performed in a spectacle that had captured journalistic imagination nationwide: the arrival of Marcoule's first heat exchanger, manufactured near Paris and transported to the site by a convoy of trucks. The national press followed this journey south for weeks, dubbing the exchanger the "atomic millipede" because of its appearance and its slow pace. As the millipede neared Marcoule on the final day of its journey, residents flocked to the sides of the road to catch a glimpse: "In Bagnols, we saw an eighty year old woman postpone an urgent trip so as not to miss the passage of the engine. 'I want to see the atomic bomb,' she said to whomever would listen. And when she learned that there was no bomb, and that there was no need to fear an explosion, she appeared horribly disappointed."[35] The atomic millipede approached the bridge of Codolet, the final hurdle before the site. The bridge, more than a hundred years old, had not been designed to accommodate a monstrous modern machine; to make matters worse, there was a tight curve in the road just before the bridge. Taking several hours, the convoy driver inched the millipede through the turn and across the bridge, making a grand entrance into the village of Codolet. There, a charming young couple came forward to great them. Bursting with eagerness, Parisian journalists dubbed them "the atomic fiancés." This episode symbolized the way that traditional structures such as the Codolet bridge accommodated modern technology. Parents could encourage children to reenact this accommodation by purchasing a toy atomic millipede, painted in the colors of the French flag.[36]

The drama of Marcoule also rehearsed and updated the centuries-old tropes of young, male, modern France versus old, female, traditional France—also manifested as civilized Paris versus the savage provinces.[37] According to regional geographer Alfred Chabaud, such contrasts were evident every day in the villages around Marcoule:

Saint-Nazaire well illustrates this evolution that associates village with new neighborhood, past with present. Next to the aging, predominantly female indigenous population there is now a foreign element, composed primarily of young men. This transformation of the village into a suburb brings change: faced by the city, the field retreats; the rural world gets submerged by the white-collar workers. These transformations occur without resistance by the primitive environment and without aggression on the part of the newcomers.[38]

Just as French colonialists had gone on a civilizing mission to Africa, so contemporary industrial missionaries brought modernity to the "indigenous," "primitive" population of rural France.[39] This encounter occurred with perfect harmony: "[We see] a curious juxtaposition in this society: a rural mass, groups of blue and white-collar workers, and an elite of technologists. This encounter of such diverse elements in the same place can secure the link between scientific thinking and work in the fields, attach the factory to the earth, and perhaps renovate the soul of the village."[40] The persistence of an agricultural society would root the new factories in the soil of tradition. But harmony also came because peasants embraced modernity. Chabaud claimed that the progressive spirit brought by the atom had begun to permeate agricultural thinking. Proof of this, for him, lay in the dramatic increase in the number of tractors in the region and in the greater attention being paid to irrigation and other technologies: "Everywhere in this environment still prisoner of its routines and structures, a will to act is born that brings with it movement and life."[41] Thus, Chabaud concluded, modern industry could free rural France from the fetters of primitive routines. Agriculture would undergo a transformative experience, a massive rejuvenation. Modernity, at last, was within the reach of the French peasant.

The story of Marcoule was most often produced as a drama, but this drama also incorporated elements of display. As objects on display, the reactors provided a breathtaking spectacle. One local journalist appeared to have a transcendental experience upon beholding the reactors for the first time:

From the very first glance, the enormous dimensions of the buildings that contain the [atomic] piles strike us dumb. These are two cathedrals of steel, 60 meters high, 40 meters wide, and 72 meters long—in other words, about the same dimensions as the nave of Notre Dame. Inside, one could easily fit three Arcs de Triomphe. . . . The pile itself is in the form of an enormous cylinder of prestressed concrete, 20 meters in diameter and 34 meters long. It is held together by steel cables. . . . Each cable is capable of supporting a weight comparable to that of the Eiffel Tower.[42]

This passage was written as though these thoughts had occurred to the author as he gazed upon these magnificent machines. The national press, however, used the same language. From the daily *Le Figaro* to the monthly popular science magazine *Science et Vie*, journalists described Marcoule's reactors as cathedrals and compared them to national monuments in the same terms.[43] We can surmise that the CEA itself had suggested these comparisons to journalists, either in a press release or at one of the few press conferences held at the site.[44]

Regional papers, scholars, and officials appropriated these metaphors in an utterly un-self-conscious, non-ironic manner in order to re-imagine the relationship of their region to the nation.[45] Alfred Chabaud offered this description in his analysis of Marcoule's socio-economic effects:

> As of today, France has realized a cyclopean achievement that prefigures the immense possibilities of the atomic era. . . . Built as an amphitheater, very picturesque and imposing to see from afar, the factory launches itself forward like a *hymn* to the *glory* of industrial creation.
> It would be up to an engineer to describe the G1 pile housed in an immense concrete *cathedral*. . . . Nothing can match the view of the cooling tower, shooting straight up in one 95 meter bound, *haloed* by its *shimmering* collar in the intense luminosity of the sky.[46]

Spiritual metaphors thus acquired local meaning. The Gard now housed a monument to surpass all monuments. Marcoule could compare to the most resplendent Parisian symbols of the nation. Notre Dame, the Arc de Triomphe, and the Eiffel Tower—traditional monuments to French religious, military, and cultural glory—lent their symbolic power and legitimacy to the nuclear reactors. Far from negating France's glorious history, these modern monuments represented the next logical step in that history. The quintessence of Frenchness no longer resided exclusively in Paris; it now existed in the Gard as well. The department would bathe in the light cast by the French atom: "Marcoule will give the Gard of tomorrow a national radiance that coal could have never brought," proclaimed the president of a local chamber of commerce, explaining that, whereas reliance on coal had led to severe economic problems, the presence of the atomic site would guarantee the industrialization of the entire region.[47] Over and over again, the nuclear industry appeared in the regional press of the mid to late 1950s as the potential or actual savior of the region.[48]

The quintessential manifestation of this reconfiguration lay in the expected transformation—thanks to Marcoule—of the region into a tourist destination. The site itself had esthetic qualities well worth viewing. *Le Midi Libre* assured its readers that the CEA "intends to prove that a big modern factory is not a prototype of ugliness, of drab uniformity that generates boredom. . . . The main evidence for this concern to flatter the eye is the choice of bright colors for the outside walls—green for the plutonium factory, ochre for the G2 and G3 piles—the installation of modern street lights with curious, conical lampshades, [and] the installation of a tourist belvedere on the Dent de Marcoule."[49] Indeed, although visitors could not tour Marcoule as they could other French monuments, the belvedere so thoughtfully built by the CEA provided a spectacular view of the region. This belvedere was celebrated not only in pamphlets put out by the CEA but also in the local press and in locally produced tourist

Figure 6.2
The Dent de Marcoule in the early 1950s, before the arrival of the nuclear site.
Source: CEA/MAH.

Figure 6.3
The view from the belvedere in 1968. Source: CEA/MAH/Jahan.

Figure 6.4
Scenes from the comic book *Bruno et Sophie au pays de l'atome* (*Bruno and Sophie in the Land of the Atom*). Ordinarily, tourists and residents could not take an actual tour of Marcoule. At best, they might experience the virtual tour offered in this comic book. In the comic, Bruno and Sophie (both around twelve years old) mischievously break into Marcoule to satisfy Bruno's curiosity. ("Seeing that amazing factory every morning without knowing what goes on there is really getting to me!") They are caught by a man who looks remarkably like Francis Perrin (the CEA's scientific head, not normally at Marcoule), who takes pity on them. He assigns Mr. Timoléon, a clumsy and absent-minded scientist with big glasses and a beard, to show them around and explain the mysteries of the atom. Timoléon takes them on a tour of the site, showing them everything except the plutonium factory (off limits even in the virtual tour) and asserting that radiation is not inherently dangerous. Timoléon's clumsiness and the children's antics are evidently intended to provide humor, but most of the dialogue in fact supplies a rather serious (if highly glossed over) explanation of Marcoule's activities, together with earnest reassurances about its safety. Source: J. Castan, *Bruno et Sophie au pays de l'atome* (no date or publisher listed). Courtesy of Jacques Bonnaud.

Figure 6.5
More scenes from *Bruno et Sophie au pays de l'atome*. These images accompany Mr.
Timoléon's explanation of why radiation, properly handled, is safe. The basic gist
of the explanation is that radioactivity exists in nature. In these panels, archetypal
French male figures represent inhabitants and consumers who receive differing
amounts of radioactivity. The first panel contrasts a resident of the Seine (a
Parisian) with a resident of Brittany, the second a man who consumes mineral
water with one who drinks milk, and the third a man who eats fish from the sea
with one who eats fish from rivers. In all three cases, the panels proclaim, both
men "live normally." Courtesy of Jacques Bonnaud.

brochures.[50] Thus a modern landscape of the Gard was constructed. From
the new vantage point, the site of Marcoule dominated a panorama
bordered on one side by the Rhône and peppered throughout with quaint
villages and fine vineyards. The new landscape provided yet another way for
the modern and the traditional to coexist in esthetic harmony.

Local officials hoped that the distinctiveness imparted to their region by
Marcoule would attract visitors. A glossy booklet published by the depart-

mental services of the Gard in 1964 recommended, in its description of Bagnols, an itinerary on which tourists could view Marcoule from the outside.[51] (Except for the occasional press conference or the even rarer tour for local officials, the site was off limits to all but those who worked there.) Those who wanted a special souvenir of their visit to atomic France could even purchase a bottle of *Cuvée de Marcoule* Côtes-du-Rhône, the crowning symbol of a perfect fusion of tradition and modernity, commissioned from the vintners of Chusclan by Marcoule's administrators.[52]

Spectacles that displayed the harmonious marriage of tradition and modernity did more than simply offer an interpretation of regional modernization. They also became a political and economic tool for local officials. For example, Mayor Boulot evoked the narrative whenever he asked the Bagnols town council to approve tax increases to pay for new, modern facilities. Such evocations drew upon the dramatic themes elaborated earlier. In 1962, for example, Boulot reminded the councilors that Bagnols represented the perfect reconciliation of tradition with modernity:

This evening, your Finance Commission presents you with a budget that pretends to nothing except a wish to be reasonable.

It is indeed difficult, you may say, to know where reason lies. Some would seek it in the tranquil and comfortable place of the man who, after a life of labor, calmly awaits the end of his days. It is no longer time for him to plant, nor even to build. He is content to keep up the old furniture that he received from his family, and to give the trees he planted in his youth the care they need in order to produce as much fruit as possible before his death.

Doubtless his house is comfortable and his orchard old, but what does he care? Those who come after him can work things out for themselves.

Others, on the contrary, feel that one must think about those who come later. This would be the idea of the old industrialist who only manages his business for his son; he abandons his old factory, which he deliberately let fall to pieces, in order to construct new workshops, equipped with the latest machines. Little does he care that the familial house has a few rotten beams and peeling paint, since it will no longer suit the next generation, and since for him reason only lies in what is projected for the future.

"In medio Stat Vortus" [*sic*] said the wisdom of the ancients, and it is in this happy medium that we have tried to find reason.[53]

Boulot used these fables to argue that what was right for Bagnols was the "happy medium," the perfect harmony of modern and traditional values. His budget, he said, requested only those amenities needed to produce such harmony. He intended to carry Bagnols forward on the path cleared by the forces of industry and modernity without giving way to excess or sacrificing the values that they all held so dearly. Throughout the 1960s,

Figure 6.6
The mayors of the towns and villages around Marcoule on a rare visit to the site.
Courtesy of Mireille Justamond, Bagnols-sur-Cèze municipal archives.

Boulot continued to draw upon these themes in order to push through his development programs.[54]

Châteaux for the Twentieth Century

The spectacle of nuclear development in the Gard took the form of dramatic narratives in which modernity became reconciled with tradition in order to save the region from decline. As objects of display, the reactors were to be gazed upon with awe from afar. In the Touraine, however, the spectacle of nuclear development primarily took the form of a display that could—at least on the surface—be examined more closely than its Gardois counterpart. The producers of this show in the Touraine appeared most concerned with harmonizing modernity and tradition in a visual esthetic.

Tourangeaux officials initially expressed more caution than their Gardois colleagues at the prospect of a nuclear site in their region. They too felt that their region was in decline. They particularly deplored the state of housing, the decline in population, the "decrepit" telephone system, and the "seriously neglected" development of television. The only major economic event near the town of Chinon since the end of World War II had been the construction of an American military base. The base had hired local residents for construction and service jobs, but officials feared a surge in unemployment with the completion of the facilities. Still, the Touraine did not face the strikes and massive unemployment problems that the Gard experienced due to the decline of its coal mines. Most local leaders expressed only cautious enthusiasm for proposals to lure industry into the region.[55] In departmental meetings and in the regional press, the traditional qualities of life in the Touraine—the gentle landscape, the abundant produce, the wonderful wines—were repeatedly lauded, even while pointing to their uneasy fit with a modernizing nation. For example, while the regional newspaper *La Nouvelle République du Centre-Ouest* worried about the decline in population and its implications for the economic resurgence of the Véron (the area delineated by Avoine, Beaumont, and Savigny), it also idealized the ways of life there. As they had in the Gard, female inhabitants of the Véron incarnated traditional France: "The Véronaise woman is self-sufficient. She has her meat, her milk, her vegetables, her grains, her fruit, her fowl, her rabbits, her potatoes, her fodder, her roots, her wood, and her wine"—all foodstuffs of which François Rabelais, a native son of nearby Chinon, would have approved.[56] Yet despite this self-sufficiency, young people continued to flee the land. The traditional lifestyle could not endure much longer. The very diversity of the produce made the mechanization of agriculture difficult, and the scarcity of roads and lack of a railway station inhibited contact with the outside world. Unless something were done, modernity would pass the region by and tradition would disappear—by sheer attrition—into oblivion.

The Indre-et-Loire's general council created a committee to investigate and implement plans to lure industry to the region in the hopes of addressing such issues.[57] Yet the continued discomfort of departmental representatives with industrial development became evident when the prefect announced, in early 1956, that EDF was considering the Véron for its first nuclear site. Most officials reacted with tempered interest. The site promised to provide the kind of economic boost they sought, but they expressed concern about the potential dangers from radiation (an issue which Gardois officials did not raise). The prefect sought to comfort them on this score, reporting on two recent information sessions held by EDF. Specifically, he noted, "we have been amply assured that the water . . . needed for the cooling operations, will not, upon its evacuation into the

Vienne (or the Loire), have damaging effects either on water use or on the fishing folk. Fishermen will even appreciate the slight heating of the water."[58] He went on to assure the representatives that EDF wanted to establish a regional development plan in conjunction with local officials. He concluded with what he evidently intended to be the coup de grâce: "We might even think that this development could be the source for a new kind of tourism, since it will [attract] . . . researchers and developers from many neighboring nations." Auguste Correch, Chinon's mayor, expressed the greatest enthusiasm. He immediately lent his support to the prefect's proposal, adding only his fervent desire that steps be taken to ensure that the new plant would "not detract from the beauty of the site."[59]

As Correch's comment suggests, the desire to achieve a certain harmony between tradition and modernity had different manifestations in the Touraine than in the Gard. The differences, I believe, stemmed from the fact that tradition itself had different meanings in the Touraine. In both regions, the notion of tradition evoked village lifestyles and ties to the land. In addition, however, the Tourangeaux tied tradition to national history. The Touraine's history derived much of its meaning from its role in national history, thanks to the presence of numerous landmark châteaux. Commemorating this history had economic as well as cultural significance, since tourism to these châteaux provided a significant source of income. Even smaller châteaux, such as the one at Chinon (site of the legendary scene in which Joan of Arc recognized the legitimate king, who stood disguised among the members of his court), received a steady 70,000 visitors a year. And less ostentatious sites such as La Devinière (Rabelais's birthplace) and the ruins of the château de Bonaventure (site of the famous love affair between Charles VII and Agnès Sorel) attracted more adventurous tourists.[60] Community leaders proudly noted that France's medieval and renaissance monarchs had prized their region's mild climate and gentle landscape. Preserving this landscape was thus a priority. In sum, esthetics and tourism provided the fundamental parameters by which Tourangeaux officials understood and described their region. These, consequently, were the primary terms they applied to the arrival of nuclear technology in their midst.

Ultimately, esthetic considerations concerned local officials more than safety issues. EDF's initial intent to locate the site at the confluence of the Loire and the Vienne alarmed people who felt that location afforded one of the most beautiful views in the region—one that on no account should it be marred.[61] Many breathed a sigh of relief when EDF announced that it would select a site a few kilometers downstream instead. Clearly, EDF had begun to understand the value of addressing local concerns. Though the move had been prompted by the results of a geological survey, in a meeting with local officials an EDF envoy also cited esthetic considerations.[62]

The move did not assuage all esthetic anxieties, however. The new site, in the territory of Avoine, might not have the same view, but officials still worried about the plant's effect on that landscape. Michel Debré (who would later become de Gaulle's prime minister) remarked in one meeting that it was essential to "protect the traditional face of sites and villages around [the site]."[63] The architect of the department's siting commission cautioned EDF against a hasty design project. He noted carefully that though he did not oppose modern edifices in principle, he wanted to ensure that the site would blend harmoniously with the landscape. The new building, he wrote, "should look like a very twentieth-century building, functional and esthetic at the same time. Indeed, it is conceivable that this installation will attract tourists more strongly than many other buildings and that its general appearance, while taking economics into account, should be adapted to the new and somewhat sensational interest raised by the use of nuclear energy. There is, in this regard, a very avant-garde project to establish which I will be very interested in examining. . . ."[64] The architect did not get approval rights on the site's design, but he and others were doubtless relieved when they learned that the housing developments for site employees would be designed by a Touraine architect using traditional materials, such as white stone and slate.[65]

As the reactor slowly took shape, community leaders appeared to accept the new industrial esthetic. In 1959 one journalist commented that "the department is very rich in tourist sites, and would not accommodate unesthetic installations with chimneys spitting out black smoke." The "sober lines" and "neat layout" of the site worked well in the landscape. And, fortunately, a landscape architect had "planted magnolias and linden trees in great quantities in order to ease the transition between nature and machines." Indeed, the journalist went so far as to call the reactor a true "twentieth-century château," an "exalting spectacle," and "100% French."[66] Marcoule may have been a cathedral (even *La Nouvelle République* described it as such[67]), but in the Touraine no metaphor could signal appropriation better than that of the château.[68]

Local officials and journalists conceptualized the Chinon reactors as châteaux first. Eventually the site's managers realized that they could turn this metaphor to their advantage: the nuclear site, like a château, could become a tourist destination, complete with guided tours. Beginning in 1958, tours were conducted on Sunday mornings. A

Figure 6.7
A small, museum-like display that greeted visitors who came to tour the Chinon nuclear site. Beginning in the mid 1960s, guides used this simple model of EDF1 to explain how the reactor worked. Photograph by H. Baranger. Source: EDF Photothèque.

Nouvelle République journalist who had taken the tour told prospective visitors what they could expect[69]:

You stand in line to go inside. While waiting, the public goes to look at the large poster that depicts the finished plant: the ball is EDF1, . . . the cylinder is the water reservoir. . . . A mother hesitates, and then tells her son, "you go ahead, you're the only one who can understand any of that." In the end, the mother, the father, and the son go inside; the daughter will wait in the car. Like certain movies, though for other reasons, the Plant is off-limits to those under 16. And together with the adolescents, you have to leave cameras at the door.

Once inside, a guide took the tour group around the site, explaining how the reactor would work and patiently defining technical terms. Despite the analogy to other tourist destinations, this monument clearly elicited different responses from visitors:

The tourist-students do not yet ask any questions. Shyness, or fear of ridicule? It's one thing to ask the name of a painter or the style of an arm-chair in a château. It's quite another thing to venture into enriched uranium or the role of CO_2!

A big guy, who probably works with a monkey wrench during the week, raises his hand: "How many kilowatts? And when will the construction be over?" "First pile [reactor]—end of 1959, 60,000 kW; Second pile—1961, 170,000 kW; Third pile—1965, 250,000 kW." He acknowledges the response with his cap, and turns to his wife: "Talk about a big job!"[70]

As they proceeded around the site, several more visitors asked questions. The tour guide volunteered some information about protection against radioactivity, calling particular attention to a concrete barrier nearly 3 meters thick—"a wall," notes the article, "which evokes the Middle Ages." The second reactor under construction, though it did not have the same appeal as the spherical EDF1, was still pleasing to look at, with "modern architecture, large windows, and brightly colored panels." As they left, members of the first tour group were accosted by a second group eagerly waiting outside—just like at the châteaux.

Making EDF1 and EDF2 into the Touraine's twentieth-century châteaux served to endow the nuclear site with a regional flavor. In so doing, local officials and the press made the site local. At the same time, they redefined and updated the relationship of their region to the nation. One article put the matter succinctly: "The Touraine, already proud of having outlined on its soil a large part of the History of France, writes another grandiose page [of this history] with the birth of EDF1, the first thermonuclear plant on the banks of the Loire."[71]

EDF, for its part, fully cooperated with this appropriation. Not only did the utility offer site tours; it also sponsored (in conjunction with departmental authorities) an atomic exhibit at the region's annual fair, the Grande Semaine de Tours. In 1958, an enormous mural erected in front of the town hall in Tours publicized the exhibit, depicting an abstract representation of an atom with huge circles and spheres jutting out at different angles. Inside, the public could examine scale models of French gas-graphite reactors, gaze at posters describing the extraction and processing of uranium, learn the basic principles of fission and fusion, and (of course) read about their country's contributions to nuclear physics and technology, from Becquerel on.

EDF willingly went along with the display of nuclear power as a tourist attraction; however, it also tried to get the Tourangeaux to think in terms of the drama of technological salvation and liberation, and to consider French technological radiance more generally. The promotional literature on the Chinon site emphasized energy production, industrialization, and modernization. In its contribution to a brochure for the regional fair,[72] EDF sternly informed visitors that France had become the world's fourth nuclear power. Atomic energy would "safeguard our economic independence and our power." EDF was working toward this end at full speed, planning an atomic plant every 18 months. The new industry would provide a splendid opportunity for young people, who could thereby make for themselves "a place of choice in the world of tomorrow." Such were the terms in which EDF wanted the region to see itself. "France is entering the atomic era, with the Indre-et-Loire at the forefront, and it is with confidence and lucidity that our youth, and our entire department, turns toward the future." The concluding words of the pamphlet abandoned all restraint: "It is up to the young generations, to the future engineers, technicians, and scientists, to build a new civilization: it is their luminary value, their desire for peace and social progress, on which the future of the world depends." In EDF's drama, therefore, France's youth, by acquiring scientific and technological training, would save not just the future of their nation, but that of all humanity.

In a parallel vein, EDF representatives reminded residents that their region was poor in energy. An "investigative report" by EDF, published in *La Nouvelle République* a few months before EDF1 went on line in 1963, showed, on a map of France, which energy sources came from which regions. Mountainous areas provided "water" (hydroelectric dams); other regions provided "fire" (coal). Only western and west central France provided little to no energy. Fortunately, however, "the plains [of western central France] with their energetically underdeveloped rivers are propitious for these plants of our times. . . . Thus the natural energy void of western central France will be filled. In a family, it is the custom to coddle the baby; let us follow this lead [and proceed] under the emblem of EDF, which is setting the course for the energetic expansion of this region of France." The nation's baby in terms of energy and modernization, the region was thus being "coddled" by receiving the best and most modern energy source:

> Everything points to the conclusion that central western France, allergic to industrialization for so long, is about to blossom. Bitter voices will say that it's about time, after the progress of heavy industry last century and that of hydroelectric energy recently.
>
> They are wrong. Nothing can be compared to the past. The future of [the region] sparkles. . . . EDF is setting the example in research, financing, investment, and development.

The reactors at Chinon were particularly laudable, "fine example[s] of Cartesianism, . . . the magnificent fruit of reason."[73] By the time the third reactor went on line, a few years later, the region would actually export energy to the rest of the nation. *La Nouvelle République du Centre-Ouest* asked "Is this not a revolution in the Hexagon?"[74]

Local journalists and officials imagined the Chinon reactors as twentieth-century châteaux that would restore the Touraine to its rightful place within the French nation. Though EDF administrators were happy to cooperate in this display, they also sought to generate enthusiasm for nuclear technology as a form of salvation and revolution. This too involved resituating the Touraine's place within the nation, but in a different way. According to EDF, the Touraine, once dependent on the rest of France for energy, would now be in a position to export electricity to other regions. This change constituted nothing short of a revolution. But how much impact would this revolution have on local life? The question remained unposed.

Residents of the Gard and the Touraine were thus offered two somewhat different spectacles of technological change. In each, nuclear technology appeared as a force that would reconfigure the relationship between the nation and the region in a variety of ways. But the promise of technology for local life differed in the two regions. Marcoule promised widespread modernization that would harmonize with traditional lifestyles, offering a kind of salvation. Chinon promised esthetic harmony with the past and with the natural environment, and little else. These promises shaped the expectations that local residents had of the nuclear sites. And those expectations helped to shape residents' responses to the arrival of nuclear reactors in their region.

The Critics: "Two Steps Away Is the Abyss"

The critics of the spectacle of French technological radiance addressed both components of the spectacle: drama and display. Some did not accept the notion that nuclear technology represented salvation. Where the producers of the spectacle saw redemption, they saw "apocalypse." Others offered more secular evaluations of technological displays. The critics included Catholic intellectuals, local activists, communists, Poujadistes, journalists, and writers.

At the national level, these critics were quite visible, though not particularly powerful and certainly not coordinated with one another. Indeed, the critiques emanating from Catholics, communists, and Poujadistes usually referred back to the central ideologies of their respective communities. Catholics argued in terms of Christian morality and against materialism; communists blamed capitalist governments for the arms race; Poujadistes excoriated Marcoule as an agent of the techno-cratic state. These ideological associations probably helped to prevent the critiques themselves from being accepted at face value.

In the Gard the critics were even more marginal, though still visible. In the Touraine they seemed almost completely invisible. Nonetheless, the existence of these voices helps to delineate the spectrum of attitudes toward technological change in France. They are worth examining, if for no other reason as a reminder that the spectacle of technological radiance, dominant as it was, did not always have a completely rapt audience. More concretely, some of these critical representations—however fleeting their appearance in the Gard—made enough of an impression to be drawn into a locally produced counter-spectacle.

Starting with the first bomb test at Alamogordo and Robert Oppenheimer's legendary quotation from the Bhagavad-gita ("I am become death, the conqueror of worlds"), commentators everywhere used apocalyptic imagery to describe the terrifying destructiveness of atomic weapons.[75] Catholic writers interpreted the atomic bomb as evidence of the moral corruption of science. Nuclear explosions were said to signal the return of the "the old thunder of the Bible."[76] As David Pace has noted, "Hiroshima made ancient apocalyptic images concrete, and the new threat of destruction soon became intertwined with the church's moral crusade against materialism."[77] Just as advocates of large-scale technological development evoked the religious language of salvation and redemption to give their plans higher moral purpose, Pace argues, Catholic conservatives placed the destructive potential of science and technology in an apocalyptic framework—in part to assert their own moral rectitude.

Catholics were not the only ones to evoke the apocalypse when contemplating atomic energy. Nuclear technology was the focus of real and widespread existential anxiety about humanity's future, particularly in the late 1940s. Writers expressed their concerns in books with titles such as *Atomic Energy or Calamity?*, *Atomic Bomb: Toward Total Destruction or Heaven on Earth*, and *The End of the World or the Golden Age?*[78] Although such publications grew less popular in the 1950s, apocalyptic imagery continued to appear, particularly in the increasingly rare publications on the dangers of radioactivity. In 1957 *France-soir* ran a series titled "Is atomic radiation preparing our collective suicide?"[79] The first few articles discussed the personal risks run by the heroes (including the scientists Henri Becquerel and Pierre Curie) and the victims (such as workers who adorned watch dials with radioactive paint and residents downstream from a uranium mine who fished radioactive pike from their rivers).[80] Some of these people appeared as martyrs—including a man who prepared samples of medical radium for years until he himself became radioactive (at least according to *France-soir*). The series expressed continued ambivalence as to whether the benefits of the civilian atom outweighed the risks. But the language it used to describe the military atom was unequivocal: "We are at the threshold of the Apocalypse."[81]

The few articles about atomic risks in *France-soir* and other mainstream publications did not appear to be connected to a general editorial policy. As we have seen, *France-soir* published far more articles evoking the salvational powers of modern technology, and its coverage of Reggane was obsequiously reassuring about the utter harmlessness of the bomb test. The popular science monthly *Science et Vie* seemed to have a more consistent editorial policy, condemning foreign and French nuclear weapons alike. It too used doomsday language ("apocalypse in one-one-hundred-millionth of a second"), but it also gave its readers precise parameters for the final catastrophe. "The smallest [bomb] is already a monster," read the heading of a table that showed bomb tonnage along one axis and radius of destruction along the other. Four well-placed 100-megaton "superbombs" would annihilate France.[82]

In articles and drawings, the satirical weekly *Le Canard enchaîné* expressed outrage and fury while offering comic relief. In 1957, a special supplement was devoted to atomic matters. One cartoon featured a group of men sitting around a conference table perched atop the earth. The southern hemisphere had been blown off, and the northern half of the planet was floating through space. One man announced to the others: "Gentlemen, I have the honor of informing you that according to this referendum, 13% of humanity is in favor of a thermonuclear truce, 12% are against it, and 50% have no opinion." In the accompanying article, the commentator lambasted science, religion, and politicians in typical *Canard* prose:

The sources of spirituality have run dry. The scarecrow God, barely good enough to scare the canaries, has been knocked down while the crows have gorged themselves. Falsehood has been banished, but no Truth has come to take its place. The necessary demolition of superstition, absurdity, and fanaticism has been accomplished, but it has left us face to face with the infinite, the inexplicable, the incomprehensible. Man, the thinking robot, cured of his visions, has lifted himself up again, blind, in the middle of a hodge-podge of knowledge that brings him not an ounce of certainty. The saints have failed. . . . To whom, to what can we turn? The stupidity of the other God made us laugh; the intelligence of yours makes us fearful, you cardinals of Hate. All science, in your paws, becomes the weapon of a crime. You draw a thunderbolt from all light, agony from all energy. . . . You have tied humanity to the atomic chair, and your child-ish executioners are playing with the throttle: "do you dare us?" Well. WE DARE YOU! SHIT! Let it all blow up!. . .

Let's erase everything, but above all let's not start over! And let the cold, blood-less planet Earth finally roll without life, without thought, head cut off from the great Everything, in the basket of silence.[83]

Prophets denounced the false salvation of atomic energy at the regional level too, though written traces of their words and actions are extremely rare. Residents of the Gard remember Lanza del Vasto, a mysterious self-proclaimed acolyte of Gandhi and Hinduism who had founded a commune dedicated to nonviolence. Even before the arrival of Marcoule, del Vasto had preached against the domination of technology over nature. In 1956 he began to stage regular protests against Marcoule.[84] In June 1958 these culminated in a fifteen-day hunger strike protesting the French atomic bomb decision. The flyer del Vasto and his seventeen supporters passed out during this strike evoked spiritual passion equivalent to what we have seen at the national level:

Our fasting is the waiting and the suffering of the whole world in front of these buildings in which the life and death of all gets discussed, in which the death of the whole world gets premeditated and prepared.

The next nuclear conflagration: hundreds of millions of victims, some of whom will be annihilated in one instant, others of whom will see them-selves consumed over a slow fire for dozens of years. As for which people will suffer the greatest blows, *"that,"* say the experts, *"will depend on the direction of the wind."* . . .

Atomic testing is war against our born and unborn children. . . . Given this truth, it does not matter whether one is right or wrong, strong or weak, victor or vanquished.

All that matters is that we open our eyes onto this evidence: *In front of us, two steps away, is the Abyss.* [85]

Lanza del Vasto may have been Hindu, but surely most of the Gardois who read this flyer envisaged Catholic representations of hell and purgatory. Perhaps the language of this flyer responded ironically to representations of the reactors as cathedrals; perhaps del Vasto deliberately meant to suggest that nuclear knowledge led to the same fate as Catholic zeal. Or perhaps he merely meant to scare his audience. With such a scant written record, it is difficult to know. In any case, most Gardois apparently dismissed del Vasto and his ideas, even as they remembered him with fond-ness as a local nut who had provided a good measure of entertainment. The Touraine also had a religious figure who denounced the local nuclear site. His story is even more elusive than that of Lanza del Vasto: some say he was a priest, others a monk.[86] Local residents did not seem eager to remember this type of opposition.

France's technological spectacle had secular critics too. Of these, only the attacks published by science and technology reporters in *Le Monde* and *Science et Vie* really worried the technologists of the nuclear program. Starting in the mid 1960s, a journalist named Nicolas Vichney began publishing extensive critiques of French nuclear development in *Le Monde.* In particular, he attacked the CEA's development policy and the technical capabilities of private contractors. Meanwhile, *Science et Vie* eschewed the nationalist language favored by so many others in the media. While it certainly painted most science and technology in a positive light, it also frequently criticized the patterns of atomic energy development in general and France's program in particular. It did not necessarily oppose civilian nuclear energy—in the late 1960s, for example, its journalists expressed high hopes for the potential of breeder reactors. But its writers did think that scientists and engineers had made rash promises about the potential of atomic energy. On the occasion of the second Geneva conference for the peaceful uses of atomic energy in 1958, for example, Jean Boiset noted acidly in *Science et Vie* that thus far nuclear reactors had only produced extravagantly priced "caviar electricity."[87] The CEA's Marcoule reactors had proved particularly disappointing. Boiset echoed the monumental language used by the CEA and much of the media, but gave it a derisive spin. G1's cooling tower was "a 95-meter minaret topped by a lampshade." G2 may have been housed in a cathedral that could contain three Arcs of Triumph. But so what? All it produced was plutonium. G2 was "a marvelous stove which you only turn on to gather some precious slag (which isn't good for anything anyway) and which only incidentally heats things up a tiny little bit." It was in response to such articles that the CEA stepped up its public relations efforts.[88]

The weekly *Canard enchaîné* showed little mercy toward Charles de Gaulle and his displays of grandeur. The front page of the first issue after the Reggane explosion was covered with amusing barbs. One cartoon, entitled "le champignon de Paris" (a reference to a type of mushroom), showed de Gaulle handing a mushroom cloud to a group of men dressed as chefs but identified as the nation's top "technocrats" (and including Pierre Guillaumat): "Make a whole dish of these," says the general imperiously.[89] One headline punned on the infamous "l'état, c'est moi": "L'éclat, c'est moi!"[90] The irresistible accompanying text was even more cutting:

This bomb has liberated France—what am I saying—it has liberated the French from a complex. Better still, [it has liberated] the old Gallic rooster that we all carry in our hearts and which hasn't dared to show itself since 1940. . . . This bomb, oh Frenchmen, this bomb is the most beautiful day of our lives. Saturday February 13 marks the beginning of a new era. . . . Do you not feel completely different since that day, since that minute, since that second? You do, don't you? Before, in the eyes of the world, we were only a people like any other, neither better nor worse. After: we are, in our own eyes, a superior people. Superior to how we'd imagined ourselves. Before, we were only the first non-atomic power. After: we are the fourth atomic power! Before: our good American allies refused to share their secrets. After: it's our turn to have secrets. Nah nah nah nah nah![91]

Le Canard enchaîné also regularly poked fun at the historical continuities that de Gaulle and others drew between the regal French monuments of the past and the technological prowess of the present. In one 1966 cartoon, de Gaulle appears dressed like Louis XIV, gazing down a long esplanade of manicured trees. At the end of the esplanade, where Versailles would have a fountain, stands a mushroom cloud. The "king" tells his minions: "Le Nôtre, my dear architects, did not foresee this grandiose perspective; it is up to you to design it."[92] Some commentators, at least, had a sense of humor about French technological radiance.

More sober political critiques of France's technological spectacle came from the communist daily *l'Humanité*. Opposing nuclear weapons—particularly, though by no means exclusively, French ones—was a major priority for the Communist Party from the 1940s on. Hélène Langevin, daughter of Irène and Frédéric Joliot-Curie and herself a scientist, followed in her parents' footsteps after their deaths and spoke out against nuclear weapons at numerous rallies and congresses sponsored by the Communist Party.[93] As we saw in chapter 4, communists challenged the equation of nuclear weapons and national grandeur. They argued that French grandeur would be better served by making the nation the foremost developer of peaceful atomic energy. In addition to asserting this frequently in the pages of *l'Humanité*, the party also staged its own spectacles to convey this message, including one in Bagnols-sur-Cèze. According to *l'Humanité*, 5000 people gathered in Bagnols's amphitheater on a cool fall day in October 1959 to hear scientists, party leaders, and even Lanza del Vasto condemn French military nuclear development. Even *l'Humanité* admitted, however, that most of the audience had traveled to Bagnols from elsewhere.[94]

As I noted in chapter 1, more criticism of the technological spectacle came from the opposite end of the political spectrum: the extreme-right Poujadiste movement. Nîmes, the capital of the Gard, had an active

Poujadiste group, which published the weekly newspaper *L'Echo du Midi*. Since *L'Echo* specifically targeted Marcoule in its tirades against state power and technocratic elites, it requires a closer look.

L'Echo opposed Marcoule not so much because of its nuclear features as because the site embodied the evils of state intervention. In mid 1957 the paper began to run a regular column, entitled "La Tribune de Marcoule," that regularly attacked the site and its administrators. In an otherwise discordant political confluence, some stories paralleled revelations by the CFTC/CFDT labor union—such as the articles which revealed that the CEA refused to let the Ministry of Labor inspect Marcoule.[95] Other articles denounced the drama of regional salvation performed by local and site officials. The claim that Marcoule would lure other industries to the Gard was a hoax; the region derived no special advantage from housing the nuclear site.[96] Several columnists worried that the site would emit harmful radioactivity.[97] One writer noticed that, contrary to earlier announcements, the reactors there produced not electricity but plutonium.[98] All these stories carried the same punch line: the state and its technocrats had duped the people.

L'Echo expressed outrage not only about Marcoule itself but also about the ways in which other local journalists wrote about the site and about their cozy relationship with the CEA technocrats. One columnist sharply criticized the opening of Marcoule's belvedere as a tourist event. Better places to admire the countryside existed than the viewpoint of an atomic factory, he noted, deriding one mainstream journalist's suggestion that the belvedere be awarded three Michelin stars. Moreover, the comparison made between Marcoule and cathedrals deeply offended religious men such as himself (most Poujadistes were Catholics).[99] The journalists of *L'Echo* attributed the fanfare about Marcoule to the CEA's courting of the local press. Marcoule's administrators had held a sumptuous dinner for regional and national journalists. Apparently, however, *L'Echo*'s journalists had not rated an invitation. No matter: "This only makes us more free in this column, where our information will never be gathered between the fruit course and the cheese course, in the euphoria of Tavel gathered from the deep recesses of the wine cellar."[100]

For *L'Echo du Midi*, Marcoule, its technocrats, and the bureaucrats who supported the project represented not the salvational power of modern technology but the evils of a state-directed economy and industry.[101] Apparently, however, the attitudes held by the paper did not make it into the mainstream of regional discourse about Marcoule. The other local papers never cited *L'Echo*, and its name did not appear in archival material.

Local residents remembered neither the paper nor its attitudes. Although this absence is not conclusive, it does suggest that the Poujadistes had little direct influence on how most local residents conceptualized Marcoule.

Nonetheless, the critiques made by *L'Echo*—along with the apocalyptic imagery of Lanza del Vasto and the Catholic critics—found echoes in the representations of local citizens. Nowhere is this clearer than in a play produced by the Gardois in 1957.

Counter-Spectacle: "When the Tale of Marcoule Is Told"

In 1957, the townspeople of Bagnols-sur-Cèze produced their own spectacle about the arrival of nuclear technology in their midst: a musical pageant in five scenes entitled "When the Tale of Marcoule Is Told."[102] In many respects, the technological spectacle produced by CEA administrators and local officials was the central reference point for the play. But rather than simply reiterating the messages of salvation and redemption conveyed there, the townspeople's pageant reacted to it and retold it, producing a counter-spectacle. The play emphasized different themes and characters from the original show. Some of these themes and characters strongly recall the spectacle's critics—for example, the false lure of technological progress, and the apocalyptic danger lurking within the atom. Others seem more grounded in local encounters between peasants and engineers: the distance between those with technical expertise and those without, the overarching power of the state. The play's scenes not only drew upon episodes that had already occurred at the time of performance; they also foreshadowed incidents to come.

The prologue features an engineer, two aides, a peasant, and two journalists, all contemplating the landscape of the Gard. The engineer launches into a disquisition on France's pressing energy needs, the importance of plutonium, and why the Gard was chosen to host the first industrial-scale atomic site. As this reiteration of the CEA's standard narrative unfolds, the two aides gaze disdainfully at the fields and mutter "Quel bled!" ("What a dump!"). The peasant informs the engineer that a levee of 3 meters will not suffice to control the flooding of the river, but the engineer dismisses his remark: "Go plant your cabbage, my good man!" This interaction sets up a question taken up later in the play, and also central to other stories locals told about Marcoule: who outsmarts whom—the peasant or the engineer?

The first scene takes place inside the home of villagers whose house will be expropriated for the site. The family mourns the loss of their home and sings a few nostalgic songs. Characters express confusion about the events that befall them and repeat a rumor that an Eiffel Tower would be built on the site.[103] Just as they reach the depths of despair, the local nobleman arrives to cheer them up:

Stuff and nonsense. Don't get sentimental. This is progress, and we can't do anything about it, do you hear? In the time of our Kings everything was justified by the *raison d'état*. In your Republic, it's become Urban Planning . . . Social Progress . . . etc. What do I know? These are all worthy things. But from there to giving up my property for a morsel of bread, stop right there, my dear gentlemen. You do not know the Maître de Gicon. I will not cede my property for anything less than 50 million.

At the prospect of making millions off their land, everyone cheers up considerably. The scene develops the story of the peasant and the engineer by suggesting that, with a little noble help, the peasant-villagers can squeeze good money out of the CEA. In a twist that gets abandoned in the rest of the play (but whose message re-emerges elsewhere), the aristocrat suggests that urbanization and social progress have become the legitimators of the republic: clearly, the French state always has justification for imposing its will on the citizenry.

The second scene, which takes place outside Marcoule's construction site, includes a hodgepodge of characters and episodes. The story of the peasant and the engineer reaches a climax here. It turns out that the peasant was right: the site flooded because the levee was not high enough. The peasant has returned to laugh at the engineer and to present him with a cabbage. On the vicissitudes of nature, the peasant has thus proved wilier than the engineer, who was too proud and foolish to listen to his wise counsel. The rest of the scene features a procession of people knocking on Marcoule's door: people looking for work or clamoring for a tour. In a particularly silly twist, Prince Rainier and his new bride Grace Kelly show up in a Rolls-Royce. They alone are offered a tour of Marcoule (which they refuse because they wish to maintain their royal innocence of worldly, technological matters). Clearly, it did not escape local residents that, despite attempts to put Marcoule on the tourist map, only special privilege would get people into the site.

The third scene pokes fun at the enthusiasm of local officials for the site. It takes place in the office of the French president. He is so busy that he cannot receive the Queen of England or take a call from the president of the United States. But when Boulot, mayor of "Bagnols-Marcoule" arrives, the president greets him with open arms, punning "Quel bon boulot vous faites, Monsieur le Maire" ("You're doing a great job,

Mr. Mayor"). (*Boulot* is slang for "job" or "work.") The mayor asks for 8 million francs to help with new construction; he gets 6 billion. The scene finishes in joyous celebration and singing. Several young women in traditional costume have accompanied the mayor, bearing a giant bottle of *Cuvée de Marcoule* and a huge cake in the shape of the G1 reactor. The scene suggests that residents of Bagnols maintain a certain ironic distance from their political leaders. Marcoule takes precedence over all else at the national level, but the scene satirically questions this priority: should the president (representing the state) not have more important things to do? The conflation of town and nuclear site ("Bagnols-Marcoule") suggests that the mayor no longer has the sole interests of his town at heart. Finally, the sheer size of the wine bottle and the cake mocks the iconography of Marcoule.

The fourth scene depicts tensions between local residents and newcomers most explicitly, exposing the fallacy of the harmony promised by the technological spectacle. This scene takes place in Bagnols's public wash house, where a group of women are doing their laundry. In the foreground we meet Robert and Marion, a charming young Bagnols couple contemplating marriage. Marion fantasizes about a new modern house, which would come complete with "refrigerator, washing machine, pretty dresses, a small car, and intelligent, well-dressed children." Robert sighs: he will never nab her, since he can only plant cabbage. In strolls Jacques, a dapper young Parisian. He flirts shamelessly with Marion, who simpers. Robert cries "These Marcoulins have taken everything from us—our roofs, our vines, our fields and our gardens—and now they're taking our girls!" and stomps out. But it turns out that Jacques does not plan to linger: his time at Marcoule has ended, and he has no plans to return. Her chances for a modern lifestyle ruined, Marion slumps off stage. The scene shifts to the washerwomen, who discuss the changes they have witnessed in their community. They complain that they no longer recognize anyone at the cinema or in church—their town no longer belongs to them. But, they admit, they will get a post office, a high school, and a sewage system, and they "needed the Marcoulins to get all that." They express their own technological hopes: "With the atom, all we'll have to do is press a button, and everything will come out washed and ironed." At the end of the scene, Robert and Marion reappear, reconciled.

This scene makes clear what women can expect from the atom: not glory, nor jobs, nor money, but rather technology-induced domestic bliss. At the same time, it mocks this version of modernity by portraying Marion's expectations that "intelligent, well-dressed children" come in the same package as new commodities.[104] It also clearly articulates ten-

Figure 6.8
The living room of a Marcoulin apartment. Presumably, it was a place such as this
that Marion thought she might live in if she married Jacques. Courtesy of Mireille
Justamond, Archives municipales, Bagnols-sur-Cèze.

sions between local residents and the "Marcoulins," as the new arrivals are
derisively called. Residents experience the Marcoulins as invasive: they
have taken over sites of sociability and places of worship. Newcomers have
robbed local men of their land, their homes, and their livelihood. In the
ultimate insult to their masculinity, the nuclear men threaten to take
"their" women away too—thereby corrupting the archetype of traditional,
female France. Jacques represents not only the selfish, acquisitive
Parisian/Marcoulin but also the false seduction of a flighty, commodified
modernity. Marion hopes to reap the harvest of progress by marrying him,
only to have those hopes slip through her fingers as he dashes off to his
next technological rendezvous. Her final reconciliation with Robert suggests that in the end, Bagnolaises do better by sticking to traditional ways.

The last scene—"Finale or Apotheosis"—conveys a mystical, dreadful
sense of the atom itself, strongly evocative of the apocalyptic language
used by some of the critics of the technological spectacle. Abandoning
local characters and caricatures, the scene begins with an offstage voice
stage intoning: "In the beginning was the Word, and the Word was God.
And this God created the Heavens and the Earth, the World and the

Atom." On stage, the company—now dressed to represent Earth, Cain and Abel, Man, and the human race—sing words from *Faust*. Earth recites a somewhat plaintive poem, to which Atom—still off stage—responds:

I share in your pain, o dear and ancient Earth

For ignored by all during millennia

I tasted in the very stuff of matter a peaceful rest

God created me everywhere

I am in the petal of the rose, and in the blackest coal

I am in the air, the sun, and space

And in the live flesh of animated beings

. . . Then, ancient Earth, hate was born on your soil and war . . . submerged the world of the living . . .

And your children, no longer happy with searching your veins, discovered my power. This immeasurable power that I myself did not know I had.

And to serve this hate they managed to capture me, to domesticate me like cattle, to subordinate me to their needs and to assuage their folly

Like Satan in his kingdom of Hell

Where will they stop?

Do they realize that one day I may take revenge?

By unleashing upon them this destructive force that they revealed to me.

Terrified, Earth begs for mercy, and asks what sacrifice Atom wants. No sacrifice, Atom responds. Treasures? No. Power and glory? Neither. How about intelligence and harmony, the qualities aroused by scientists and artists? In a transport of biblical passion, Atom responds:

You stray, ancient Earth; that which my heart longs for is PEACE and Love. That infinite love of man for his brother man, whether he be great or miserable. . . . Whether he lives in icy deserts or in the sunniest of places. . . .Whether his face be black or white or yellow. The day when I see all men love each other, I will leash my power to their service. And I will bury my vengeance when I see that PEACE reigns among them.

Earth appears to grasp this message. In the grand finale, the company sings the "Ode to Joy" from Beethoven's Ninth Symphony.

This scene clearly expresses fear and ambivalence about the prospects represented by the atom. In the spectacle of technological progress, the atom is never detached from the technology needed to control it; together, the atom and the reactors appear as savior and redeemer. In this scene, however, the atom exists independently, as a pure force of nature, a creation of God. The biblical language parallels that of Lanza del Vasto and the Catholic critics. The dominant spectacle proposed nuclear technology (including and especially the bomb) as France's redeemer and savior. The counter-spectacle suggests that the atom will redeem nothing as long as peace does not reign. Indeed, man (the only female presence in this scene is plaintive, confused Earth) may have erred mightily in trying to enslave the atom to his violent impulses. Someday the atom might escape human control and wreak vengeance upon all the earth. Peace and love are more important than treasures, power, glory, intelligence, or harmony—that is, more important than anything Marcoule could bring to the region. For those who identified with the story told in this scene, technological progress clearly did not mark the telos of human existence.

∗

Local political and intellectual elites in the Gard and in the Touraine treated the nuclear sites as symbolic mediators between their region and the nation. The sites brought the nation into the region. Appropriating the nuclear sites through regional metaphors and dramatic narratives made the sites—and therefore, in some ways, the nation—local. In invoking the nation, local elites (as well as technologists) endowed the sites—and, by extension, local modernization projects—with a higher moral purpose. At the same time, their appropriations resituated the regions within the nation and defined a role for them in the emerging technological France. In the scenarios imagined by the producers of the technological spectacle, regional history, national destiny, and technological development all worked together. These scenarios cultivated national and local historical consciousness as a way of defining modernity, tradition, and the relationship between them.

Frank opposition to Marcoule and to the state that built it shows that it was at least conceivable to reject the story told by this spectacle. Other meanings were possible. But the marginality of this opposition on a local level suggests that most residents did not consider outright rejection a serious option. For the residents of the Gard and the Touraine, nuclear technology was not an abstract issue. Nor were nuclear plants radically separate from other aspects of modern life. Still, the representations proposed by the critics added to the vocabulary created by the technological spectacle. Together, the show and its critics provided a set of concepts through which their audience might imagine and interpret technological France.

"When the Tale of Marcoule Is Told" indicates how some of that audience used that vocabulary. It shows that the spectacle of technological progress was powerful but not monolithic. Performing or watching the play gave locals an opportunity to grapple with representations of technological progress and with their experience of modernization. The play's lack of a single plot line appropriately reflected how residents constructed their understanding of Marcoule. Whereas the technological spectacle tried to impose a unified interpretation of Marcoule, the counter-spectacle has a more fragmented quality and offers multiple meanings. These meanings center not around abstract concepts such as progress, but around human emotions and experiences: hope, fear, betrayal, arrogance, cunning, ambition, and invasion.

7

Atomic Vintage

The villagers gathered along the roads of Codolet to watch the "atomic millipede" and the queue for guided tours of Chinon demonstrate that, at least in some respects, residents of the Gard and the Touraine did experience nuclear development as a spectacle. Indeed, their willingness to behave as enthusiastic audiences and eager tourists contributed mightily to the creation of the spectacle. After all, a show without an audience is a not a spectacle but a flop. But were Gardois, Tourangeaux, and other French citizens *merely* audiences for a grandiose pageant? How did ordinary people feel about nuclear technology in the 1950s and the 1960s? What impact did the nuclear sites have on the Gardois and the Tourangeaux?

Results of public opinion polls make it possible to draw a rough sketch of public responses to nuclear technology. A small number of polls taken in the 1950s and the 1960s asked adults what they thought about the prospect and the reality of a French atomic arsenal. One poll also tried to determine how people felt about nuclear power and what beliefs they held about the dangers of radioactivity. The responses to such poll questions provide some indication of how "the French" (often an undifferentiated category in these polls) thought about nuclear development. We must be careful, however, to keep in mind the many limitations inherent in these sources. Polls—particularly polls with multiple-choice questions—can make categories of opinion appear where none might exist otherwise. (To what extent, for example, did citizens really think of themselves as having "an opinion" about a French atomic bomb?) Furthermore, multiple-choice poll questions obscure potential diversity by forcing respondents to choose "yes" or "no," "for" or "against." The categories of opinion supposedly revealed by polls may in fact be simply those of the pollsters or of the groups that commissioned them.[1]

Nonetheless, judicious reading of poll results can provide a useful entrée into public interpretations of nuclear development. For example, these results hint at ways in which ordinary citizens might have distrusted or ignored representations of technological prowess. They also show that some people did indeed embrace the basic concept of French technological radiance. Still, the aggregated opinions depicted in polls can, at best, reflect only the outer layer of experience with technological development. Going beyond polls requires moving to the local level. Juxtaposing the experiences of local residents in the Gard and the Touraine shows that significant variations existed both between and within these two regions. Were it feasible to do a more comprehensive study of local experience, we could expect even bigger differences between these two regions, which confronted large-scale technological development directly, and other regions that did not.

After a brief discussion of polls, therefore, I focus on the local level in the rest of the chapter. I attempt to capture how Gardois and Tourangeaux experienced the construction and expansion of the nuclear sites and the influx of large numbers of outsiders into their regions. Readers will see that in the 1950s and the 1960s—before the development of organized opposition to nuclear power—most residents of these regions were neither "for" nor "against" nuclear power. Pollsters may have tried to classify citizens in stark categories, but for the most part such categories did not adequately reflect how residents thought and lived. For the Gardois and the Tourangeaux, nuclear technology was not a hypothetical issue about which they had abstract opinions. Reactors were an increasingly dominant feature of their natural landscape, and nuclear workers were an increasingly large presence in their social landscape. Residents had complex responses to the sites, which arose from the mundane interactions they had with site workers, from the economic impact of this influx, and from the expectations raised by the producers of the technological spectacle.

Though I have used all the written sources available, a substantial part of my analysis rests on oral interviews. These interviews provide important insights not just into people's experience but also into the construction of local memory. The relationship between the history narrated here and the evocation of local memory can come through more clearly once the reader has a sense of the stories told by residents. Reflections on differences in the construction of memory in the two regions therefore appear after an examination of local experience in the Gard, but before a parallel discussion of the Touraine.

Representations of Public Opinion

Even if we were to take public opinion polls at face value, interpreting their results would pose a tricky problem, particularly when (as is the case for the polls discussed here) we know nothing about poll design and execution. Without such information we cannot evaluate the representativeness of the responses, nor do we have access to the reasons behind them. Did people base their answers on what they read in the paper? On the opinions of their parents, their spouses, or their colleagues? At best we can only speculate. The importance of these issues notwithstanding, poll results can at least suggest how ordinary citizens might have thought about nuclear technology.

The technological subject that most occupied French pollsters in the two and a half decades immediately after World War II was nuclear weapons. During that period at least thirteen nationwide polls asked French citizens what they thought about their nation's developing its own military nuclear capability (table 7.1).

The poll questions are as interesting as the responses. In 1946 pollsters simply asked "Should France have its own atomic bomb?" It is not too surprising that in the aftermath of the war—and given the efforts to write

Table 7.1
Results of public opinion polls asking whether France should develop its own nuclear military capability.

	Yes (%)	No (%)	No answer (%)
January 1946	56	32	12
January 1955	33	49	18
July 1956	27	51	22
December 1957	41	28	31
July 1959	37	38	25
March 1960	67	21	12
July 1962	39	27	34
January 1963	42	31	27
July 1963	37	38	25
August 1963	34	37	29
November 1963	39	37	24
April 1964	39	40	21
January 1967	23	50	27

France into the history of the two American bombs—more than half of the respondents replied "Yes." Still, nearly a third responded "No"—not a negligible minority. Nine years later, the same question generated nearly the inverse result.

In July 1956, the question was changed in a subtle but significant manner: "As you know, France conducts atomic science research but does not build atomic bombs. Do you think France should build atomic bombs?" The question hints at ways in which the very act of taking a poll may have itself shaped public understanding. Poll designers almost certainly did not know that the CEA had begun work on a French bomb. Innocently or not, however, their question reinforced the false notion that France was not preparing its own military nuclear capability. That poll also provided a tiny glimpse into the reasoning behind some of the responses. The sample reasons French should not build a bomb included "We have no money, it would mean losing money stupidly, it would mean new taxes" and "If we make a bomb, it's to use on our neighbors [c'est pour foutre chez le voisin], and I don't agree." The sample reasons in favor of a French bomb mostly referred to national prestige ("If foreign countries know that France has atomic bombs, they'll respect her"; "It's symbolic to keep the rank of great power"; "Simply to show that she's as strong as the others").[2] When examined along with the data, these samples suggest that, although those who supported the idea of a French bomb did so for nationalist reasons, such arguments had little consequence for at least half of those polled.

In the month immediately after the Reggane test, pollsters asked "Do you think France should have its own atomic *force de frappe?*" The issue was now no longer one bomb but a whole arsenal. The technical success of Reggane apparently fired people's imaginations. Enthusiasm for French bombs peaked in March 1960. But it waned over the course of the 1960s. In 1967, a poll taken just before a parliamentary election asked voters "What is your opinion of the *force de frappe?*" Respondents could reply "satisfied" or "unsatisfied."[3] Overall support for de Gaulle had hit an all-time low; support for his arsenal apparently followed suit.

Much of the time, 20–30 percent of people did not respond to poll questions. Possible reasons for this include indifference, confusion, and surliness. It is also likely that many people simply did not think in the categories proposed by the pollsters. Whatever the case, these polls made no room for them to express such thoughts. All we can take away from the results is that, despite the best rhetorical efforts of politicians, technologists, Catholic protesters, and the Communist Party, about one-fourth of

Table 7.2
Responses to the question "Have you heard of 'Zoé' and can you say what you
know about it?" Source: poll conducted in early 1949, published in *Sondages*, 1
March 1949 (cited on pp. 10–11 of Fourgous et al. 1980).

Poll's categorization of responses	Poll's sample answers	Percentage of responses
Correct	"First French atomic pile"; "atomic pile built by the Joliot-Dautry team"	36
Fairly close	"An atomic clock"; "I read something in the paper. It's an atomic thing."; "Yes, the first manifestation of the French atomic bluff"	3
Vague	"A French pile"; "A weird engine"; "Yes, it can blow us up"	4
Atomic bomb	"Name of the first French atomic bomb"; "Miniature atomic bomb"	2
Other	(none)	2
No answer		53

French citizens usually declined to express a black-and-white opinion of
nuclear weapons.

The pollsters apparently surmised that indifference and ignorance
went together. They occasionally conducted polls to determine how well
informed people were on the subject of their nation's nuclear achievements. Table 7.2 shows the results of one such poll taken in 1949, along
with indications of what kinds of responses qualified people as "well
informed." A similar poll conducted eight years later suggested that the
overall level of information had dropped: in 1949, 36 percent of respondents gave answers that pollsters qualified as accurate; in 1957, only 18
percent did. The later poll categorized answers less specifically than its
predecessor, noting only that well-informed people included those
"capable of providing at least an approximate definition of Marcoule,
Saclay, Zoé, and of ranking France fourth or fifth among the great
atomic powers."[4] The ability to correlate geopolitical prestige and
nuclear achievement apparently counted as accurate knowledge.[5]

In 1957–58, one government council commissioned a poll to obtain a
more detailed picture of how the French thought about atomic energy.[6]
The poll began by asking "When you hear about the atom, what uses
come to mind?" From the nine categories offered to them, respondents
chose as follows:

bombs, weapons	35%
destruction, war	20%
energy source	17%
peaceful uses	14%
industrial progress	10%
medical uses	7%
scientific progress	3%
interplanetary rocket	3%
means of transportation	3%

Respondents could choose more than one category, but we do not know where the overlaps occurred. Thirteen percent gave no response. According to the pollsters' breakdown, men, younger adults, white-collar workers, professionals, and more highly educated people seemed more likely to respond "energy source," "industrial progress," or "peaceful uses." Women seemed slightly more likely to think of military uses or give no response at all, as did agricultural and blue-collar workers and less educated respondents. Another question in the same poll tried to determine how people understood the dangers of atomic energy (table 7.3). Women appeared somewhat more likely than men to find atomic energy dangerous. Forty percent of respondents with little to no education thought atomic energy was dangerous for everyone, versus 25 percent of those with university-level education. Less educated people were also less likely to respond to the questions. When asked to be more specific about danger, answers were fairly uniform across the board: "radioactivity" and "accidents" appeared as the most likely sources of danger; "health" the most likely effect. Again, though, more than one-third of those polled declined to specify.

Table 7.3
Responses to various questions beginning "Is it correct that the use of atomic energy for peaceful purposes is . . ." (cited on p. 29 of Fourgous et al. 1980).

	Yes (%)	No (%)	No answer (%)
"dangerous for those who work in atomic plants?"	62	9	29
"dangerous for the inhabitants of the region?"	52	16	32
"dangerous for everyone?"	36	27	37
"not dangerous at all?"	7	53	40

The poll also tried to determine how the French compared civilian and military uses of atomic energy: "Is it more urgent for France to intensify the development of peaceful uses for atomic energy, or to proceed to make atomic weapons?" Sixty-four percent answered that peaceful uses were more important; 15 percent said that weapons mattered more; 3 percent said both were important; 18 percent said they didn't know. According to the pollsters' analysis, 20 percent of those who voted in favor of peaceful uses thought it "better to work for peace than for war." Sample responses cast the matter less in high moral terms than in terms of national prestige: "It would be a great revenge if scientists in France were at the avant-garde of progress while everyone else cares only about weapons"; "France has always been a peaceful nation"; "France's destiny is to impose itself in intelligence and peace." Such arguments echoed those made by the Communist Party; the poll, however, offered no correlation between these responses and political affiliations. Another 20 percent of respondents did not want to make weapons because they did not want to use them: "If we make weapons we'll have to use them sooner or later"; "The USSR-USA antagonism is bad enough without France getting involved." Twenty-five percent cited the importance of peaceful atomic energy for better living conditions and overall prosperity, in some cases relating prosperity to prestige: "It can create well being for all, in our generation and in future generations"; "It would allow the modernization of our industry"; "It would allow us to attain economic independence"; "Greater industrial development would automatically make us a great power." Pollsters did not provide sample responses for those in favor of military over civilian pursuits, commenting only that "people who feel it more urgent to pursue atomic weapons all say that it's the only way to be equal to the other powers, to protect oneself, and to get respect."[7]

Finally, and most directly pertinent to the rest of this chapter, pollsters asked residents in various regions across France how they felt about the installation of a nuclear plant in their region (table 7.4). Only Normandy and Brittany had more people in favor of than opposed to a local plant. Residents of Paris, its suburbs, and Center-West France (where Chinon was located) came out most strongly against a plant. Still, fully one-third of the Center-West residents polled declared themselves "indifferent." In the other regions, most responses were split between opposition and indifference.

These numbers certainly suggest that people would not embrace the arrival of a nuclear plant in their immediate vicinity. And indeed, ordinary residents of the Gard and the Touraine did not manifest as much enthu-

Table 7.4
Responses, by region, to question about installation of a local atomic energy plant (cited on p. 28 of Fourgous et al. 1980). Marcoule is in the South/Mediterranean region, Chinon in the Center-West.

	In favor (%)	Opposed (%)	Indifferent (%)
North	12	43	45
Normandy and Brittany	38	13	51
Center-West	15	51	34
South-West	15	46	40
Center	17	41	42
South/Mediterranean	21	43	36
Center-East	13	43	44
East	15	42	40
Paris region	15	65	20
Paris	13	61	27

siasm for nuclear sites as did local political and intellectual elites. However, the attitudes of locals were too complex to be categorized simply as opposition, support, or indifference. Faced with the reality of a nuclear plant (as opposed to a hypothetical question about its desirability), people developed more complex interpretations of a plant's significance.

Peasants and Engineers: Bagnolais de Souche and Marcoulins

The counter-spectacle examined in chapter 6—the play "When the Tale of Marcoule Is Told"—suggested that the Gardois elaborated their own representations and interpretations of local nuclear development. Although the stories told in the play exaggerated events and personalities for the sake of humor, many also had bases in real encounters between local residents and nuclear technologists. Some of these encounters evolved into stock stories that locals told and retold about the early years of Marcoule. Of these, the most common revolved around a wily peasant (embodying the region) and a supercilious engineer (embodying the state). Rather than narrating the reconciliation of modernity and tradition, these tales evoked invasion and distance, representing local customs and ideas not as romantic traditions and preludes to great modern achievements but as real lifestyles facing unwelcome challenges from uninvited technocrats. The outcomes of these stories varied. Sometimes the peasant was a hapless victim of the engineer's desires. At other times he proved himself more cunning than the expert, sometimes gaining profit, sometimes just grim satisfaction from their encounter. Several of these tales involved the expropriation of land in the villages of Chusclan and Codolet, on whose territory the nuclear site had been built. In one story that Chusclan's mayor told a journalist, engineers appear as bizarre intruders upon the land. "Mysterious characters appeared in our little village. These strangers called themselves oil prospectors, but they weren't carrying any equipment. They did no boring. These were serious men. All day long, with empty hands and grave [expressions], they surveyed the vines. . . . In the evening, they regrouped in the village square and got back into their car without saying a word. Since they didn't go into either of the two local cafés, we knew nothing. Three days later, they disappeared just as mysteriously."[8] Here, the engineers violated the customs of the land and distanced themselves from the peasants by failing to frequent the local cafés. In retrospect, the mayor declared that he and his villagers had known all along that something strange was afoot. Not until the prefect of the Gard called him and a dozen other mayors in, however, did they know why those engineers had nosed around their land. Their sense of victimization grew further when the land expropriations began.

In another story, the expropriators preyed on one of the only female landowners, Mme. Vigié, whose property lay in the middle of Chusclan and whose vintage ranked among the very best in the region. She was paid a mere 20 million (pre-devaluation) francs—"a laughable price," as the land would have been worth much more in smaller parcels. In a 1956 interview, one *vigneron* recalled: "It was not without heavy hearts that the entire village witnessed the destruction of Mme. Vigié's vineyard. Oh! If you could have seen those bulldozers. They went right through the vines, tearing up and tossing the roots away. An hour later the ground was razed."[9] The journalist who published the interview concluded sadly: "One of the most extraordinary vineyards of France, . . . which had made Chusclan's reputation, which Louis XIV had named 'My Garden' and whose wine he demanded to have on his table every day, existed no longer."[10] In this story, modernity did not build harmoniously on tradition. Instead, progress destroyed a piece of land that had been a source of income and pride to the village. The tale thus reversed the message of the national technological spectacle.

Other stories featured shrewd villagers. According to a former aide to the mayor of Chusclan, many landowners soon realized that the longer they held off on selling their land, the more they could get for it. A particularly salient instance occurred when a representative of the state began to buy land to build a road to Marcoule. At first, the state paid relatively little for the land parcels along the road. Those who waited to sell found themselves in a better bargaining position as the path of the road became increasingly defined. The last one to sell—the mayor himself—ended up getting the best price.[11] Another example of villagers' outsmarting the CEA is recorded in the minutes of town meetings at Chusclan and Codolet (and gleefully retold by village residents). Faced with an army of state-sponsored experts who wanted to buy communal lands, the municipal councils of the two villages decided that they needed help from their own experts. Before quoting the CEA a price, they therefore asked a professor from the agricultural school in Montpellier to assess their land values. In the end, they obtained the price they demanded.[12] Villagers could thus learn to manipulate the state's system to their profit.

Not all encounters with Marcoule ended so happily, however. CEA administrators may have accepted local expertise in matters of land assessment, but they seemed unwilling to admit that residents might have valuable knowledge about their own environment. Indeed, the opening scenes of "When the Tale of Marcoule Is Told"—in which the peasant counsels the engineer to build higher levees and returns triumphantly with his cabbage after river flooding proves him right—were dramatically reenacted in the fall of 1958, this time not on a stage but on the Cèze flood plain. The villagers of Chusclan and Codolet had indeed issued ominous warnings about the flood potential of the Cèze river. But Marcoule's administrators refused to believe that such a seemingly calm, relatively small river could cause much damage. They went ahead with their plans, building roads and raising levees in ill-advised places. In one tale told to me by a lifetime Chusclan resident, Marcoule's director imperiously declared that he would feel confident laying his hat by the side of the new road, because even the worst of floods would not reach that far. In October 1958, rivers flooded all over the Gard. Codolet was badly hit. The road leading to Marcoule prevented the Cèze from flowing easily down to the Rhône. Instead, it surged straight into the heart of the village. Waters rose so high that houses had 30 centimeters of water on the second floor. Disaster repairs and reconstruction took several years.[13] Legend has it that one peasant, crossing paths with the director after the floods, inquired acerbically: "So, Mister Director, what happened to your hat?"[14]

Many of these stories appeared in several newspapers in the late 1950s. They also became part of local memory. They were all spontaneously told to me, with many of the same details, in interviews I conducted with long time residents in 1994. Clearly, the memory of these encounters and tensions was nurtured, told, and retold by residents over the decades. The central message of these tales was that, for all his expertise and sophistication, the naive engineer—in contrast to the wily peasant—could not come close to understanding the land or the vagaries of nature. The memories carried a note of bitter triumph: although state technologists might try to extend their reach deep into the life and landscape of the region, the region—through both its natural features and its people—could thwart the technologists. The state could not completely control the region without the cooperation of local residents, but when locals tried to help they were rebuffed. The memories thus conveyed a tremendous distance between local residents and state technologists—a distance repeatedly perpetuated by the newcomers.

As we saw in the tale of the mysterious land surveyors, from the very beginning residents of the Gard felt that CEA representatives told them as little as possible. The inhabitants of Chusclan, Codolet, and other villages learned about the arrival of an "atomic center" not from the CEA or department officials but from a press release. Offended by this manner of communication, the municipal council of Chusclan issued a statement in 1953 opposing the installation of the center on its territory. Such a statement had no force in the face of a determined state institution such as the CEA, however, and it was quickly forgotten.[15]

Locals felt that their sense of distance from Marcoule and its employees was reinforced by their inability to grasp what was being done at the site. Almost all of the residents I interviewed claimed that they did not understand the goings on at Marcoule, and that they never had. The former aide to the mayor of Chusclan insisted that most people simply did not care about such details. Over the course of nearly forty years, he himself had visited the site in his official capacity a dozen times, and had repeatedly heard the basic description of how the reactors and plutonium factory worked; still he could not repeat such explanations. But who bore the blame for this lack of comprehension? In a rare critique of Marcoule, the newspaper *Le Provençal* blamed the experts rather than popular ignorance. Reporting on the CEA's pavilion at the 1957 Nîmes town fair, one reporter wrote:

Without question—and here we are only expressing the opinion of the man on the street, which is to say the vast majority of visitors—the CEA Pavilion was a disappointment.

It is not the CEA's effort which is at fault, but rather its conception. . . . The explanations offered were too technical, too dry—in a word, too scientific. Nuclear science is brand new, and therefore practically unknown by the masses. It was indispensable, therefore, to try to popularize [this information] in the clearest, simplest manner. The means of education were distinctly inadequate: a few diagrams, some photographs, four or five models, and some equipment which may have been easy to manipulate but which, in the end, did not help any bit of actual knowledge penetrate the mind of the visitor.[16]

The problem was not that people could not learn, but that experts could not teach.

A kind of socio-political distance deepened this technical gulf. In the 1950s, the CEA refused to hire anyone with communist affiliations to work at Marcoule. Although officially classified, the procedures that eliminated suspect applicants were, in fact, well known.[17] Not that the region was teeming with communists—most residents of this part of the Gard considered themselves centrist or apolitical.[18] But anyone in contact with CEA agents quickly realized (well before the French bomb project became official news) that secrecy characterized Marcoule's activities. No matter how understandable the circumstances, the inability of CEA employees to talk about their work made conversations awkward.[19] Residents conveyed a sense of fatalism about this distance: just as knowledge of the atom and its technologies lay beyond reach, so did knowledge of the state and its purposes.

For locals, the state and its technologies were deceptive as well as inscrutable. The false lure of modernity, personified by the seductive but elusive Parisian researcher in the play, was a theme in stories and memories about the modernization of the region. First to experience disappointment were Chusclan and Codolet. Villagers had believed that Marcoule's presence would compensate for their land sacrifices by bringing modern amenities such as sewage and sanitation. In 1956—three years after the land expropriations—they still had nothing. Frustrated, the municipal council of Chusclan wrote to departmental officials protesting that, even though they had not impeded the march of progress, progress had not yet marched into their village. Would Marcoule need so much land that it would force them out of the village altogether? The council requested departmental money to modernize the village, arguing that dilapidated Chusclan made a poor impression on outsiders visiting Marcoule's "ultra-modern" installations.[20] Other villages made similar requests, perhaps hoping to capitalize on elite efforts to make nuclear development a national spectacle.[21] Codolet residents grew particularly frustrated after the 1958 floods. Apparently the CEA expected municipalities to take care of such matters, while the municipalities expected either the CEA or the state—from their point of view essentially interchangeable institutions—to do so. French industries were supposed to pay a *patente* to the villages whose land they used—a kind of tax based on their production figures. But Marcoule had been exempted from the *patente* because it was both a national defense site and a research center. After considerable agitation, Codolet and Chusclan realized in the mid 1960s that the only way to extract money from the CEA was by imposing a municipal tax on its electricity consumption.[22] The villages did eventually get new schoolrooms, better roads, running water, and sewers—but they had to pay a significant portion of the costs themselves, and getting these amenities took a decade. The minutes of village council meetings make clear that residents felt cheated by the failure to deliver on promises of modernization.[23]

The town of Bagnols fared better than the villages, mostly because the CEA and other state agencies, having decided to house most of Marcoule's employees there, played an active role in urbanization. The town received the modern facilities it expected. But the Bagnolais had been promised more: they had been assured that tradition and modernity would blend in peaceful harmony. And many felt the betrayal of this promise in countless small ways. The new and old parts of town, for example, supposedly blended seamlessly in mutual esthetic reflection, the high-rises of the new town echoing the spires of the old town. Yet the Bagnolais felt that the edge of the new town formed a barrier between the two parts. Rather than extending an extant boulevard, urban planners had placed a shopping center and a building perpendicular to it, thereby creating a physical boundary and a visual barrier between the two parts of town.[24] Local merchants feared that the shopping center threatened their livelihood.[25] Established townspeople—who called themselves Bagnolais de Souche— complained that the newcomers flaunted their greater income; certainly the washing machines and refrigerators that began to appear in stores in the 1960s were beyond the means of most old-timers.[26] Even the town council felt that the newcomers were greedily trying to extract special financial favors (such as reduced water rates) from the council without granting equivalent favors in return (for example, the CEA's sporting leagues charged townspeople a membership fee).[27] According to a sociologist who did field work in Bagnols in the late 1960s, tradition and modernity had only succeeded in grating against each other:

While there is no will for isolation, the old-time Bagnolais associate with each other and constitute a closed society for the newcomers who, in turn, do feel the difficulties of forming relationships in the town but who also don't realize the upheaval that they created in the small town of 5000 inhabitants. The way of life of the Bagnolais had to change somewhat to face this sudden growth. Thus, faced with the heavy traffic that grew so quickly, certain customs have disappeared; during nice weather in the peaceful neighborhoods of Bagnols, residents used to bring chairs to their doorsteps to savor the cool evening and chat with their neighbors or with passers-by—for everyone knew each other. Now, with the ever increasing traffic and the large numbers of "strangers," . . . the town natives no longer feel quite at home, and the custom of sitting in front of their doorways has disappeared in most neighborhoods.[28]

In these stories and memories of everyday life, the harmony between tradition and modernity promised by the technological spectacle disappeared, revealing instead a stark division between old and new.

Interlude: Reflections on Local Memory

The stories and memories of the Tourangeaux have quite a different flavor from those of the Gardois. Passion and discord propel local narratives about Marcoule. Narratives about the Chinon nuclear site speak more of indifference, acquiescence, or satisfaction. Natives of the areas might attribute the contrast to regional temperament: local stereotypes hold that people from the south of France are fiery, while those from central France are even-keeled. I believe the contrast in memory derives not only from differences in local expectations and experiences in the 1950s and the 1960s but also from how each region handled subsequent nuclear development and how the locals situate themselves in France's nuclear history. This interlude addresses these last two points.

The region around Marcoule experienced its most dramatic changes in the years covered in this book. Though Marcoule acquired several new research facilities after 1970, these did not produce comparable upheavals. But if people remember the 1950s and the 1960s as the time of most significant change, it may not be due entirely to lived experience. An additional source for this perception may be the scholarly attention devoted to this period of the region's history. Sociologists, economists, geographers, and ethnographers (both professionals and students) have devoted considerable research to the impact of the nuclear site on the region.[29]

Interacting with these researchers over several decades has helped to give Gardois a strong sense that this period of change forged a unique place for them in contemporary French history. Everyone I approached either had been interviewed before or knew someone who had. They knew themselves to be interesting scholarly subjects. They told well-rehearsed stories about the arrival of Marcoule. They even talked back to researchers. Consider this blurb, which appeared on July 26, 1994 in the Bagnols edition of the *Midi-Libre*:

Intrigued by an American research mission focused on Bagnols. Californian academics . . . have come to observe the fabric of local economic life, and especially to learn how people from here reacted to the implantation of large enterprises near their homes. Of course one thinks of the Marcoule site and its surroundings. Now we just need to wait for the conclusions. Then we will see whether the Americans have succeeded in understanding us.[30]

I probably would have missed this paragraph had not a gleeful employee of Bagnols's municipal archives brought it to my attention. I could only presume, from his mischievous smile and his refusal to comment, that he was the one who had told the local press about my work. Of course I was amused to see that I had become an entire research mission. But I also felt an odd sense of disquiet upon realizing that this blurb transformed me from an observer to an object of observation. Accustomed to social science investigations, the Bagnolais saw me as simply the latest in a long line of researchers. If I was interesting at all, it was by virtue of my being American. Could an American understand the Bagnolais? Only time would tell. And only they could judge.

This strong sense of historical subjectivity and ownership contrasted sharply with the attitudes I encountered in the Touraine. While most people I solicited for interviews there responded cordially, many also expressed puzzlement. Why should I want to talk to them? Did I realize that they themselves had not worked at the plant? One woman told me sharply "I have nothing to say on the subject" and hung up the phone. Another gently and humbly insisted that her opinions did not matter. In the end, most people agreed to an interview. But the dynamic of the interviews made it clear that the experience (indeed, the whole concept) was new to many of them. Most did not have well-rehearsed tales to repeat. They tended to reflect more before answering questions. With the exception of the elected officials, they tended to express more deference toward me than the Gardois. In the Gard I was just another researcher, interesting only because I was American. Four years earlier in the Touraine, my nationality had barely registered. (There had been an American military base near Chinon until the mid 1960s, when de Gaulle evicted the US military from France. Between the camp and the steady influx of tourists since then, locals appeared well used to Americans' traipsing about their region.) Instead, people appeared impressed by my status as a researcher—despite the fact that I was a graduate student at the time.

The Touraine's nuclear history since 1970 also helps explain some of the variation in how locals remembered the 1960s. In the 1970s, Chinon became the site for a series of light-water reactors, which covered much more acreage (because of their cooling towers) than the original gas-graphite reactors. People experienced this construction phase differently from the first one. The new reactors brought a second wave of EDF workers and their families. According to some residents, these newcomers were younger, more snobbish, more numerous, and less interested in local life than the first wave. There was thus a "good old days" quality to the stories they told about the 1960s.

Somewhat paradoxically, however, the 1970s was also a decade in which local community leaders constructed an image of their region as one that accepted—even embraced—nuclear power. During that decade, waves of anti-nuclear protest rippled across the nation.[31] The residents I spoke with insisted that these protests had largely bypassed Chinon, and that what little anti-nuclear rhetoric they heard was instigated by outside activists. Many people contrasted their region with Brittany, where the "ecologist movement" (as French environmentalists refer to themselves) had successfully put anti-nuclear contestation on the local political agenda. National parliamentary deputy André Voisin expressed great pride in the excellent relationship between departmental officials and EDF's administration during the 1970s. EDF, he said, viewed Chinon as its model site. Utility managers would invite elected officials from other regions there, where they would provide a tour of the plant and a good lunch. Voisin would then help EDF persuade these officials of the benefits of a nuclear plant in their midst:

I remember some general councilors from Brittany. I had been invited to give them my perspective. They were very surprised when I said that we had had our first plants built for fifteen years already, functioning with no problem except the fact that they brought lots of money to the department! They asked me if people didn't lose hair [by being] near the plant, and I answered that I went to the plant quite often, and that I still had all my hair. The fact that I went to EDF to make sure that everything was going okay—well, this helped EDF, and it was grateful to the department.[32]

The Bretons made an impression on other local residents, too. Listen to this couple, who ran a grocery and a café in Avoine:

Woman: I don't think that people here were really scared like in Brittany. We never said that we were scared of this or that. We just accepted things as they happened. While as in Brittany—

Man: It was a political affair. And anyway, the plants here had proved themselves, but in Brittany they were opposed to the construction.

Woman: They would call us and ask "Aren't you afraid?" A man who wanted to open a campground asked me that!

Man: Because they had been told that around this nuclear plant, nothing grew, the land was arid, there were no trees or birds, and that nothing was fit for consumption. They had been inculcated with this, and they really thought that!

Woman: Whenever we went on a trip, and said that we lived near a nuclear plant, people would say: "You're not sick? You're not tired? No problems?" It had gone that far!

Man: There were one or two anti-nuclear demonstrations, with people from neighboring departments. What was strange was that people came to demonstrate to tell us, the locals, that there was danger, and they demonstrated at the gates of the plant. There were mothers with their little babies. It was political.[33]

"Political" here clearly meant irrational and interest-bound. During the 1970s, it appears, many local residents ended up viewing themselves and their region as the rational, reasonable ones who did not buy into the anti-nuclear "hysteria" that had swept the rest of the nation (especially Brittany). The continued fertility of their soil served, for them, as proof that there was nothing dangerous about living near a nuclear reactor. Personal or social conflicts with "arrogant" EDF workers went unremarked; they appeared to think of these as internal matters not to be shared with outsiders.

To the Tourangeaux, what was at issue in the 1970s was the *nuclear* quality of the site—a quality separate from politics. These nuclear characteristics, they insisted, posed no problems. With elected officials taking the lead, they began to see the Touraine as the model for nuclear development elsewhere in France. Further, they naturalized their acceptance of nuclear power by attributing it to regional temperament. In the words of another resident I interviewed: "Brittany has not accepted [nuclear power]. But the Bretons are more chauvinistic, while we are more welcoming."[34] Indeed, the Touraine cultivated an image of serenity, which locals dated back centuries. Consider this passage from a 1980s tourist brochure published in Chinon:

This is a land of balance; nothing is excessive, neither the cold of winter, nor the heat of summer, nor emotions, nor the language which remains the purest in France. This is a land of harmony: harmony of the landscape bringing a harmony of thought and of character. There is not even a local dialect; the few expressions or words that might surprise the ear date from eighteenth-century French.[35]

Accepting nuclear power fell into the natural, historical order of things— which, in turn, made it easier to forget the conflicts that did occur in the 1950s and the 1960s.

Of course not everyone forgot. A few people remembered interpersonal conflicts that had accompanied the site's initial development, and dwelled on those rather than on stories of harmonious integration. And those who had sold their land to EDF had strong memories of conflicts over land prices. But it seemed to me that the Tourangeaux—unlike the Gardois— had not developed a collective memory about the 1950s and the 1960s.

I do *not* mean that the collective memory of the Gardois has produced uniform agreement on the place of nuclear technology in their region. The termination of the gas-graphite program in 1969 created a kind of reconciliation between Gardois leaders and Marcoulins, in which the two groups affirmed a common set of economic interests. In the 1970s and later decades, a budding sense of solidarity coexisted with periodic anti-nuclear demonstrations. I encountered a wide spectrum of attitudes among the Gardois with whom I spoke in 1994, ranging from enthusiasm to tolerance to opposition. Many of the people who spoke of conflicts in the 1950s and the 1960s had since become strong supporters of Marcoule.

Therein lies my point: In the Gard, collective memory made room for conflict, regardless of current attitudes. In the Touraine, conflict appeared to remain mostly hidden, an uncertain part of the nuclear story. The Tourangeaux had evolved an identity that involved nuclear acceptance. A collective memory that made room for early conflict had not emerged. Those who spoke about problems did so hesitantly, and they did not have stock anecdotes to illustrate their general points. Perhaps in part because the Touraine's early nuclear history had not been a focus of much scholarly attention, locals had not spent much time rehearsing the details of that history and interpreting its significance.

With these reflections in mind, let us now turn to the tales of the Tourangeaux.

The Little Kuwait of the Indre-et-Loire

"Land was sacred then," the mayor of Savigny-en-Véron told me in an interview.[36] Like Chusclan and Codolet, the villages of Avoine, Savigny, and Beaumont were dominated by agriculture before the arrival of the nuclear site. Wine grapes and asparagus prevailed, but most farmers also grew smaller quantities of other produce. Not surprisingly, they were reluctant to sell the land that ensured their livelihood.

Considerable disagreement existed, both in the 1950s and in 1990, over whether EDF had adequately compensated landowners for their property. According to the mayors of Savigny and Avoine, it had not. The price offered by EDF had been "laughable." The utility had "acquired parcels dirt cheap" and "paid five times less than what they were worth."[37]

In contrast, parliamentary deputy Voisin (who had been a member of the department's general council in the early nuclear decades) asserted that EDF had been generous: "If a [parcel] was worth 10,000 francs, EDF didn't hesitate to pay 12 or 13,000 francs." In fact, while some landowners sold their parcels easily, others forced the utility to expropriate their land.[38] Not all the parcels were for the site. Some went toward EDF's housing developments, and still others were for roads. Some Avoine residents expressed reluctance to sell their land for a mere road (a considerably less prestigious development than a nuclear site). Others did not want large roads running past their homes. But Avoine's municipal councilors had little patience with such reluctance. Recalcitrant owners, they opined, were motivated by greed or other "personal" reasons, and the municipal council declined to support them.[39] By the end of 1956, EDF had acquired much of the land it would need for the next decade.

Some local officials hoped that this rapid rate of land acquisition meant that the site would immediately lead to a wave of hiring. In the fall of 1956, however, EDF had still not hired any local residents to work on the site. People began to wonder about the economic benefits of the new project. Most disturbing to local officials, EDF had begun construction without keeping them updated on its plans. Rumors and grumblings drove Auguste Correch, Chinon's mayor and general councilor, to write the site's top engineer asking for explanations. Did EDF plan to extend its site beyond the foreseen limits? The engineer responded "no." Did EDF plan to hire local labor? The utility itself, explained the engineer, would not hire anyone until the reactor was built, but the private companies in charge of construction would surely hire locals. Would the "factory" be dangerous? No, categorically not. And finally, did the utility intend to zone land specifically for the construction companies, and to create a zone where no construction would be permitted? EDF had not yet considered this matter, admitted the engineer. Correch seemed satisfied with these answers, which he presented to his fellow general councilors. But the council expressed its displeasure with the utility. "Surprised," it declared "by the secret conditions under which the activities preliminary to implanting the nuclear factory are occurring," it asked that the utility keep the local authorities and population informed about events as they occurred.[40]

Site administrators eventually understood that things would go more smoothly if they kept locals at least nominally involved in the site's development. In this respect, they demonstrated more savvy than Marcoule's administrators had. EDF made a bigger and more sustained effort to court local officials. Site directors regularly invited officials to lunch with them.[41] In 1961, the utility invited elected officials from the whole region to a four-day seminar on atomic energy, held at a civil defense training school near Paris and led by scientists, medical doctors, and engineers from the CEA and EDF. The program included lectures on the basic principles of fission, descriptions of how reactors worked, and explanations of the safety measures EDF took to protect workers and the local populace. Not many councilors accepted the invitation, but those who did returned to their municipalities armed with a huge volume containing transcripts of the lectures.[42] Jean Chamboissier, a young pharmacist recently elected to Bourgeuil's municipal council, returned brimming with enthusiasm for nuclear physics and eagerly shared his new knowledge with his fellow councilors.[43] Thirty years later (during which time he had served as Bourgeuil's mayor), Chamboissier's enthusiasm had not waned:

They explained everything they were going to do, what the potential dangers were, and all the safety measures taken. The lectures were at a very advanced level, and very interesting. . . .This is to tell you that nothing was left out in [efforts] to inform area residents in the most objective manner possible. I insist on this point. The information was transmitted. This is why I've always been scandalized to hear that EDF doesn't inform people.[44]

He went on to denounce more recent accusations to that effect in *Le Monde*.

As we saw in the last chapter, other efforts to inform the population took the shape of displays and lectures at the annual fair in Tours. *La Nouvelle République du Centre-Ouest* had a much better opinion of these than the *Midi-Libre* had had of their CEA equivalent in Nîmes. Describing the lectures given by two engineers (one from the CEA, the other from EDF), one journalist said: "With rare talent, they gave us fascinating glimpses into the prospects opened up by the terrible and magnificent secret of atomic energy, both in this waning century, and in the ones to come."[45] Not everyone was so articulate; the scientist who spoke the following day "tried hard to make the history and application of radio-elements accessible to his audience"[46] (apparently without much success). Perhaps EDF had learned from the CEA's mistakes.

Site tours also appeared to be a hit, at least with some locals. Tourangeaux certainly found EDF1 (known locally simply as *la boule*—the ball) to be every bit the promised spectacle. "You could see people welding the ball in the evenings. You felt like you were watching a fireworks display."[47] Equally spectacular were the "thousands of cubic meters of earth" that were moved to make the land suitable for the reactors.[48] Some residents went only once. For others, it provided a regular excursion, as they became tourists in their own region. Few, though, manifested as much enthusiasm as Jean Chamboissier, a frequent member of the Sunday morning tours:

You didn't have to sign up. You just showed up, and then they said "the tour will start now"—just like in a Loire Valley château. . . . I went very often. I remember going with my father, who was a medical doctor and who was always very interested in science. . . . Mostly it was people from the area [who went on the tours] . . . and often it was also friends or relatives of site workers.[49]

For those who were interested, then, the site could function as the display and tourist attraction its supporters had dreamed of.

Villagers in Avoine seemed enthusiastic about the prospects—touristic and otherwise—the site offered their municipality. They also wanted nominal credit for housing the site. They had assumed that, since the plant lay completely within their municipal territory, it would be called the Avoine nuclear site. When EDF began calling it the Chinon nuclear plant, Avoine's residents expressed outrage. Recognizing that the town of Chinon "is naturally known for its historic reputation and [thus] serves as a geographical landmark," the municipal council nonetheless saw no reason to name the plant after a town located 7 kilometers away. "Precedents have already been established in which small communities [have housed] dams or electric plants . . . and seen the name of their municipality on that of the installations built on their territory."[50] Backed by all the neighboring towns and villages except Chinon, the mayor asked EDF to include "Avoine" in the plant's name. But the utility refused to budge. Its first nuclear accomplishment had to have a widely known, historically significant name. Avoine would just have to swallow its pride.

As money began to flow into the village, this grew easier to do. Besides, Avoine could take credit unofficially, if not officially. In the 1960s, for example, the general store began to sell postcards depicting the site, some of them labeled "the nuclear plant of Avoine-Chinon." Local residents found "the ball" and its water tower compelling. Images of "the ball" appeared on wine labels and on the paper the butcher used to wrap meat. Some, like Avoine's mayor, endowed the shapes with deep modern significance:

Do you know why it was a ball? The ball is one billion times as big as an atom. I learned this from the engineers. EDF asked the advice of an engineer when it built the first plant, and he answered that he envisaged it as an atom and a candle. In fact, when I was mayor of Avoine, I had a [postal cancellation] stamp made with that ball and that candle.[51]

Others preferred local historical metaphors. One man called the water tower "Gargantua's cigarette."[52] Another, referring to the mounds on either

Figure 7.1
Postcards sold at Avoine's general store. Cards courtesy of M. & Mme. Georges Arrault.

side of the reactor, remembered that "people used to say that Gargantua must have scraped his boots there."[53] Still others saw advertising potential in this iconography. Upon noticing a marked increase in sales since the arrival of the plant, for example, vintners in Bourgueil created a "tourist circuit" that included guided cellar tours and wine tastings. The nuclear site figured as a prominent landmark on the accompanying brochure and on the billboard advertising the circuit at the village entrance.[54] Perhaps the example set by the production of tourist souvenirs relating to Loire Valley châteaux in other Touraine communities

APPELLATION CONTROLÉE

DISTRICT RURAL DU VÉRON

PHOTO: VUE DES VIGNES PRÈS DE LA CENTRALE   IMP. F. MORON & FILS CHINON

Figure 7.2
A wine label featuring a photograph of the Chinon reactors. The label was initially made by one of the local winemaking cooperatives. Courtesy of M. Raffault.

helped spawn this grassroots production of nuclear iconography. Whatever the case, the contrast between the Gard (where the CEA had commissioned the cuvée de Marcoule) and the Touraine (where vintners initially produced a label depicting "the ball" of their own accord) was striking.

Clearly, many merchants and vintners—particularly in Avoine and Bourgeuil—embraced the economic opportunities afforded by the Chinon site and its steadily increasing influx of construction workers and EDF employees. In addition, the site provided a huge income to Avoine through the *patente* tax. As a research and national defense facility, Marcoule had been exonerated from the *patente*. EDF's plant, however, was a production facility. Even before EDF1 went on line in 1963, Avoine collected regular taxes as well as small *patentes* from the construction companies. This already represented a substantial increase in income for the village. When EDF began to pay its *patente* in 1963, the village budget skyrocketed. Avoine rapidly acquired an enduring nickname: "the little Kuwait of the Indre-et-Loire."

Avoine spent the first francs of its newfound wealth on basic public works: running potable water, a sewage and drainage system, complete electrification for the entire village, and repaved roads. The whole village seemed to get into the spirit of modernization. In 1959 it elected several new people to the municipal council, including the first female councilor (the young wife of a prominent farmer) and a new mayor (a young businessman who had moved to Avoine in 1945 to open a small canning factory). These young people shepherded changes in village life that were themselves nothing short of spectacular. Avoine acquired a new school, complete with its own gymnasium. Sidewalks began to line roads in the village center. The main square was remodeled, with a new town hall at one end and access to a new sporting green at the other. Plans were laid for a new cultural center. And, as if to celebrate the end of the village's decline and the new era of youth and wealth, the cemetery moved from the center to the edge of town.[55] In addition to these community projects, the municipal council spread the wealth to individuals and groups. The woman who delivered telegrams received a 70 percent pay raise. The athletic league and the hunting group got larger subsidies. Even community groups based in other towns and villages received money from Avoine's newly beneficent municipal council. The tremendous influx of workers also generated income for residents who were not merchants. Transient workers who had come from all over the country to build the reactors needed lodging, and villagers provided it. Anyone with a spare room could find a tenant willing to pay an impressive rent. In short, the nuclear site made Avoine and its residents rich—at least compared to their previous incomes.[56]

Departmental officials worried that modernization would be haphazard and wanted to take a hand in managing change. In 1957, they encouraged Avoine and nine nearby towns and villages to form an urbanization consortium to manage regional development. The proposal was novel in that it asked townships on both sides of the Loire to work together. (Traditionally, people on one side of the river did not even socialize with those on the other.) The consortium would ensure that construction companies spread their headquarters throughout the region, oversee rural infrastructural development (including electrification, water, and sewage systems), manage housing development, and formulate plans to attract other businesses and industries to the region.[57] In principle, the municipalities agreed that the consortium was a good idea. In practice, however, intercommunal disputes thwarted the plan's implementation. Chinon residents found the urbanization plan too restrictive.[58] Bourgueil's municipal council resented having to obtain approval from the consortium for every development project it proposed.[59] In the end, only Avoine appeared happy with the arrangement, and the plan faded.

The other townships thus ended up managing the arrival of the nuclear site on their own, which led to considerable variation across the region. Most villages appeared content to let modernization happen haphazardly. Individual landlords in Beaumont, Savigny, and Huismes rented rooms to transient workers, and some farmers left their land to work at the site. EDF built a housing development for its employees in Beaumont, which required water and drainage. But the municipal councils in these villages made no concerted effort to court industries, and they continued to focus on the traditional agricultural concerns of their constituents.[60] One inhabitant of Beaumont explained: "At that time Beaumont was headed by a mayor who was a good peasant from the township, but who just found it all to be too much for him to handle."[61] Chinon, the largest town in the area, appeared no more interested in fostering modernization than its village neighbors. It already had potable running water and other such infrastructural amenities, and it appeared content to bask in the glory of having the site named after itself.[62] The town's modernization and expansion efforts before the 1970s remained modest.

The leaders and residents of Bourgueil formed the exception to this rule. Numbering 3000 in 1962, the Bourgueuillois considered themselves a town more than a village.[63] They too hosted an EDF housing development, and they hoped to make the most out of the nuclear site. They could not do so directly—though the municipal council tried to give its town amenities similar to Avoine's, for example, it could not convince EDF to finance a municipal swimming pool, and it had to take out a large loan to build a gymnasium. Instead, therefore, Bourgueil's leaders focused on luring other industries to their township. Reasoning that the presence of the reactors would, with time, attract more business, they wanted to hasten and manage change. They withdrew from the regional urbanization consortium and sponsored a smaller collective that included only townships on their side of the river. They zoned part of their territory for industrial development and passed a decree that would exempt any industry generating twenty or more new jobs from 50 percent of the *patente* for five years.[64] The municipal council also implemented several advertising campaigns to attract the newcomers to its wineries. Perhaps the most significant symbol of Bourgueil's eagerness to embrace modernity was the election of a plant engineer to its municipal council in 1965.[65]

The process of modernization thus occurred unevenly in this part of the Touraine in the 1960s. For Avoine, it happened easily and rapidly, almost entirely through the influx of industrial taxes, with not a murmur of complaint once it became clear that the village could have everything but the name of the plant. The surrounding communities were not so lucky, and their experience depended much more on the initiative of municipal officials.

Like economic and technological modernization, personal accommodation was also uneven. The first wave of workers in the region were not EDF workers but construction workers. At Marcoule, construction workers had stayed in temporary barracks and tents, and only sometimes in local homes; residents remembered very little about them. In the Touraine, construction workers lived in people's homes. From the start, then, locals and newcomers came into close contact. Much depended on the individual, of course, but generally residents had fond memories of these transient workers, who would often keep them informed of events at the site. One couple, who lived in Beaumont at the time, recalled:

Man: Yes, there were some welders near us. The first thing that was built was the ball. We knew what was going on through them, we could more or less follow the development of the site that way. They came with wife and kids.

Me: And they talked to you readily?

Woman: Oh yes, they weren't proud. And I would go get water in their courtyard.

Man: That led us to talk with them. They were used to working outside their homes, so they opened up to everybody. They were from the Midi, and I think they open up more easily, maybe because of the sun.[66]

Transient workers from the Midi (i.e., southern France, including the Gard) left a strong memory in the minds of several residents, who attributed their sunny dispositions to their climate of origin.[67] Some residents also noted, though, that if the transient workers got such a good reception, it was also because locals were themselves friendly and welcoming. Jean Chamboissier attributed this to winemaking: "In viticultural regions, you know, like Bourgueil—Bourgueil has a good reputation—in viticultural regions there is always a much more open mentality than in other agricultural professions."[68] Residents thus explained the good relations between transient workers and locals by referring to essential, universally "known" regional temperaments.

Relations with the first wave of EDF employees took somewhat longer to establish, since they resided with their families in separate housing developments. Nonetheless, many appeared to assimilate to local life easily enough. Consider the participation of several workers in Chinon's annual Gargantua festival. Although the town of Chinon had long cherished the memory of Rabelais, this festival was a recent addition to local culture. Instituted in 1959, it consisted of "two days of jollity,"[69] culminating in an eating contest whose winner would be crowned Gargantua of the year. In 1964 there were five finalists for the competition: a butcher, a plumber, a cook who worked in the plant's cafeteria, and two plant workers (both veterans of the contest). Dressed in homespun monks' frocks, these five men had one hour to consume as much food as possible. Each sat down to 6 kilograms of Rabelaisian food, which included ten slices of pork snout, half a kilo of chitterling sausages, a goose, a huge camembert cheese, a salad, a baguette, and all the wine they could drink. One of the plant workers won the big prize: several liters of wine and a box of tripe.[70]

Figure 7.3
The Gargantua festival. This photograph is from a 1980s tourist brochure for Chinon. Courtesy of Chinon municipal archives.

EDF workers also mixed with local men regularly by playing in the same sporting leagues. The women appeared to mix well too. "The first EDF agents we had here, we knew them all, we knew the whole family, we would always ask after each other."[71] One resident contrasted this first wave with the workers who came in the 1970s: "After it was very different. The other generation hired by EDF, to make EDF agents, were trained in schools. While as the first ones [got their training] by experience, through manual skill. The young agents were much less friendly, more distant."[72]

Despite the reasonably friendly relationships between local and EDF employees in the 1960s, despite the tours offered by EDF, and despite the seminars given for local officials, some local residents were clearly anxious about the safety of the site. But the traces of this anxiety—both in the historical record and in memory—are faint. In 1962, two general councilors from districts 20 to 30 kilometers away from the site urged the department to engage in independent and systematic monitoring of radiation around the plant. In 1963, an article in the *Nouvelle République* mentioned that the local population had expressed concern about radioactive waste and about the water that EDF1 would reject into the Loire, but dismissed such concerns as unfounded.[73]

Only one couple acknowledged their anxieties about radioactivity in the interviews I conducted:

Woman: When they said "nuclear," "radioactivity," that made you think of Hiroshima. People talked about it, they were afraid that the rivers would get polluted. At first, some precautions were taken, I don't know if they're still in place, but I remember that at that time . . . they went every week to sample water from the Loire. There was a staircase that had been fitted to give access to the water. And they regularly took samples of milk from farms in Savigny to analyze it.

Me: And what effect did that have on you?

Woman: I was afraid. But I didn't analyze things, we didn't know much.

Man: But really, even after that . . . when we left at 4 in the morning for work, in '64–'65 . . . we would see these things that looked like mile markers on the side of the road, and those were radiation detectors. And the EDF people would pick them up before dawn, so that people wouldn't see them. Few people know this, but we were outside at night, so we saw them. But anyway, if you leave the area, you'll just find another nuclear plant 300 kilometers further away. And if it's not that it's something else.[74]

The apparently clandestine nature of the radiation monitoring made these people suspect that if something serious happened at the plant they would not be informed. At the same time, they did not expect to understand any explanations that might have been offered to them. Everyone else I interviewed affirmed that no one had been worried about the radiation. A few people mentioned the Chernobyl accident; they said that people worried more as a result of that than they ever had before.[75]

With such limited information, it is difficult to draw any conclusions about the degree to which residents worried about the dangers posed by the Chinon reactors. Though scant, available evidence does suggest that people worried more than they generally admitted. When asked about their fears, however, they almost invariably answered—like a mantra— that nuclear power had always been well accepted in their region. The 1970s had left an indelible mark on their memories.

*

Clearly, at least at one level, the Tourangeaux had a better experience with the implantation of their nuclear site in the 1960s than the Gardois had with Marcoule. Some of the reasons are easy enough to decipher, and have to do with economics and demographics. Chinon brought more money, with fewer strings attached, than Marcoule did. At Chinon, both the construction companies and EDF paid Avoine a *patente*. This meant that the village did not need to raise municipal taxes or take out a loan to finance its modernization projects. This contrasted with Chusclan and Codolet, which suffered from the lack of a *patente* from Marcoule. Bourgueil (on the other side of the Loire from the Chinon site) did require loans to modernize its public facilities, but this did not appear to strain the municipality. Meanwhile, Chinon and nearby villages seemed content to let modernization occur at a leisurely pace. In and of themselves, the special-rate state loans obtained by Bagnols for its urbanization projects did not appear to strain the townspeople (though taxes did rise). But the pace of change in the town was overwhelming for many Bagnolais. The sense of invasion they experienced was heightened by the fact that the vast majority of CEA employees at that time lived in the town of Bagnols. In the Touraine, by contrast, EDF employees were spread out over Avoine, Beaumont, and Bourgueil.

Another set of explanations for why the Tourangeaux had an easier time than the Gardois in the 1950s and the 1960s lies in the expectations harbored by inhabitants of each region. The spectacle of technological progress in the Gard took the form of a drama of salvation. This drama promised a reconciliation between modernity and tradition. The fact that the producers of the spectacle claimed (at least nominally) to include tradition in the drama of modernization may have raised expectations among the Gardois that they would be assigned an active role in this drama. Instead, they felt shut out. Marcoule was largely off limits, the work there a secret. And the alleged harmony was nowhere to be found. Instead, as the locally produced counter-spectacle and many other stories suggested, modern technological France—incarnated by CEA employees and Bagnols's urban projects—seemed to want to dominate and destroy, not harmonize with, existing knowledge, customs, and values.

The technological spectacle advertised to the Tourangeaux—more in the nature of a display than a drama—made less extravagant claims. Certainly people were promised jobs and modern facilities. But the emphasis was more on the esthetic qualities of the site; the operative metaphor was that of the château. In this spectacle, residents should not expect to be actors—instead, their place was clearly designated as audience and tourists. Site tours thus played a double role: on the one hand, they kept locals informed and made nuclear technology seem more accessible; on the other hand, they provided a way for residents to perform their assigned roles as spectator-tourists. Ultimately it was much easier for Chinon to meet the expectations set up by the producers of the technological spectacle than for Marcoule to do the same. In the Touraine, the new technological France appeared, above all, as a source of income; the state might have invaded, but it brought so many gifts that few people seemed to mind much. Tension around the Chinon nuclear site existed in the 1950s and the 1960s. But with a few exceptions (notably, land expropriations), there appeared to be little space to discuss these tensions, either then or now. Ironically, then, although the Tourangeaux appeared, in principle, better informed than the Gardois about nuclear matters, the silence around their history is greater.

8

Warring Systems

In November 1969, hundreds of CEA employees around the country went on strike to protest the demise of the gas-graphite program and the imminent purchase of an American license for the construction of light-water reactors. Workers, technicians, scientists, and engineers marched through the streets of Paris and staged sit-down strikes at Marcoule and Saclay. "We are in the process of losing our national independence," they cried. "We are on the path to underdevelopment and colonization."[1] They also feared that they would lose their jobs: rumor had it that the government would soon announce several thousand layoffs. The French public followed the strikes in newspapers, on radio, and on television.[2] In southern France, the Bagnolais suddenly became alarmed that the Marcoulins might have to move away, leaving them stuck with the large debts they had incurred for their new public facilities. Gas-graphite engineers and workers at EDF also became angry, but they had more immediate worries which prevented them from staging anything more extensive than a few protests: the day after the termination of the gas-graphite program, the new reactor at Saint-Laurent underwent a partial meltdown. EDF employees thus had to contend simultaneously with the demise of their program and the cleanup of the most serious accident they had ever faced.

Dramatic though these events were, they quickly faded from the official history of the nuclear program. Men who had participated in the strikes or the accident cleanup remembered well, but those who hadn't soon forgot. Some remembered if I jogged their memories, others did not: one former CEA scientist was not only surprised but also skeptical when I assured him that the demise of gas-graphite had loomed large in the 1969 strikes—all he remembered were protests over the layoffs of cleaning ladies at Saclay.

Such lapses in memory, I believe, stem partly from the fact that the narrative of the *guerre des filières*—the war of the systems—has been transformed over the years from the story of the demise of the gas-graphite program to that of the birth of the pressurized-water program. The standard version of the story which circulates in French industrial circles goes something like this: In the beginning, there were two nuclear systems: one centered around gas-graphite reactors, the other around light-water reactors. In the late 1960s France had to choose between these two systems. A battle ensued. The nationalist CEA wanted to pursue the gas-graphite system for irrational political reasons, while the eternally reasonable EDF wanted to switch to the light-water system for rational economic reasons. Rivalry among engineers aggravated the quarrel. De Gaulle supported the CEA because he trusted it better, and because top CEA officials had his ear and could plead their case directly. De Gaulle thus became mistakenly and unfortunately convinced that only the gas-graphite system was compatible with national independence. Once de Gaulle stepped down, his better-informed successor, Georges Pompidou, could make the obviously correct choice. France could finally abandon the inferior gas-graphite system and build the superior light-water reactors under license. EDF's current nuclear program was born. Later, the light-water system became *francisé* (Frenchified), thereby providing the ever-coveted national energy independence as well as a source of national pride.

Like many origin stories, this one has served to erase events and circumstances crucial to understanding the process, the outcome, and the meaning of the *guerre des filières*. Some scholars have begun to unravel this history, showing that the positions for and against the gas-graphite system did not divide so neatly along institutional lines.[3] The light-water system had serious supporters in the upper echelons of the CEA, while many middle-ranking EDF engineers defended the gas-graphite system. But other aspects of the process remain unclear. How were the comparisons between the two systems carried out? How did the light-water system emerge as the rational, "apolitical" solution? How, indeed, did a nominally apolitical solution even become desirable? And what did the *guerre des filières* mean to gas-graphite engineers, workers, and technicians? Ironically, this is both the most studied episode in the history of the gas-graphite program, and the least well understood.

The *guerre des filières* weaves together the diverse thematic threads which I have teased out so far. Most of the historical actors we have encountered make an appearance: technologists, engineers, labor militants, technicians, workers, journalists, politicians, and Gardois (only Tourangeaux are absent). One of the central questions debated during this episode concerned French radiance: could the nation not only preserve its autonomy but also export reactors if it abandoned the gas-graphite system? In the roughly three years over which this war extended (1966–1969), engineers, managers, and union leaders proposed different technological scenarios and corresponding conceptions of France and its future. In advocating the light-water system, EDF's economist-managers (whom we last encountered in chapter 3) sought to reformulate the utility's technopolitical regime. Their reformulation efforts made the relationship between technology and politics a central arena of struggle in the battle. Ensuing debates were not only about the features of each system but also about the appropriate selection criteria. Should the choice be based on economics or politics? To what extent could these be intertwined? In an attempt to define selection criteria and defend the gas-graphite system, labor militants at both the CEA and EDF presented their own economic and political analyses. CEA employees further responded to the threat to gas-graphite by staging a series of strikes. These called attention to the specter of layoffs at Marcoule, which prompted a reconciliation of sorts between Bagnolais de Souche and Marcoulins. Finally, EDF employees had to contend with the Saint-Laurent accident. The cleanup provided them with a means of responding not only to the technical threat posed by the accident, but also to the cultural and political threat posed by the termination of the gas-graphite program. The *guerre des filières* thus provides an appropriate finale for my story.

Preliminaries to the War: Public Relations and Technological Mishaps

The idea that France might pursue other reactor technologies did not appear out of the blue. The research and military branches of the CEA had been investigating alternative designs for some time. These included a small light-water submarine reactor as well as heavy-water, high-temperature, and breeder prototypes. EDF helped the CEA with some of these prototype efforts—especially for the heavy-water and breeder reactors. EDF's nuclear division also sought reactor projects not tied to the CEA. In the late 1950s the United States concluded an agreement with Euratom that favored the importation of American reactor designs into Europe; reluctantly, the French government allowed EDF to cooperate with Belgium in the construction of the first such reactor in 1960.[4] These efforts notwithstanding, support for the gas-graphite system held fast in both technopolitical regimes through the mid 1960s.[5]

What happened in 1966 to shake this consensus? Most scholars agree that the trigger for the *guerre des filières* came from outside. In 1965, American reactor manufacturers embarked on an aggressive marketing campaign based on extremely optimistic capital cost estimates. (Recall that the capital cost of a reactor is the amount of money required to build it.) Persuaded by these attractive numbers, American utilities jumped on the nuclear bandwagon, ordering 49 reactors—destined to produce nearly 40,000 megawatts of electricity—in 1966 and 1967. Not until the mid 1970s did utilities discover that actual costs were more than twice the original estimates.[6] In 1966, however, the American estimates presented a serious temptation to French program leaders.

Of course, these figures did not, by themselves, mandate a change in policy. Had harmony reigned in the nuclear program, the *guerre des filières* may not have taken place at all. As we have seen, however, relations between the technopolitical regimes of the CEA and EDF had worsened as their collaboration deepened. Conflicts had also emerged within each regime. The possibilities raised by the American capital-cost estimates deepened and rearranged existing fault lines.

The first official intimation that EDF leaders were seriously considering abandoning the gas-graphite system came in March 1966 in a letter from André Decelle, EDF's director-general, to Robert Hirsch, the CEA's administrator-general.[7] Affirming that gas-graphite reactors were competitive with conventional power plants, Decelle stated that they were nonetheless "significantly more expensive" than light-water reactors.[8] A system based on natural uranium did promote French autarky, Decelle observed, but in the long term this advantage might not be worth the price difference. Noncommittally, Hirsch agreed that the matter required further study. In May the two leaders created a committee to study the various reactor systems operating in Europe and America. The committee was jointly headed by Jules Horowitz of the CEA and Jean Cabanius of EDF.[9]

Much as Decelle and Hirsch may have wanted these investigations to maintain a low profile, this quickly proved impossible. The press learned about the committee, and a wave of articles asserted that the current nuclear program would soon be terminated. These articles flustered regime leaders and worried labor militants. CGT representative Claude Tourgeron demanded an explanation at the May 1966 meeting of EDF's board of directors. Was EDF indeed abandoning the gas-graphite system? Decelle insisted that this was not the case, and that the media had, as usual, gone overboard. EDF and the CEA had simply decided to study both breeder and light-water reactors. Hirsch (who represented the CEA to the board) backed Decelle, declaring firmly: "This information . . . is more spectacular than it is well-founded, since of course the question of abandoning the natural uranium, gas-graphite system has never been raised anywhere."[10]

Clearly the directors hoped that the publicity problem would simply disappear. However, a series of technical mishaps at EDF3 not only aggravated the utility's public relations problem but also intensified ongoing debates within the establishment over industrial contracting and EDF's relationship with private companies.

EDF engineers and managers had nourished high hopes that EDF3, the latest Chinon reactor, could compete economically with conventional power plants. Construction delays had dampened these hopes, but in late 1966 all appeared ready. As engineers began to power up the reactor, however, more difficulties arose. Two of these were serious: the heat exchangers developed numerous leaks, giving engineers reason to doubt the integrity of the entire exchanger ensemble, and the system intended to detect rupturing of fuel rods failed to function adequately.[11]

Nicholas Vichney at *Le Monde* and staff writers at *Le Canard Enchaîné* quickly publicized these difficulties. Vichney blamed both the private builders and EDF. French technology could not meet the high standards demanded by nuclear plants, he wrote, and furthermore EDF had flawed industrial contracting practices.[12] The utility had tried to build something too complicated too fast, and the technical abilities of its personnel could not rise to the occasion. The CEA, Vichney continued, should stop distancing itself from EDF's difficulties and make more serious efforts to help. Predictably, *Le Canard* interpreted events more bluntly: "In short, our home-grown nuclear equipment doesn't hold up." *Le Canard* also gleefully noted de Gaulle's displeasure at the incidents: "Heads will roll, citizens!"[13]

These accusations of incompetence disturbed EDF's board. Hirsch tried to console board members and mitigate these harsh judgments by praising EDF design teams for their dynamism.[14] EDF's management noted indignantly that the press had neglected to mention the complexity of the reactor, the difficult trial periods undergone by all new technologies, and the troubles experienced by foreign reactors.[15] Union militants joined management in expressing outrage at the accusations leveled against EDF's technical know-how. Tourgeron insisted that the personnel was experiencing a deep "malaise," which he attributed to the "harm done to the prestige of [our] Establishment."[16] The personnel urgently needed reassurance, he said; the board had to counterattack.[17] But CEA and EDF management declined to engage in an active battle with the press, instead issuing a brief statement that simply averred the intimate collaboration between the two institutions.[18]

As time went on, media attacks on the nuclear program—particularly from these two newspapers—continued to escalate. Relations between Vichney and the CEA became openly hostile. Suggestions that French nuclear engineers lacked technical expertise angered both union and non-union employees at the CEA.[19] In April 1968, Hirsch accused *Le Monde*'s editor of lacking patriotism.[20] The paper's recent report of an EDF2 incident lacked objectivity, he said (thereby associating disinterestedness with nationalism), and demonstrated Vichney's "customary lack of good will in the face of the difficulties inherent in the development of nuclear energy." Hirsch continued: "Such publicity over an incident that even the article characterized as minor can only complicate the task of French industry, currently in charge of constructing the same kind of reactor abroad." In a subtle insult, he contrasted the unhelpful attitude of *Le Monde* with the discretion of the German press during similar incidents in Germany's power plants. Vichney retorted that he was the one acting in the public interest. He called the recent technical incidents "'serious' to the extent that they cast doubt both on the competence of French industry and on the competence of engineers in the different institutions called on to build these different plants."[21] "*It is therefore a matter of national interest,*" he concluded. By evoking journalism's claim to speak in the public interest, Vichney challenged the CEA's claim that it acted inherently in the national interest.

On the surface, such media attacks led the two regimes to close ranks, both internally and with respect to each other. EDF's unions defended the utility as an institution, and EDF and the CEA affirmed their solidarity. Officially, technical difficulties were just a normal part of technological innovation. Internally, however, the technical difficulties—transformed by the media into a failure of French technology—triggered renewed battles over the EDF's technopolitical regime. As in the past, the debates centered on the role of private industry in nuclear development. Repairing EDF3, for example, would be costly and time consuming. Who should bear the responsibility? There seemed to be two options. In the first, EDF's Direction de l'Equipement could accept the contractors' suggestion simply to repair the existing leaks and restart the reactor to find out whether the problems were systemic. But if that turned out to be the case, the plant would suffer frequent shutdowns, which would entail heavy "psychological" consequences.[22] The threat of more negative press, in other words, militated against this course. The second solution involved having contractors rebuild all the potentially problematic components from scratch. But the builders would undoubtedly object, leading to "a lawsuit whose outcome could only be uncertain and far away."[23] This too, then, entailed technopolitical costs.

EDF's management proposed a third solution, which it hoped would minimize technopolitical costs. This scheme called for the complete replacement of EDF3's heat exchangers. The contractors would pay for the faulty exchanger components. (This cost was estimated at 13 million francs.) EDF would pay for the rest of the new exchangers, and would also pay the cost of dismantling, rebuilding, and improving the exchangers (estimated at 29 million francs). This solution would allow EDF3 to restart more rapidly with better equipment and would avoid a long and difficult quarrel with the manufacturers. But several board members objected that this solution was too easy on private industry. Tourgeron protested that the builders had had plenty of opportunities to test the exchanger prototypes and that the problem stemmed quite simply from manufacturing defects. Management's solution would "reward mediocrity."[24]

Though most of the board members would have liked contractors to take greater financial responsibility, they ended up voting for management's solution. The government's representative to the board, Charles Chevrier, took the opportunity to make a little speech:

[EDF's] policy of rapidly increasing reactor power . . . enabled [us] to reach competitiveness with four reactors, while our English friends have not attained this with 16 reactors, even though for the moment they have a higher capacity factor [i.e., proportion of time that the reactor is actually on line and producing electricity] than we do.

The EDF3 incidents . . . do not throw this competitiveness into question. Indeed, what they will cost does not represent a very considerable percentage of the expenses initially foreseen.

Now we need to increase the capacity factor of the equipment, and for this it is necessary to make a big effort to improve the technology. This above all is the lesson to learn from this unfortunate affair.

In the future . . . we must find a way to link the manufacturers to [reactor downtime], not in order to make them pay for all the expenses, but . . . to develop their solidarity and their conscience with respect to the equipment they build.[25]

EDF3's difficulties thus had heterogeneous origins (including media scrutiny) and demanded a hybrid solution. This was not simply a matter of fixing a few leaks. Instead, a host of heterogeneous issues clamored for attention: the reactor's capacity factor, the relationship between EDF and private industry, institutional prestige, the components of the heat exchangers, France's competition with Britain, and the public image of the nuclear program.

Tensions over EDF3's leaky heat exchangers were also tensions over the changes that EDF's new managers wanted to make in the institution's technopolitical regime. Management's willingness to foot most of the bill was not just about saving time and repairing EDF's public image. Managers also sought to redefine the meaning of "public service" for EDF. For these new economist-managers, the utility could serve the public not only by making electricity but also by creating a climate congenial to the development of private industry. This outlook worried labor militants, who wanted to preserve a regime in which EDF would *control* industrial development. For example, the Confédération Française Démocratique du Travail worried that private industry would gradually take over making programmatic decisions about nuclear development at the expense of EDF employees. It also worried about management's internal changes, claiming that rank-and-file employees had recently lost a great deal of decision-making responsibility. EDF's original mission as the model for a new society was being betrayed. "EDF's ambiguous and weak attitude toward the Manufacturers in applying contracts" exemplified these complaints.[26] The CFDT did not oppose management's desire to help industry ready itself for international competition. But these efforts could not come at the expense of salaried employees or of "nationalized firms, which represent a social and economic achievement which we all value."[27]

EDF3's leaky heat exchangers thus became technopolitical tools in a growing struggle over the utility's regime. Would nationalization continue to mean the contractual and technological subordination of private industry? Or would it acquire a new, more ambiguous meaning in which EDF would make national energy policy by supporting rather than dominating private industry? As we shall see, the *guerre des filières* would provide ample terrain for this struggle. All the issues raised in the course of negotiating the repair of EDF3—industrial relations, reactor capacity factors, public relations, EDF's technopolitical regime, France's international position—would be played out in the debates over the future of the nuclear program and its relationship to the future of France.

The War Starts in Earnest: The Horowitz-Cabanius Report

In late January 1967, the Horowitz-Cabanius committee, charged with comparing available nuclear systems, presented its results. It immediately became obvious that the effort to smooth over quarrels had failed. Though Horowitz and Cabanius agreed on some basic numbers (including the capital costs for various reactors), they strongly differed on the technopolitical meanings of those numbers—so much so that they produced separate reports, each written as though the other did not exist. The points of contention formed a dense weave composed of

- the responsibility of private industry in guaranteeing reactor reliability,
- the meaning of public service for the two regimes, particularly with respect to their role in fostering the growth of private industry,
- the proper criteria for evaluating the performance of the gas-graphite system,
- the development status of gas-graphite reactors (were they fully functioning industrial systems, or mere prototypes?),
- the relevance of the French context to the final choice, and the nature of that "context,"
- uncertainty in the data, and how to handle it, and
- the "political" dimensions of the comparison, and who had the right or responsibility to analyze these dimensions.

In some respects, the last point is the most important. While the CEA's report maintained the tight links between technology and politics that had characterized both regimes during the first two decades of the nuclear program, EDF's report separated technology and politics. Horowitz maintained these links in order to justify the gas-graphite system. Only by severing these links, however, could Cabanius advocate the switch to light-water.

Cabanius, the director of EDF's Direction de l'Equipement (which designed reactors, coordinated industrial contracting, and supervised construction), stated that his goal was simply to compare the cost of the kilowatt-hour produced by different reactor systems. Politics was not his affair: "Political considerations, in particular those relating to the acquisition or manufacturing . . . of enriched uranium are up to the public authorities and will not be raised in this study. The rapporteur has strictly limited himself to the industrial, technical, and economic side [of the comparison]."[28] In contrast, Horowitz, the director of the CEA's Direction des Piles Atomiques (which engaged in research and development related to experimental and industrial reactors), stated that his goals were to "reveal the lessons to be learned from the current program," to "appraise once more the possibilities of the gas-graphite system," and to examine the future orientation of the French program.[29] While Cabanius sought to distance himself from the "politics" of system choice, Horowitz aimed to address "political" issues directly.

Cabanius began by briefly summarizing the state of French industry. This summary alone clearly reflected the desire of EDF's management to change its technopolitical regime by redefining public service. EDF, wrote Cabanius, played the dual role of customer and supplier to French industry. As a supplier, it had to offer manufacturers cheap electricity in order to help them compete with foreign companies. As a customer, it had to help companies reorganize themselves into large consortia capable of taking on the massive investments required by reactor manufacturing. Encouraging these consortia to work under a foreign license would further help French industry because the dynamism and success of the licensers would provide important financial and technical support for the licensees.[30] Ostensibly, then, EDF's first priority should be to strengthen French industry.

Horowitz focused his definition of public service on the gas-graphite system. He proudly noted that the system had already exceeded expectations in several ways: the price of natural uranium fuel had dropped faster than anticipated, the fuel rods had proved technically reliable, and the reactor cores had performed well. He argued that the system costs would have been much lower without the many difficulties that plagued the construction and startup of Chinon's reactors. He blamed EDF's technical incompetence and inconsistent attitude toward private contractors for this poor performance, rather than the system itself. Marcoule, after all, had proved that French reactors could maintain a high capacity factor.[31]

Cabanius blamed the mishaps not on EDF but on the technology. The complexity of the gas-graphite system, he observed, posed particularly delicate construction problems. Yes, Marcoule had performed well, but its reactors were smaller and less complex than EDF's, and the difficulties of building gas-graphite reactors increased dramatically with the scale of the reactor. "The natural uranium-gas-graphite system," wrote Cabanius, "therefore contains a source of incidents which could have serious consequences not so much for the safety of people as for the length and frequency of stoppages and therefore for the capacity factor of a series of plants which are integrated into an energy system."[32] Thus the gas-graphite system was inherently flawed. The light-water system, however, was not. Cabanius described this system with considerable enthusiasm. The capital costs were low. With many reactors on order, manufacturing could be standardized (leading to even lower costs and greater reliability than the gas-graphite system). Best of all, General Electric would price reactor fuel as a function of its energy production, thereby guaranteeing a performance standard for the fuel rods. "This formula," Cabanius asserted, "leaves a powerful manufacturer [GE] with the costs of technological uncertainties. Surely its acceptance is based on tremendous confidence in the technological quality of the supplies. This trend will probably be irreversible."[33] General Electric's confidence justified France's confidence. Cabanius portrayed the spread of light-water reactors as inexorable. He thus planted the seeds of technological determinism among EDF's economist-managers, simultaneously seeking to separate technology and politics.

Horowitz admitted no such determinism. Yes, American utilities had ordered a remarkable number of light-water megawatts in the last two years. Even more amazing, he noted snidely, this enthusiasm was based on the actual performance of just two 200-MW reactors! The performance data, therefore, were hardly statistically significant. True, the American program would probably succeed, thanks to its technical rigor and the vast resources of its manufacturers. *But this did not mean that the same program would succeed in France.* National context mattered deeply. Insufficient data made good predictions nearly impossible: "The catalogue [listings] for 'nuclear boilers' . . . do not give a breakdown of the price per [boiler] component; they do not, therefore, enable one to make a detailed techno-economic analysis. And in any event, as the promoters themselves admit, these prices do not correspond to the cost of a few isolated plants; it is only by anticipating the effect of [building] a whole series [of reactors] that nuclear power has been able to penetrate the market in the United States."[34] While Cabanius treated the American figures as reasonably accurate characteristics of the *technology*, Horowitz treated them as rough estimates based on the *context*. In Cabanius's analysis, the technology, abstracted from its context, was the most important variable. In Horowitz's analysis, the functioning and cost of technologies could not be separated from their contexts. What worked for the United States would not necessarily work for France.

Handling Uncertainty

Despite his skepticism about the American numbers, Horowitz used them in his calculations; they were, after all, the only ones available. Therefore, like Cabanius, he estimated the cost of a conventional kilowatt-hour to be between 3.95 and 3.35 centimes (depending on the plant's capacity

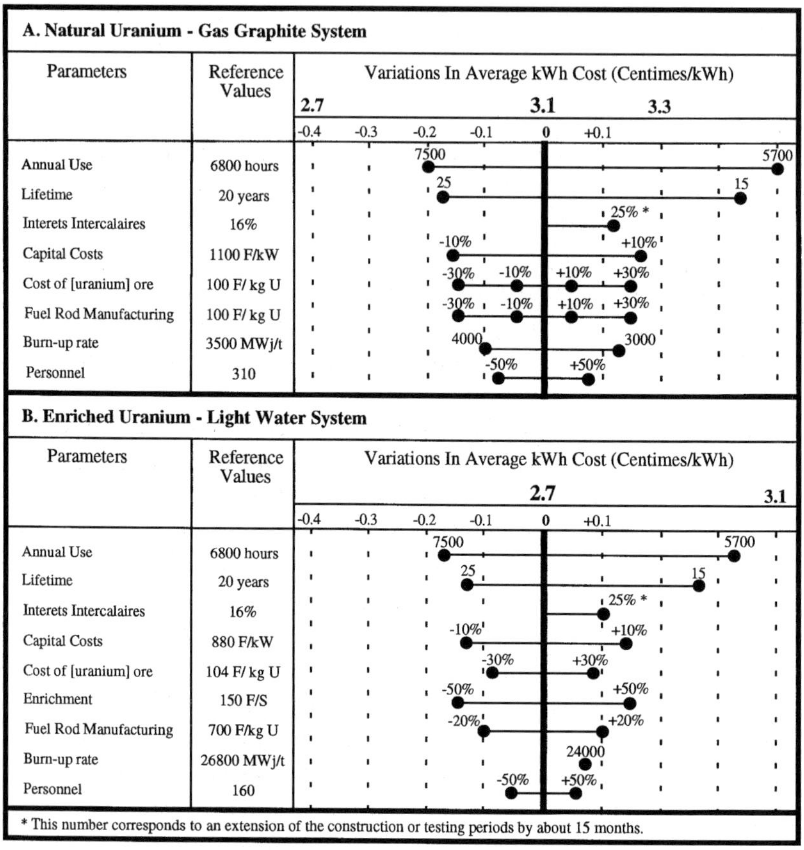

A. Natural Uranium - Gas Graphite System

Parameters	Reference Values	Variations In Average kWh Cost (Centimes/kWh)
		2.7 ··· 3.1 ··· 3.3

		-0.4	-0.3	-0.2	-0.1	0	+0.1
Annual Use	6800 hours			7500			5700
Lifetime	20 years			25			15
Interets Intercalaires	16%				25% *		
Capital Costs	1100 F/kW			-10%	+10%		
Cost of [uranium] ore	100 F/ kg U		-30% -10%	+10% +30%		
Fuel Rod Manufacturing	100 F/ kg U		-30% -10%	+10% +30%		
Burn-up rate	3500 MWj/t			4000	3000		
Personnel	310			-50%	+50%		

B. Enriched Uranium - Light Water System

Parameters	Reference Values	Variations In Average kWh Cost (Centimes/kWh)
		2.7 ··· 3.1

		-0.4	-0.3	-0.2	-0.1	0	+0.1
Annual Use	6800 hours			7500			5700
Lifetime	20 years			25			15
Interets Intercalaires	16%				25% *		
Capital Costs	880 F/kW			-10%	+10%		
Cost of [uranium] ore	104 F/ kg U			-30%	+30%		
Enrichment	150 F/S			-50%	+50%		
Fuel Rod Manufacturing	700 F/kg U			-20%	+20%		
Burn-up rate	26800 MWj/t				24000		
Personnel	160			-50%	+50%		

* This number corresponds to an extension of the construction or testing periods by about 15 months.

Figure 8.1
Variation in cost of a kilowatt-hour with variations in base parameters. This diagram (shown here as reconstructed by Carlos Martín) originally appeared in Cabanius's "Rapport du Groupe de Travail."

factor) and that of a kilowatt-hour produced by a pressurized-water reactor at 2.67 centimes.[35] The two men differed only on the cost of a gasgraphite kilowatt-hour: Cabanius priced it at 3.14 centimes, while Horowitz priced it at 3.04 centimes. Horowitz hoped, on the basis of experimental data, that the CEA's new fuel rods would reduce the cost, whereas Cabanius refused to rely on the experimental data. Horowitz thus emphasized the paucity of data on actual operating light-water reactors while expressing great confidence in the CEA's equally unconfirmed estimates. Cabanius took the opposite approach, expressing confidence in the American estimates and skepticism toward the CEA's.[36]

The two men also handled the uncertainty in the data differently. Cabanius, admitting that the parameters used to calculate costs were subject to change, constructed a table describing the "sensitivity of the cost of the kilowatt-hour" (figure 8.1). This table showed how the costs of the two nuclear kilowatt-hours would change with variations in parameters such as fuel burnup rate, capacity factor, and capital costs. Plotting uncertainty offered a sense of control, suggesting that because uncertainty was quantifiable it was manageable. Naming and describing uncertainty, in other words, eliminated the need for qualitative assessment.

For Horowitz, however, uncertainty *required* qualitative judgment. In the absence of hard "facts," political acumen had to guide the choices. This conviction came through most clearly in Horowitz's discussion of enriched uranium supplies. Cabanius had dismissed the topic in a single short paragraph, suggesting that, although light-water plants would initially rely on foreign supplies, eventually France or Euratom would build enrichment plants. He relegated any further discussion to the realm of "politics," which defined as beyond his mandate. Horowitz too saw this as a political matter, but he understood it to fall well *within* his mandate. Where Cabanius had written of foreign "supplies," Horowitz wrote of foreign "dependence." "Political reasons" (which he left unspecified) would make a European enrichment plant impossible, and France could never afford one on its own. Furthermore, the enriched uranium produced in France or Europe would cost considerably more than American uranium, thereby negating the cost advantage of the light-water system.

And Horowitz found other reasons not to plunge headfirst into the light-water system. Foremost among these was the need to capitalize on the time, money, and knowledge already invested in France's existing technologies. In addition, France's plutonium needs would increase as its breeder-reactor program took off. Buying plutonium abroad would increase France's dependence on foreign sources—a compelling reason to continue with plutonium-producing gas-graphite reactors.

Nonetheless, Horowitz concluded that France should probably acquire some experience with light-water reactors. But France could not afford to pursue both types of light-water reactors (pressurized-water and boiling-water), and the choice between them had to take political considerations into account. "General technological and economic arguments do not provide enough information to choose between pressurized and boiling water reactors; this [choice] must therefore follow from considerations proper to the French context, and in particular from the consequences in either case of the recourse to American licenses. We must ensure that, pending a decision, French industry does not get involved in . . . costly relationships that may later prove useless or premature."[37] Further, whatever the choice, it should not come at the expense of the gas-graphite program, which was a major source not only of France's political security but also of its financial security.[38]

Cabanius conceded some of the advantages of the gas-graphite system and agreed that the plans to build two gas-graphite reactors at Fessenheim should proceed. But he concentrated his efforts on subverting the connection between gas-graphite reactors and national glory by transferring France's national interest onto the light-water system. This system, he wrote, would "allow our manufacturers, grouped into consortia, to assert their technological value, acquire references that will carry great weight both for export purposes and for agreements with other European manufacturers, and participate in the great industrial confrontation of the next decade."[39] Thus light-water reactors, even though they were not made in France, would still contribute to one aspect of France's radiance: they would enhance the nation's ability to export technology. That the light-water system was operated under an American license was relevant only because it meant that the French would get American technical support for their endeavors. What mattered in terms of national interest was that French companies would be manufacturing parts for the world's leading nuclear system.

Independence or Interdependence?

Thus, both Cabanius and Horowitz constructed their arguments in terms of France's national interest. Horowitz's argument conformed to the familiar CEA association of the gas-graphite system with French independence. Cabanius, rather than deny this notion, sidestepped it. His arguments became part of a larger effort to reform EDF's regime. The national interest remained foremost, but its definition revolved around a vision of France in a set of interdependent international relations. He centered French national interest on the economy, which in turn depended on the growth and competitiveness of private industry, which in turn were measured in international terms. The corollary conception of EDF's regime reflected a subtle but significant shift: EDF no longer commanded private industry; instead, it helped to reshape industrial structures so that France could compete in world markets. In this regime, pursuing light-water technologies under a foreign license made national sense. Horowitz did not repudiate the goal of helping French companies compete internationally; indeed, as we saw in chapters 2 and 3, this had been a goal of the CEA's "policy of champions." However, for the CEA industrial "champions" had always meant French companies building French technology. Abandoning this pursuit now, Horowitz asserted, was folly. In arguing for different systems, the two men promoted two different versions of France. Choosing a system was also about shaping the nation's future. Technology and context were thus inseparably entwined for both men, even if only one of them admitted it.

The differences in the two reports signaled divisions that would sharpen over time. Gas-graphite technology had succeeded as a technopolitical system. Its developers, having consistently invested its technical features with political meaning, promoted the resulting hybrid as the best choice for the nation. The success of this practice (particularly with de Gaulle) had made it difficult to argue against the gas-graphite system, for to do so could appear unpatriotic. Promoting the use of a foreign license, therefore, required a different strategy—one that involved rhetorically separating technology from politics. Cabanius claimed repeatedly that his responsibilities did not include political analysis. For him, politics included anything that had to do with fuel supply. He construed resituating EDF with respect to private industry not as politics but as economic good sense. This position necessarily involved a reconstruction of EDF's role in such a way that the utility would move from the political to the economic epicenter of the nation. EDF's initial technopolitical regime had deliberately conflated politics, economics, and technology. In advocating a new regime, management sought a rhetorical separation of the first element from the latter two. From the viewpoint of this book, of course, rhetorical separation did not mean actual separation; this strategy itself constituted a political quest to change not only the identity of EDF but also that of France.[40]

Although EDF's economist-managers sought to exclude politics from their analysis, they could not exclude politics from their world. Indeed, their arguments in favor of economic criteria held little sway with de Gaulle, who ultimately had the final say on this matter. De Gaulle held fast in his commitment to French independence and glory, and his close advisors assured him that these were synonymous with the gas-graphite system. The historical record becomes murky here, especially because appropriate documentation remains inaccessible. In retrospect, the light-water victors argue that had de Gaulle truly understood the technological and economic aspects he might not have supported gas graphite technology so fervently. But de Gaulle (an enduring icon who even today remains above serious criticism from all but the most ardent leftists) was "ill-advised." One man usually emerges as the nefarious advisor: Maurice Schumann, the minister of scientific research and atomic questions in 1967–68. Proponents of the light-water system claim that Schumann maneuvered de Gaulle into an intransigent position in favor of gas-graphite. One commented disparagingly: "Schumann wasn't an economist, but a typical *homo politicus.* . . . [He] did not reason in terms of international industry. He did not recognize the international situation."[41]

Of course, Schumann's version of the story differs somewhat. In a 1981 interview, he proudly admitted that he had defended the gas-graphite system. But, speaking twelve years after the launch of France's light-water program, he claimed more nuanced reasons for this defense. "After studying the file very attentively, I gave absolute priority to the breeder."[42] Breeder reactors would ensure French independence in the future; in the meantime, "the simple and pure abandonment of the French system was not justified; certainly not before the *francisation* of the light-water systems."[43] He portrayed his position not as a simple knee-jerk reaction against light-water technology but as a reasoned argument that prioritized French independence and that saw in gas-graphite reactors the technological bridge to a future of breeders. He had to ensure that de Gaulle understood the implications of each option. Proponents of the light-water system, he noted, "had advocates—I was about to say 'accomplices'— inside the CEA, who invoked the dangers inherent in the French system as arguments against it," but "the studies that I commissioned showed that foreign systems had at least as many accidents and delays."[44] Had de Gaulle remained president, Schumann concluded, the gas-graphite program would have continued.

Whatever the case, all agree that Charles de Gaulle had a formidable will. André Decelle, EDF's director-general, passionately desired a solution to the frustrating impasse. He had tried to persuade various ministers to change de Gaulle's mind. But, he said in a later interview, Pierre Massé (then EDF's president) would not back him—not because he disagreed with Decelle's position, but because he disagreed with his strategy.[45] Depressed and discouraged, Decelle resigned in September 1967, citing health and personal reasons.[46] Apparently, he and Massé had agreed not to mention Decelle's advocacy of the light-water system in the resignation statement because doing so would only make it more difficult to pursue that American system later on.

This precaution failed completely. The very day that Decelle announced his resignation to the board, *Le Figaro* proclaimed: "André Decelle, director-general of EDF, resigns. Partisan of the enriched uranium system, he disagreed with the government."[47] Pierre Massé, backed by Robert Hirsch, denounced the headlines as "serious counter-truths."[48] They objected to claims that Decelle and others had blindly "championed" particular technologies. Like others, they said, Decelle had simply aimed at "determining, with maximum objectivity and in the spirit of science, where the interest of the country lies in this matter." Hirsch was "particularly shocked to see this effort transformed for public opinion into some kind of passionate conflict."

Regardless of such disclaimers, Decelle's resignation did not signal the end of the debate; quite the contrary. As Marcel Boiteux replaced him at the end of 1967, it became increasingly clear that positions with regard to various reactor systems did not fall neatly along institutional lines. The advocates of light-water included EDF's top management, private manufacturers, and a handful of CEA top officials (including Robert Hirsch). Some CEA employees adopted a middle position, arguing that, should it become necessary or expedient to build light-water reactors, France should develop these itself on the basis of the CEA's submarine prototype. According to this middle position, reactors built in France should remain French, regardless of type.[49] The advocates of gas-graphite included labor unions and rank-and-file employees in both EDF and the CEA who had devoted their careers to that technology. All attempts at negotiation having failed, the impasse was referred to the PEON commission.

PEON: Defining the Context for Technological Development

The Commission pour la Production d'Electricité d'Origine Nucléaire (PEON), founded in 1955, was a government-appointed commission composed of top EDF and CEA leaders, ministerial officials, and a few industrialists. Its ostensible purpose was to advise the government on matters nuclear. It was not a decision-making body. At least until the late 1960s, programmatic decisions were negotiated within and between the CEA and EDF. PEON did little more than discuss and bless such agreements.[50]

The PEON commission's role grew more subtle and complex during the *guerre des filières*. In the contentious climate fueled by technopolitical uncertainty, meetings of this commission provided a place for constructing notions of objective arbitration. The commission's discussions and reports provided a stage on which members could play a hybrid role: although they were there to represent specific institutions, their membership in PEON symbolically separated them from their institution and gave them a larger constituency—the nation. This hybridity conflated the self-proclaimed disinterestedness of state technologists with that of the nation. At least in principle, any PEON conclusion or document represented an arbitrated negotiation for the greater good of the nation among otherwise competing interests. When reporting back to their home institutions, PEON's members carefully nurtured the commission's status as objective arbiter. The same policy conclusion would carry more weight all around if reached by PEON.[51]

PEON inherited the Horowitz-Cabanius mission: to investigate the ramifications of each reactor system and provide a rational, objective basis for short-term and medium-term programmatic choices. Accordingly, in late 1967 its members produced reports on a variety of issues[52]:

- the current technological state of each system
- national and international fuel sources and their costs
- the comparative costs of various energy systems (not only gas-graphite and light-water reactors but also advanced gas-cooled and heavy-water reactors and conventional—i.e., non-nuclear—power systems)
- industrial organizations and contracting
- licensing agreements.

The criteria for comparing different reactor types were heterogeneous:

- the reliability and longevity of reactors
- capital, fuel, and operational costs
- construction times
- dependence on foreign countries
- export potential
- existing industrial infrastructures
- the shape of foreign licensing agreements.[53]

On the surface, these reports appeared to meet expectations for an objective and consensual conclusion, particularly in the domain of cost calculations. PEON's cost calculations favored the light-water system, though its numbers differed somewhat from those of Horowitz and Cabanius.[54] Further (and this conclusion appeared especially objective, since PEON was supposed to be fundamentally pro-nuclear), the commission did *not* find nuclear power competitive with conventional sources—largely, it seemed, because the price of conventional fuels had dropped significantly.[55]

Once again, however, numbers did not tell the whole story. To begin with, many numbers were missing, uncertain, or incalculable. Comparing system costs raised the same problems for PEON as for the Horowitz-Cabanius commission: the data corresponded to very different economic contexts, and the prices offered by American and German companies did not necessarily reflect the real building costs. Commenting on the PEON discussions, Jules Horowitz noted bitterly that "the variation in American prices, the sacrifices that AEG and Siemens admit having made recently in order to obtain their first large orders, and the difficulties encountered by Belgian industry in the Doël and Tihange affairs all illustrate well the distinction that must be made between the real cost of an undertaking and the price that must be conceded in order to succeed in certain markets."[56] Another CEA commentator reached similar conclusions:

After reading [the PEON reports] one can see, as indeed is well known:

1. that the differences between the systems are the same order of magnitude as the uncertainties.

2. that light-water is being "pushed" and gas-graphite is being "jinxed" on the following economic bases: what is gained on the investment front will very certainly be lost on the fuel front, and the only parameter that tips the balance is a lower operational cost for light-water (21.7 F/kW-year, which is to say 0.3 c/kW-h instead of 33.9 F/kW-year or 0.5 c/kW-h).

One could ask oneself whether the real decision-making point is not simply a big difference in the reliability of the two systems (what comes from abroad always seems more attractive to French minds, but watch out for painful awakenings).[57]

EDF members of PEON were just as aware of the uncertainty in the numbers. Later, Pierre Massé acknowledged that the cost difference between the two systems was less than the margin of error in the data used to calculate that difference.[58]

The numbers were uncertain, and calculations indicated that conventional power might prove a wiser course. But PEON did not recommend abandoning the entire nuclear endeavor. Instead, PEON members attempted to define and describe, and therefore shape, not the artifacts directly, but the *context* in which they would operate. For industrial leaders, this context was the Common Market—a context that demanded the pursuit of nuclear technology regardless of the current price of fuel. Offering familiar arguments about competitiveness, the structure of French industry, and the increasing worldwide dominance of this technology, one industrialist added slyly: "Just imagine the position of French industry in a Common Market in which, nuclear power having succeeded, German industry dominated this sector."[59] Since light-water reactors dominated the world markets, he continued, the light-water system was the only plausible choice for this context. Industrialists added weight to their argument in favor of light-water by stipulating that they would offer warranties for that technology, but not for gas-graphite technology.[60] The mere act of offering warranties transformed the light-water system into a more reliable technology than the gas-graphite system, without doing any technological work per se.

CEA representatives to the PEON commission sought to limit the context to France (rather than Europe). Here they met with stubborn resistance from the industrialists, who apparently refused to discuss matters in these terms:

It has been practically impossible to get [PEON] to concretely consider the national context, technological continuity, the dangers of dispersion and oversupply in a market that will remain fairly narrow for a long time—in short, the real cost for the country, not to mention the concern to create a truly major French nuclear industry that could negotiate on equal terms with the largest European companies. I tried several times to provoke a discussion about these important industrial problems; the Industry representatives to the Commission remain prudently reserved.[61]

When the question of French economic independence arose separately from that of the Common Market, said another CEA member, "opposition came both from the industrialists, who refused to provide the smallest piece of data, and from the Planning Commission, which as always preferred multiple abstract schemes."[62]

In April 1968 these disagreements were glossed over by PEON's formal report. The report concentrated on two elements: the outcome of the cost calculations once the uncertainties were factored out, and the need to base decisions on "objective" economic criteria rather than on political considerations. This second item reflected efforts to redefine the French context: "It is pointless to hope for total independence. . . . The potential for economic independence can be defined as the capacity to maintain economic competitiveness over the long term and on the international front. . . ."[63] The numbers and the context, in turn, pointed to a clear set of recommendations:

- France should immediately build an American-style reactor.

- Pending a reevaluation in 1970, no new gas-graphite reactors should be ordered in the next two years.

- The Canadian heavy-water design might deserve further consideration.[64]

One industry periodical joyfully proclaimed these conclusions the result of a "profound unanimity." And where did this unanimity come from? Quite simply, from the separation of technology from politics:

The essential reason for this unanimity comes, we believe, from the fact that men in good faith, from the most diverse origins, were eventually bound to agree over the analysis of such a complex question from the moment that this [question] was entirely depoliticized and subjected to the objective analysis of the real problems involved.[65]

The important point was that politics had not dominated the debate. This, in turn, provided the government with a clear basis for action.

The main achievement of PEON's 1968 report was, therefore, to legitimate two key strategies of light-water's supporters: the separation of technology from politics and the redefinition of the context of nuclear development as Common Market economics.

Breeder Reactors: Flexibility and Consensus

Nonetheless, turmoil continued to lurk beneath PEON's facade of consensus. De Gaulle continued to favor the French system. Within both EDF and the CEA, employees remained split. Not everyone agreed that technology and politics should be separated, or that the context for the nuclear program should be primarily economic.

A different source of consensus emerged in discussions outside PEON: the breeder reactor. As a technology that still existed primarily on paper (only one prototype existed: the CEA's Rapsodie), the breeder was still flexible enough to fulfill a broad spectrum of technopolitical scenarios. As we saw in the cases of Jules Horowitz and Robert Schumann, gas-graphite enthusiasts had already begun to endow breeders with the power to carry France's technological glory. In the wake of the PEON report, proponents of gas-graphite reactors focused increasingly on a future of breeders. Light-water's proponents, meanwhile, used that future to build a stronger constituency for the American solution.

Some engineers and labor militants at EDF maintained that the breeder future demanded further pursuit of the gas-graphite system. In July 1968, for example, Claude Bienvenu, the leading project engineer for Saint-Laurent 1, lambasted recent decisions that jeopardized the gas-graphite program. He was angry because an impasse over industrial contracting methods had stalled the construction of a gas-graphite unit at Fessenheim. Worse, the companies in charge of construction were trying to reinvent the pressure vessel, the heat exchangers, the command and control systems, and nearly everything else. "Saint Laurent will have been useless!" exclaimed Bienvenu. "The gas-graphite system, which had been ready to derive maximum profit from the experience accumulated and perhaps even to battle with some chance of success against the American system, will find itself blown away like a straw in the wind."[66] Breeders could return France to a more rational path. They also provided the best reason for maintaining the gas-graphite system, which could supply both the plutonium and the experience required by breeder development. Such a course would ultimately allow France to surpass the United States, which had no breeder experience.[67]

The CGT militant Claude Tourgeron also saw a future of breeder reactors. His, however, was a socialist future. Tourgeron juxtaposed his argument for breeders with an argument for the "formation of nationalized companies that would free this industry from the joint pressure of large capitalist monopolies and military management."[68] These nationalized companies would provide the basis for a true socialist democracy, which could only lead to national economic growth. Breeder technology would take some time to mature, though, so France had to pursue an intermediate system in order to maintain its nuclear knowledge. Only a system based on natural uranium would allow France both to escape the clutches of American imperialism and to produce plutonium for the breeder future. Cost calculations that disadvantaged the gas-graphite system resulted from nefarious capitalist practices. The Fessenheim estimates, for example, had been inflated by capitalist monopolies in their thirst for profit and their desire to tip the balance in favor of the American design. Thus, successful gas-graphite reactors, breeders, and a socialist order were mutually dependent.[69] The technopolitical circle was complete.

Though their visions differed, Bienvenu, Tourgeron, Horowitz, and Schumann all saw a future of breeders. This consensus was remarkable, since aside from their enthusiasm for gas-graphite and breeder reactors the four men had little in common. Proponents of the American system seized on this consensus to propose a different path to that future. For example, in February 1969 EDF's top management sent a memo to the prime minister arguing that France should make every effort to research and develop breeders ("the system of the future"). But it contended that the main road to that future went through the American system. Using an American license would allow France to recover from the disappointment of the gas-graphite experience and to "catch its breath while waiting for a new breakthrough—that of the breeders—to which it will devote all its research and development efforts."[70] Not even the CEA's experience in designing a light-water reactor for submarines would go to waste. Instead, this experience would help French teams "mix French intelligence with American experience to build a Frenchified reactor."[71] Thus they too transferred the burden of French grandeur to the breeders. Further, the nebulous "Frenchifying" of American reactors would preserve French nuclear know-how (and, presumably, pride). In April 1969, Marcel Boiteux and Robert Hirsch proposed a "plan of action" that essentially reframed the proposals and arguments of the PEON report to suit the logic of a breeder future.[72]

In presenting this "plan of action" to EDF's board, Hirsch and Boiteux emphasized that the plan prudently kept the natural-uranium option open. They stressed that "the realization of the first light-water reactor will take place in the framework of a general license in order to draw from the Americans a maximum of amount of knowledge about the chosen system."[73] The French would derive maximum benefit out of the partnership while leaving the Americans responsibility for the technical warranties. Paul Delouvrier, Pierre Massé's successor as president of EDF, waxed enthusiastic about the plan. Although light-water reactors were more expensive than oil-fired plants, he affirmed that this was the price that France had to pay to keep up to date on matters nuclear. "It is not without some sadness," he said, "that one sees AEG and Siemens put a plant in Holland, given that the nuclear industry got a much later start in Germany than in France. It is definitely time for the country to get hold of itself in order not to be surpassed and dominated."[74] Once again, then, light-water appeared to provide the path to French radiance. Delouvrier gave the plan of action his blessing. With Tourgeron absent from the board meeting, no one raised any objections. PEON approved the plan in May 1969. Meanwhile, EDF's managers had already begun to prepare for the first light-water reactor.[75]

In the mid 1950s, the CEA capitalized on the ambiguity of the gas-graphite design to advance the French bomb program. In 1969, light-water advocates capitalized on a variety of ambiguities to move forward with plans to buy an American license. Each successive report tightened the case for the light-water system, using a combination of three techno-political strategies:

• managing technological and economic uncertainty, either by quantifying and plotting potential data fluctuations or by pronouncing on the relevance and function of different uncertainties

• defining the context in which future nuclear development would occur, notably by renegotiating the meaning of "national independence"

• constructing a new logic for light-water development in which that development would contribute to French radiance.

Embedded in successive reports, these strategies created a narrative teleology of nuclear development. As Philippe Simmonot argues in *Les nucléocrates*,[76] each report further instantiated a logic of technological determinism. With each successive refinement, American light-water reactors became more and more necessary for the future of France.

Unions Strike Back

As 1969 wore on, opposition to light-water became increasingly difficult to orchestrate. Advocates of the American system had developed their plans incrementally, carefully reframing ambiguities that could have argued for either system in their favor. Their stated goals—to give France cheap energy and to make breeder reactors the new symbols of French technological glory and independence—were irreproachable. Further, they had not actually proposed terminating the gas-graphite program yet. Finally, these advocates occupied the top administrative positions in the CEA, in EDF, and in private industry.

The case was not closed—the government had not yet made a decision on the choice of system. But things looked bad for gas-graphite. Clearly EDF's management, private industry, and top CEA officials were poised to buy American. Equally clearly, buying American would come at the expense of the French system. Furthermore, by then the man seen as a guarantee against the purchase of a foreign license—Charles de Gaulle—had resigned the presidency and had been replaced by Georges Pompidou, who had distinct sympathy for the American system.[77]

In an effort to obstruct the growing forces in favor of light-water, labor militants began to contest the economic analyses constructed by program leaders. Unions offered alternative figures, calculations, and interpretations. These efforts began when CGT representative Claude Tourgeron registered an official protest at EDF's June 1969 board meeting.

Focusing on the uncertainty in the light-water data, Tourgeron's protest contested the notion of light-water's worldwide dominance, resurrected the issue of national independence, and challenged the push for "purely" economic selection criteria. He noted that in the United States orders for light-water reactors had dropped dramatically between 1967 and 1969. Tourgeron attributed this dropoff to an increase in the capital costs of these reactors (now up to 1000–1100 francs per kilowatt, in the same range as the gas-graphite reactor Saint-Laurent 2). He also argued that American utilities had lowered their predictions for the capacity factor of light-water reactors and were even building "rustic" thermal plants to take over when reactors had to go off line. He reiterated familiar arguments about the threat that reliance on enriched uranium would pose to France's independence. He also added a new twist: American enriched uranium was inexpensive primarily because isotope separation plants "had been financed a long time ago [presumably during World War II] by taxpayers."[78] Neither France nor Europe could ever hope to approach American prices. Finally, he argued that economic criteria alone did not suffice for making decisions about the future of the French program: prices fluctuated too much to provide a reliable foundation for decision making. Both technical and political considerations militated in favor of more gas-graphite reactors to link the present with the breeder future.

Several board members countered Tourgeron's claims. Robert Hirsch attributed the decline in American reactor orders to market cycles. Others denied the validity of Tourgeron's calculations by simply reiterating PEON's economic estimates. Marcel Boiteux closed the discussion by insisting that consensus existed on two matters. First, the future belonged to breeders, and France should do everything to preserve its lead in this domain. Second, the country had to engage in some kind of intermediate program to ensure that French industry would maintain its mastery over nuclear technology. The only two truly viable candidates for this intermediate program, Boiteux continued smoothly, were the gas-graphite system and the light-water system. Boiteux then completely ignored Tourgeron's estimates by asserting: "All the numbers cited in this discussion—which are based on experimental results, developments, and recent requests for bids—proved that the light-water system was the most economically viable and the least capital intensive. This is the reason it was chosen."[79] Had a choice been made? Was this last statement a slip of the tongue, or a reference to American decisions? It was not clear. Nor were there any significant new data about the economics of light-water reactors. Boiteux elided these points and hastened to add that, for the moment, the natural-uranium option had not been closed. He postponed that decision for another twelve or eighteen months, pending further data. Other board members murmured their assent, and the matter was temporarily tabled.

Meanwhile, CEA union leaders had begun objecting to the emerging plan, which they felt threatened both their jobs and the future of the French nation. Initially, they protested that these plans had been drawn up not by the government—the ultimate representative of the people, however objectionable it might be—but by institutional leaders. They added that, contrary to allegations in the news media, political rather than technical weakness had caused the current problems. A CFDT publication asserted that "the difficulties faced by the [CEA] today do not come from technical failures, but from the government's lack of research policy and industrial policy."[80] The CEA's scientific and technical potential was being ignored. The government needed to establish a coherent research program addressing "technological areas in which France is especially and dangerously underdeveloped."[81] The CFDT advocated a new institution similar to that proposed by Claude Tourgeron: a state-run financial institution that could create new companies or regroup existing companies. The state could thus manage private industry and give rational direction to the nation's research and industrial development. This would also prevent Westinghouse from taking over France's electromechanical industry.[82] Finally, the CFDT—echoing the *autogestion* (self-management) demands of the May 1968 strikes, during which many CEA employees had become radicalized—asked that CEA workers (white-collar and blue-collar) be given more say in program management and in decision making.[83]

By October 1969, rumors had begun to circulate that the CEA's programs would be cut back and that layoffs would ensue. The CEA's five main unions joined forces to protest the layoffs, the introduction of American light-water reactors, the implied slurs on their technical competence, and the incoherence of French nuclear research policy. On October 10, some 800 employees staged a demonstration at Marcoule. Meanwhile, at the Saclay research center, unions avidly defended the performance of the French nuclear program, which, according to one flyer, had been "submitted to systematic . . . unfounded criticism by the press, encouraged by the eloquent silence of CEA and EDF top management."[84] The real problem, this flyer suggested, "contrary to what is written daily in the press, has nothing to do with the high price of French nuclear plants, but instead [is due to] on the one hand, the dumping prices practiced by oil companies . . . and on the other hand the current structure of the French electromechanical industry in general and the nuclear industry in particular."[85] The price of a gas-graphite kilowatt-hour was already 30 to 40 percent lower than the most optimistic estimates of the Plan several years earlier. On that basis, the Plan—which, however imperfectly, still represented popular will better than EDF and CEA leaders—had called for 2500 megawatts' worth of new reactors. Only 1300 megawatts' worth had been built.

The unions demanded a coherent nuclear program whose main criteria of success would be continuity, independence, and the development of a national electromechanical industry. This policy "must first and foremost be translated into the development of the gas-graphite system."[86] It would be "stupid" to abandon this and other national technologies. In a separate statement, the CGT called for the publication of reports that would "reestablish the truth which is indispensable to the defense of French atomic energy. . . . The CGT's engineering and white-collar worker section will not hold back in its efforts to ensure that France remains independent in the energy sector."[87] Others used even stronger language to denounce the intrusion of Westinghouse into French industry:

What some are calling the "*guerre des filières*" is a booby trap! It's really a war between international trusts orchestrated by one of them: Westinghouse. What could Westinghouse's intrusion into the closed world of bourgeois businessmen and technocrats which governs us mean, other than the brutal manifestation of American imperialism in our midst. Elsewhere, it kills by war; here, it seeks to reduce us to the state of an economic colony. Let us not be dupes: *the French government is not neutral in this affair. It's an accomplice. It's the enemy of the workers.*[88]

The government could not be counted on to produce a reasonable solution. It was colluding with private industry to orchestrate an American takeover of France.

Boiteux Declares the End of the Gas-Graphite Program

The situation finally exploded in mid October at Saint-Laurent 1, the pride and joy of the gas-graphite program. The reactor had been operating for several months and had already produced a respectable amount of electricity. On October 16, Marcel Boiteux, accompanied by Robert Hirsch and Francis Perrin, went to the site for the official inauguration of the reactor. During his press conference, Boiteux congratulated the site's teams on their success, declaring that Saint-Laurent was the best of EDF's reactors. Unfortunately, he added, the gas-graphite system was not commercially viable. From then on, he said, EDF would be building light-water reactors under an American license.[89] This announcement sent a shock wave throughout the nuclear program, the government, and the press. Everyone knew that this was the direction in

Figure 8.2
Marcel Boiteux gives a press conference at Saint-Laurent on October 16, 1969. EDF's official caption for this picture is "Inauguration of the plant at Saint-Laurent-des-Eaux." Gas-graphite supporters would have titled this picture "Boiteux announces the termination of the French system." Photograph by Michel Brigaud. Source: EDF Photothèque.

which the program was headed, but no one realized that a decision had been reached.

Reactions in the press were mixed. Nicholas Vichney of *Le Monde* was jubilant. Acknowledging the technical success of Saint-Laurent, he followed Boiteux in emphasizing its economic drawbacks; then he added several comments about the CEA's unreasonable attitude.[90] But Pierre Juin, writing in the Communist daily *L'Humanité*, was scandalized. His front-page article featured a photograph of Saint-Laurent with a caption describing the site as a "prestigious French achievement."[91] Saint-Laurent's technical success, Juin wrote, might lead one to expect that "the top brass of EDF and the CEA who piloted specialized journalists through the vast construction site of Saint-Laurent on Thursday would be overjoyed. Well no." He went on to impute the decision, not just to EDF, but to the government more generally:

In his press conference last Monday, Mr. Ortoli, minister of industrial and scientific development, had declared that France's nuclear policy would be fixed at the end of the year. . . . Mr. Boiteux, however, could not hide that the case had already been heard. . . . During a lightning interview, which only allowed for a half-dozen questions, Mr. Boiteux affirmed that all countries were now oriented toward light-water reactors and that, as a result, it would be tasteless to obstinately pursue our own technology in the restricted space of the French hexagon.

The decision, said Juin bitterly, had been the result of pressure by foreign monopolies and would seriously endanger French independence.[92] In a similar vein, the caption of one *Canard Enchaîné* cartoon read "US Go Ohm!"[93] *Le Canard* did note, however, that the decision had not emanated from the government. "Pompidou is slowly rushing to decide nothing," sneered the weekly. "For the moment, he is still in training. After all, the French system is the General's gadget. Got to treat that carefully. The dear old gentleman might take offense."[94]

Indeed, Boiteux's announcement apparently surprised Georges Pompidou, who had taken no official decision, and some of EDF's board members, who had thought that matters were still up for debate. Boiteux himself emphatically denied that he had announced any sort of decision. He had merely stated that, because the economic success of Saint-Laurent was less certain than its technical success, the future of the gas-graphite system remained uncertain. It was, he said, "regrettable that his words were given the political meaning that they were."[95] Journalists had misinterpreted his responses to their questions. He had said that "there was no reason to regret what had been done in this domain, [since] the effort poured into the 'gas-graphite' system fit into the logic of the nation's history, but that today the fact nonetheless remained that nuclear plants were too costly, and it was only right to question whether an Establishment like EDF should continue to build them."[96] He had merely indicated that EDF had a *preference* for the light-water system. The press had not mentioned that he had referred all final decisions to the government. Paul Delouvrier expressed his support for Boiteux. Claude Tourgeron and other labor union representatives reiterated their objections.

Disclaimers notwithstanding, Boiteux's statement was widely understood to signal the end of the gas-graphite program. For CEA employees, the first of the rumored layoffs confirmed this signal: the same day that Boiteux held his notorious press conference, 98 cleaning ladies subcontracted to the Saclay research center were let go. The next few days saw several more layoffs, all branded by the unions as violations of the labor agreements drawn after the 1968 protests. On October 23, Saclay's director returned from a trip to find the site's union personnel up in arms. He refused to revoke the layoffs. Four days later, 700 Saclay employees launched a series of strikes that would last for more than a month.

The CEA Strikes

In order to understand these strikes, we must briefly go back in time to 1968. During the nationwide protests that year, numerous CEA engineers and technicians had joined unions.[97] Like demonstrations elsewhere in the country, the 1968 sit-ins at the CEA focused on democratizing the workplace and loosening the institution's decision-making hierarchies. From the perspective of the protesters, the results had been somewhat mixed. They had obtained new administrative bodies that, at least in theory, made room for broader participation in managing daily workplace affairs. However, as the October 1969 layoffs indicated, not all of the CEA's directors had taken well to these new structures. Further, as the termination of the gas-graphite program showed all too clearly, the CEA's administration had no intention of including the personnel in programmatic planning, not even in the cursory style to which EDF's board of directors had devolved.

Still, the 1968 sit-ins had provided a brief opportunity for many CEA employees to experiment with solidarity among engineers, technicians, scientists, and workers. (Recall from chapter 4 that the national confederations had advocated this solidarity in their discussions about recruiting the technical elite.) The most extensive of these sit-ins had occurred at Saclay, where the working population consisted primarily of engineers, scientists, and technicians.[98] At Marcoule, 1968 apparently did little to change the relationship between workers and engineer-managers described in chapter 5. But the fact that protests occurred at many CEA sites suggested that workers there might share sentiments with engineers and technicians around the country. In 1969, this shaky alliance across multiple CEA sites resulted in protests that combined the practices and goals of labor unions with those of engineers. The ensuing strikes combined demands to halt the layoffs with calls for greater employee participation in management and challenges to the termination of the gas-graphite program. The 1969 CEA strikes, in other words, fused questions of national technological policy with concerns about social relations.

The first group of protesters at Saclay included five hunger strikers. These men demanded the revocation of all layoffs. Echoing the tones of 1968, they presented their case as a moral issue, a matter of basic social equity:

> We refuse to play the game of dividing the personnel between CEA employees (the nobility) and subcontracted employees (the pariahs) which the administra-

tion wants to impose on us and which does tempt some of the personnel. Workers, not matter who they are, have a right to a decent life. . . .

We refuse to be complicit in a hypocritical and cowardly society that always makes those pay who can defend themselves the least.

We refuse to be complicit in a repressive society that uses all means, even those that run counter to its own legal framework, to manipulate and intimidate those who in the end are the source of all wealth: the workers.[99]

On that note, the five men installed themselves in Saclay's labor union offices on the afternoon of October 29, only to be evicted a few hours later when 240 policemen stormed the site. For the next two and a half weeks, the strikers would continue their fast in a nearby church.[100]

On October 31, news leaked that the CEA's administration planned to announce another 2000 layoffs. The unions responded by broadening their demands and intensifying their strike actions. Though they continued to express outrage on behalf of the cleaning ladies and other subcontracted workers, protests now focused primarily on nuclear policy. The strikes spread to all of the CEA's research and production centers and continued through the end of November.[101]

Echoing earlier arguments, strikers denounced the termination of the gas-graphite program and the threat of an American industrial takeover.[102] They contested the assertion that gas-graphite reactors were not competitive and argued that "profitability [was] not the only important criterion."[103] National independence had to count too—especially independence from the United States. Never had the threat posed by American capitalism loomed larger. "We are," one tract warned, "in the process of losing our national independence; we are on the path to underdevelopment and colonization."[104] French plants, the protesters asserted, were equivalent in quality and cost to American ones. Nuclear research had been a source of French pride for decades. Even the British were said to have admitted that the French had a "natural flair" for nuclear technology and science.[105]

The problem, said the unions, lay in the fact that the government had not handled either industrial or research policy properly. "Such an important decision . . . should be preceded by consultations with employee representatives, not announced on the fly by a bureaucrat, no matter how highly placed he might be."[106] Only the unions had the nation's welfare firmly in sight: "Our goals are clear. We are in favor . . . of funding research which will ensure the intellectual, economic, and social future of an entire people and guarantee its independence."[107] Although nuclear weapons were not a significant subject of discussion in the strikes,

Pour faire face à la technologie américaine, le COMMISSARIAT
A L'ÉNERGIE ATOMIQUE peut et doit devenir dans l'intérêt national
un groupe puissant et diversifié où les agents seront associés à la
marche de l'entreprise et la garantie de l'emploi assurée.
En raison de la crise énergétique actuelle, le Commissariat, qui est
à l'origine du développement nucléaire en France, doit recevoir, dans
le domaine de l'électro-nucléaire, des responsabilités accrues, des
crédits suffisants et le personnel nécessaire.
Une nouvelle politique du personnel doit être définie impliquant
la reprise des recrutements.

REJOIGNEZ LA C.G.C. !

Figure 8.3
A flyer issued during the 1969 strikes by the Confédération Général des Cadres,
the white-collar workers' union. Note the alterations made to the front of the bill
(all in English): "What is good for Westinghouse is good for France," "In EDF
we trust," "Business is business," "MB" (for Marcel Boiteux), and "PWR" (for
pressurized-water reactor). The following explanation was printed on the back of
the flyer: "To confront American technology, the COMMISSARIAT A L'ENERGIE
ATOMIQUE can and must—in the national interest—become a powerful and
diversified group in which employees are involved in managing the firm and in
which employment is guaranteed. Because of the current energy crisis, the
Commissariat, which is at the origin of nuclear development in France, must
acquire greater responsibilities in the nuclear arena, as well as a sufficient budget
and the necessary personnel. A new personnel policy must be defined, one which
involves the resumption of recruitment." Flyer courtesy of Jean-Claude Zerbib.

three of the unions could not resist a jab at the military program: "Strangely,
military applications, which constitute the least important part of nuclear
research, are not in the least affected by restricted funding. The government
talks of national independence when atomic bombs are involved; at the
same time, it is liquidating our national industry, which is the measure of
true independence and the source of progress and well-being."[108]

Figure 8.4
CEA protesters march past the Eiffel Tower, a historic symbol of French technological glory. Photograph by Philippe Mousseau, Lumifilms. Courtesy of CFDT archives.

On November 14 the CEA's administration reinstated the cleaning ladies and the hunger strikers stopped their protest. If the administration hoped that this concession would end the strikes, however, it hoped in vain. That same day, President Pompidou formally announced the termination of the gas-graphite program for the foreseeable future. Although the "foreseeable future" clause was intended to leave open the possibility (at least rhetorically) that the gas-graphite system might find favor again some day, no one paid it much attention. The CEA strikes continued to intensify. On November 17, between 4000 and 6000 protesters descended on the Place des Invalides in Paris and marched past the Eiffel Tower. Strikes continued in the provinces too. According to CGT statistics, 90 percent of Marcoule's personnel were on strike between November 14 and 18.[109]

These strikes combined typical employment matters and issues of national industrial and research policy in a seamless web. The heterogeneity of the issues raised during the strike doubtless was due in part to the heterogeneity of the strikers, who ranged from the blue-collar workers at Marcoule to research scientists and engineers at Saclay. No doubt realizing that purely political tactics would have little effect in a debate whose terms were defined by its dominant participants as economic and apolitical, a group of union engineers, scientists, and technicians prepared a counter-report on the relative merits of the competing nuclear systems.

Economic Comparisons, Union-Style

One major difference between the union report and those written by Cabanius and by PEON lay in how the reports posited the relationship between technology and politics. As we have seen, advocates of the American light-water system sought to remove what they derisively called "political" considerations from the decision-making process. Union advocates of the French system, on the contrary, sought to retain such considerations. Paralleling but also extending the points Horowitz had made, the union document attributed the importance of political considerations in nuclear energy policy to the fundamental uncertainty of the data on nuclear power.

A major source of this uncertainty, according to the union report, were differences in the financial and technological conditions under which power plants operated in the United States and France. These differences gave the American system an artificial advantage in at least four ways. First, the amortization period for reactors in France was twenty years, whereas in the United States it was typically thirty. Since a shorter amortization period penalized plants with higher capital costs, this difference unnecessarily disadvantaged gas-graphite reactors. Second, the capacity factor used in the calculations differed in the two nations: 6800 hours per year in France versus 7500 in the United States. This too penalized French reactors, and there was little empirical evidence to suggest that American reactors really spent that much more time on line. Third, price comparisons between nuclear and conventional power in the two countries operated under different principles. In the United States, for example, the price of conventional fuel included the cost of transportation to the power plant. French pricing included freight costs only to the port of entry. Taking port-to-plant transportation into account would raise the price of French conventional fuel by 0.37 centime per kilowatt-hour and therefore make the gas-graphite system more attractive. Finally, were France to engage in the "draconian" precautions taken in the United States to reduce pollutant emissions, the price of conventional fuel would increase even further, perhaps by as much as an additional 0.85 c/kWh. None of these factors, said the unions, had been included in the EDF's calculations.

Indeed, the report maintained, "the capital costs announced by EDF are incomprehensible and incoherent."[110] For example, the figures used to represent the capital costs of light-water reactors did not include the fact that two such reactors in the United States had gone, respectively, 30 percent and 90 percent over budget. Combine this with the spectacular reduction in the gas-graphite costs achieved at Saint-Laurent 2 and the two reactor types had equivalent capital costs. According to calculations presented in the report's appendix, electricity from an American light-water reactor would cost between 2.93 and 3.08 c/kWh, whereas the electricity from the second Fessenheim gas-graphite reactor would have cost 2.91 c/kWh.

Still, the union report argued, such numbers had limited value: "All the plants on which current economic comparisons, and therefore decisions, are based are 'theoretical' plants."[111] Reliable numbers for fuel cost, use rate, operational costs, and amortization would come only with more extensive operational experience. Further, it was impossible to tell how the numbers used by EDF and PEON had been derived, since the actual calculations remained hidden. And finally, current economic studies were based only on the direct cost of the reactors, without taking into account the investments that either nation had already made in the technological system that supported each reactor type. (This system included fuel manufacturing plants, treatment plants for spent fuel, research infrastructure, the military functions of reactors, and more.) "The unannounced but implicit abandonment of this system, into which the CEA and EDF have poured considerable investments, is therefore completely incomprehensible."[112] The unions thus demonstrated that the decision had followed the logic of politics, not that of abstract economic rationality.

How did the unions view the politics of the situation? For them, the decision to terminate the gas-graphite program represented a capitulation to capitalism—American capitalism in particular. "Everyone is aware of the concerted offensive launched by American industrial consortia to get hold of the French electromechanical [industry]," said the union report. Pompidou's announcement merely confirmed the "Americanization of the French electronuclear [program]."[113] But the report did not argue that politics should have been left out of the decision. Instead, it argued, the *wrong* politics had guided policy makers. When the uncertainty of the economic data was taken fully into account, the resulting estimates were "sufficiently close for other criteria of choice (currency flow, capitalizing on existing investments, national independence, full employment) to be considered on the same plane."[114] Rather than base a decision purely on the politics of capitalist development, in other words, the government should have also taken the politics of social relations into account. And it should have weighted other political elements (such as national independence) differently. In conclusion, the report called for the creation of a new commission—composed of ministerial officials, EDF and CEA management, and labor unions—to reexamine the case.

On the evening of November 20, a delegation of union representatives brought this report to a meeting with Prime Minister Jacques Chaban-Delmas. He refused to revoke the layoffs, and he did not offer much hope on the programmatic front. He did not refuse outright to consider the unions' proposals, but he made it clear that he would probably turn them down. Discouraged, the CFDT, the CGT, and Force Ouvrière called for more strikes the following Monday. Those would be the last of the strikes: as it became increasingly clear that the programmatic decisions would hold, the unions lost heart.

Back to Bagnols

Although the strikes had no practical effect on national nuclear policy, they did have an important consequence at the local level. In the Gard, the strikes served to reassert a sense of common destiny among regional elected officials and Marcoulins—a sense that, as we saw in chapter 7, had begun to weaken by the end of the 1960s.

In mid 1969, as rumors about CEA layoffs began to circulate in Bagnols, members of the municipal council and shopkeepers who catered to the Marcoulins began to worry. True, the newcomers had caused unwelcome upheavals in local life, but in fifteen years they had also become tightly integrated into the region's new economic life. The departure of large numbers of Marcoulins would mean a significant loss of tax income for the municipality; without that income, the town would have an extremely difficult time paying the debts it had incurred while building its new facilities. "The state provided the town of Bagnols with large subsidies," the council fretted, and "it would be disastrous if such expenditures were approved without measures taken to ensure that they become profitable."[115] Similarly, Chusclan, Codolet, and other villages had only just begun to reap benefits from the presence of the site. Not only could the region ill afford to lose jobs; with a birth rate of 600 per year, it would soon need new sources of employment. Two hundred job cuts had been announced for Marcoule, with rumors of more to come.[116] Anxieties ran high.

Marcoule's labor militants and Bagnols's municipal council wrote a joint petition to department and state authorities explaining the gravity of the situation and demanding that "initiatives be taken in high circles in order to calm the emotions and dissipate the unease that currently weigh on the people that [we] have the honor to represent and the duty to defend."[117] Their suggestions included measures to encourage further industrial development in the region, creating new jobs, and averting the threat of job loss. Not content with writing plaintive letters, Mayor Pierre Boulot marched over to Marcoule to meet with Michel Molbert, the site's new director. Molbert, no happier than his employees about the looming unemployment, calmed Boulot's worst fears, assuring him that, in the end, not more than fifty people would be asked to leave the site. Through the prefect of the Gard, Boulot also obtained an appointment with the CEA's upper management in Paris in order to air his concerns personally.[118]

Meanwhile, unionized Marcoule workers had gone on strike to protest the demise of the gas-graphite program.[119] With the help of Mayor Boulot, a departmental official, and several social scientists from regional institutions, they prepared a document, titled *Marcoule et sa vocation dans le Languedoc-Rhodanien*[120] and more than 450 pages long, that described the site's functioning and organization and its importance for the region and for the French nation. In essence, this document rehearsed the narrative of local modernization explored in chapter 6—with one important difference: it also confronted the disillusion and difficulties that had characterized the actual experience of modernization.

The document did this by discussing Suzanne Frère's *Bagnols-sur-Cèze: Enquête Sociologique*,[121] which had concluded in 1968 that the region was irredeemably divided between Bagnolais de Souche and Marcoulins. The authors of *Marcoule et sa vocation* argued that Frère's conclusions—initially true—were now, only a year later, outdated. One had to look to "those elements that promote evolution toward the integration of the [two] populations, toward the creation of that future Society"[122]: the new schools, where the children of the two groups mixed everyday; the fact that more and more newcomers had begun to build their own houses, thereby coming into greater contact with the Bagnolais de Souche (they took up the same language); leisure facilities such as the pool or the cultural center, where the two populations increasingly had fun together; and more. Marcoule workers, the 1969 document reported, increasingly felt that Bagnols was their true home. The established population was "conscious of the step forward taken since 1954 and fear[ed] the economic consequences of recession. The first [group] has discovered a town, the second an economy."[123] The two groups now had a common "destiny."[124]

The municipal council of Bagnols supported this conclusion. CEA employees had become emotionally attached to the town; townspeople, the council felt, therefore owed these employees solidarity. Boulot argued that the council had a responsibility to maintain the regional economy. Other councilors elevated their motives to the national level: ultimately, they argued, the strike was about French energy independence, and, just as the CEA employees defended their profession, the municipality had to defend its taxpayers. Some even suggested that the municipal council go on strike if the demands of Marcoule's strikers were not met.[125] For better or worse, the two communities had to face their common destiny together.

In the Gard, therefore, the *guerre des filières* and the strikes that accompanied it brought a reconciliation between the new inhabitants and old-time local leaders. This occurred through a mutual recognition of the problems experienced by both groups. The reconciliation did not mean the end of conflict between the groups, nor did it mean that all local residents now welcomed the Marcoulins into their midst. It did, however, acknowledge that encounters between the two groups had been (and for some, would continue to be) difficult. A collective memory that made room for conflict had been born.

The *guerre des filières* had little effect on the local residents near Chinon. Site employees staged brief demonstrations protesting the termination of the gas-graphite program, but they did not worry about losing their jobs. Their initial difficulties over, the Chinon reactors appeared to have a long life ahead of them, even if they would have no more heirs. In 1973, EDF1 would be decommissioned and transformed into a museum—a fitting end, given that it had been compared to and even served as a tourist destination since its inception. That decision had not been made in 1969, however. At that point, not only did the three existing reactors function well, but it seemed likely that future light-water reactors would be constructed at Chinon. Life in the Touraine continued as usual.

The Cleanup at Saint-Laurent: Healing the Technopolitical Wound[126]

In view of what we know about the two institutions, it may seem ironic that the strikes protesting the termination of the gas-graphite program occurred at the CEA and not at EDF. But the termination of the program did not threaten EDF jobs. The EDF's labor statute guaranteed against technological unemployment—at worst, employees would have to learn new skills. In addition, both union and non-union proponents of the gas-graphite system had already had several opportunities to present counter-arguments in reports and at board meetings. Though these arguments had little effect on top management, EDF employees did not feel excluded in the same way as their CEA counterparts. Furthermore, although switching to another technology did hurt the pride of those who had labored on the gas-graphite system, it did not threaten the foundation of EDF as it did that of the CEA. Top management did use the *guerre des filières* to try to reshape EDF's technopolitical regime, but even this did not threaten unionized EDF employees as profoundly as the termination of the gas-graphite system threatened CEA employees. At EDF, unionized employees could continue to oppose the new regime by challenging its contracting practices. CEA employees had no recourse but strikes.

Those who might have protested the loudest—the designers and workers of EDF's gas-graphite reactors—had a more urgent task ahead of them. The day after Marcel Boiteux's announcement, in a strange coincidence, one of the most serious accidents the nuclear industry had yet seen caused a partial meltdown of Saint-Laurent 1. Engineers and workers from Saint-Laurent, and a few men sent over from Chinon, spent a year cleaning up the mess and putting the reactor back on line. It was in this activity, rather than through strikes, that they expressed their reactions to the termination of the program. Instead of contesting the decision by striking, they contested its *meaning* by working to repair their reactor. No discussion of the gas-graphite program's demise can be complete, therefore, without examining the cleanup of Saint-Laurent 1 in 1969–70.

As I have noted elsewhere, even proponents of the light-water system called Saint-Laurent 1 an outstanding technical success in the first few months of its operation. Its designers and workers proudly proclaimed it the most elegant and efficient of all French reactors. They insisted that its exceptionally well-planned and efficiently executed construction showed that nationalized companies should indeed lead France's technological efforts.[127] Even more proudly, they noted that Saint-Laurent 1, at 480 MW, was one of the most powerful reactors in the world. When it went on line in March 1969, it promised to help "defend the colors of gas-graphite"[128] by proving that the French system could compete not only with conventional plants but also with nuclear power in other nations. Workers were prepared to put in long hours to help it succeed. Time, said one man, did not count: he once worked 36 hours in a row without sleeping just to get a job done, and remembered his boss coming by at 2 A.M. to bring his shift "a snack and a pat on the shoulder, and to say how you guys doing?" The work atmosphere was "very friendly, very convivial. We worked hard, but for love, eh?"[129]

Saint-Laurent 1 appeared to hold the technopolitical key to the continuation of the gas-graphite system.[130] As such, the significance of its success for those who designed and operated it was both political and personal: their time, energy, and skill had made it France's most important technological achievement. It is not difficult to understand, therefore, why Saint-Laurent employees experienced Marcel Boiteux's announcement as "a stab in the back."[131]

On October 17, 1969—the day after Boiteux's press conference—loading machine operators began testing a new control tape. As far as they knew, the loading machine contained five uranium fuel rods and was about to load them into an empty channel. In fact, however, the machine contained five slightly thicker rods filled with solid graphite. Everything went smoothly until 6:32 A.M., when the last rod from the loading machine began sliding into place. The operators were puzzled when this rod protruded from the top of the channel. They thought that the difficulty might lie with the automatic control system, which had been acting up a little recently. They decided to override the automatic mechanisms manually, and by 6:58 they had managed to cram the recalcitrant rod all the way into the channel.

At 7:08 A.M. the terrifying siren of the reactor's alarm system blared. Because graphite rods were slightly thicker than uranium fuel rods, the last graphite rod had blocked the flow of cooling gas in the channel, and the uranium rods had begun to overheat. The uranium melted the metal cladding around the rods. The rods then fused together, producing a meltdown (though only in that channel). Fortunately, the operators soon realized that something had gone amiss. By shutting down the reactor quickly, they managed to avoid an accident on the scale of the one that would occur at Chernobyl 17 years later.

Figure 8.5
Saint-Laurent 1's nearly completed core from above. The circles in the floor are the openings into the core's fuel rod channels. The core was made up of nearly 3000 vertical channels, each of which contained fifteen uranium rods. Each rod was encased in a metal shield and surrounded by a graphite shell. The fission reaction took place inside the core, producing a great deal of heat. Carbon dioxide gas flowed through the channels in the core and absorbed this heat. The hot gas then flowed into the heat exchangers, where it converted water into steam; the steam powered the turbines (not shown), producing electricity. The entire reactor was encased in a concrete pressure vessel. On top of that vessel sat the loading machine. Photograph: Jacques, 1967. Source: EDF Photothèque.

Nonetheless, the reactor suffered considerable damage. After the fuel rods fused together, metal shards were blown out of the channel by a sudden burst of pressurized cooling gas. In addition to the damage caused by a melted channel, more than 100 kilograms of contaminated debris littered the structure that supported the reactor core. Furthermore, the pipes that contained the cooling gas had been exposed to radiation. Before Saint-Laurent could go back on line, the contaminated debris had to be cleaned up and the damage repaired.[132]

Figure 8.6
The fuel loading machine at Saint-Laurent 1 in 1974. Reactor operators used this machine in order to remove spent fuel from the core and load new fuel into it. The machine was guided by remote control with the aid of a huge calculator. Operators fed the calculator perforated paper control tapes which contained coded instructions that told the machine which channel to load or unload, and how many fuel rods to load it with. The machine then executed these instructions automatically. Photograph by Michel Brigaud and Marc Morceau. Source: EDF Photothèque.

For Saint-Laurent employees, the accident in their plant enacted the crisis in the gas-graphite program. Repairing the reactor became their way of handling both disasters. Doing so required a complex conflation of technological and cultural work.

Site employees routinely used the word "pollution" to describe the radioactive contamination of their reactor. On the most obvious material level, this pollution threatened the proper functioning of the reactor.[133] The pollution also posed a threat to a fundamental basis upon which Saint-Laurent employees constructed their identities as nuclear

Figure 8.7
A 1966 aerial view of Saint-Laurent 1 under construction. Photograph by Michel
Brigaud. Source: EDF Photothèque.

workers, engineers, and managers: it cast doubt on their ability to control the reactor. Meanwhile, the decision to terminate the gas-graphite
system threatened their place in the great story of French technological
glory. If the gas-graphite program was no longer at the forefront of the
French nuclear program, then they were no longer at the forefront of
high technology work and therefore no longer pioneers. Finally, the accident seemed to prove that EDF's top managers had been correct to judge
the gas-graphite program unsuitable for further development, and thus
it seemed to validate Boiteux's decision at the very moment he
announced it. The best way for workers and engineers to meet these heterogeneous threats was not to go on strike but to clean up the reactor.
The reactor was not only in technical danger; it was also defiled by the
implication that it could not perform its electricity-production duties

properly. A quick and effective cleanup would restore its functionality and its reputation. On another level, the cleanup would serve a psychological function, providing a means for employees to redeem their skills. Thus the technological and cultural dimensions of the cleanup were inseparable. If the workers failed to repair the reactor, or did so poorly or with many casualties, then the cleanup would not help them confront threats to their cultural identities. The most challenging cleanup operation in the history of nuclear power would reaffirm their solidarity as nuclear employees, restore their identities as pioneers, and make sense of the decision to build no more gas-graphite reactors.

Even before the cleanup began, site employees attempted to prescribe its meanings. When reporting on Boiteux's speech, the engineer-editor of the site's newsletter did not refer directly to the termination of the gas-graphite system. Instead, he asserted:

The incident of October 17. . . does not cast doubt on the [operational] principle of our reactors, but it does show that industrial certainty does not exist. There was much talk after the visit of our director-general and the breakdown of the reactor. Terms like design competitiveness, national independence, and foreign offensive were abundantly used. It is normal that each of us should express himself freely about in-house projects or plans, but this should be done without passion, for nothing is certain in technological or economic [matters]. At Saint-Laurent, the time has arrived for repairs, and we will be judged according to the role that we have to play there. The endeavor is sizable, but it will be useful to all regardless of which "nuclear system" is chosen.[134]

Clearly, the writer of this passage was trying to minimize the damage by suggesting that the accident did not threaten the working principles of their reactor and that employees should temper the rage they felt about the discontinuation of their design. Rage served no purpose, and now only their success in repairing the reactor mattered.

Venturing Inside the Reactor

Cleaning the debris under the core posed a particularly thorny problem for the engineers in charge of designing the cleanup operation. There was no passageway to the mezzanine, where much of the debris was located. At first the engineers thought about building a special remote-control device that they could lower down into the mezzanine through the damaged channel.[135] After considerable debate, however, they decided that such a device would cost too much and take too long to build. Instead, they decided to send people directly into the space under the reactor core to clear the debris and the contamination.[136]

The radiation level in the mezzanine was so high that engineers estimated that a single hour there would expose workers to between three and six times as much radiation as they were normally allowed in a year. Engineers decided that no single employee should spend more than 12 minutes in the contaminated zone. This limit, coupled with the extremely dangerous conditions of the work space, meant that every movement would have to be meticulously planned.

Three elements were essential in this planning: dress, motion, and space. Because of the high levels of radioactive contamination, dressing properly involved donning multiple layers of shining white garments, wearing radiation detectors on various parts of the body, and hooking up breathing and communication apparatus.[137] Even wearing all this equipment, however, men could not expect to stay in the work space longer than a few minutes. And much needed to be done while they were there. They had to remove the arm of a remote-control device that had fallen to the bottom of the channel during a previous rescue attempt, clear and scrub the flooring on which the reactor core rested, scrub the cells around the melted channel, and more.[138] Motions therefore had to be carefully choreographed and rehearsed on a replica.[139] Finally, the space in which the "intervention" would occur had to be prepared. A tunnel had to be built, and ventilation, lighting, signals, intercoms, and television cameras had to be installed. Such arrangements notwithstanding, the conditions in this space remained harsh: in addition to high radiation levels, the temperature hovered at around 35–40°C (95–104°F), and the air circulation was very poor.[140]

By April 1970, these preparations were complete and the time to begin the cleanup had arrived. Figure 8.9 shows the space at the entrance to the tunnel where workers prepared to enter the contaminated zone. From the lock chamber, a worker crawled up through the vertical tunnel, using pitons and other equipment. Once at his workplace, he spent roughly 10 minutes performing the motions he had rehearsed so carefully in the replica. These might involve removing a chunk of debris, scrubbing a surface, or any of a number of other small tasks. When his allotted time was up, he then towed whatever debris he had removed back down the tunnel with him, dropped it off in its designated spot, and removed several kilograms of clothes and equipment from his body. Once he left the tunnel, the next worker could enter to perform his tasks. In this fashion, workers succeeded one another in "interventions" which lasted two or three hours each. Each working day consisted of two such interventions. The entire operation took three weeks. Approximately 300 people participated in the operation in some capacity. About 100 actually entered the reactor.[141]

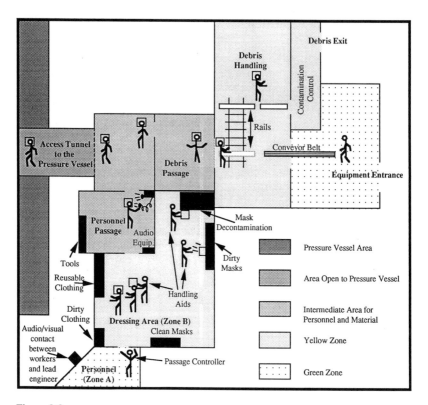

Figure 8.8
The entrance area to the tunnel. In zone A, someone monitored entrances into and exits from the reactor. Workers arrived in their standard work outfits: T-shirt, jacket, pants, socks, and tennis shoes, all made of white cotton. Here they donned additional clothing: two pairs of cotton overshoes, a pair of long-sleeved cotton gloves, a pair of long-sleeved vinyl gloves, and a pair of vinyl leg coverings that came up to the knees. They picked up two kinds of radiation detectors (dosimeters and film badges) and proceeded to zone B. There, each worker received a mask with a filter hooked up to an air supply and equipped with a microphone and a tiny speaker to allow him to communicate with the men watching him on TV monitors. He then put a white cowl over his mask and added a white overcoat with a hood that fitted over the cowl and mask. A team of dressers sealed the seams of his outfit with adhesive tape and stuck radiation detectors all over his body: two on his head, one on his chest, one on his wrist, one at his crotch, and an additional detector somewhere else on his body (which would sound an alarm if it registered a radiation dose over 2.5 rems). Thus equipped, the worker then entered a lock chamber where he got pressurized (the reactor vessel was not at atmospheric pressure). Off to the side of the lock chamber was another set of spaces through which the equipment that the worker needed entered the chamber (and through which the contaminated debris that he removed left the reactor). Sources: M. J. Grand, and M. J. Hurtiger, "Aspect de radioprotection pendant les interventions de Saint-Laurent-des-Eaux," *Bulletin de l'Association Technique pour la production et l'utilisation de l'Energie Nucléaire* 91 (1971): 38–53; Centrale de St. Laurent des Eaux (Electricité de France, GRPT C), "Etat d'avancement des études et travaux, planning au 1er juin '70," Dépannage du réacteur SL1, Rapport 13. Drawing by Carlos Martín.

The physical conditions and motions of the cleanup cannot be understood without also examining the language used to narrate and explain the process. Employees used these narratives to articulate the meanings of their motions and to assert their status in the French technological adventure. The most extensive and coherent of these narratives, titled "Great Spring Cleaning" and published in the site's newsletter, is well worth quoting in full:

> This is truly a rescue [mission], and doubtless this is why those involved in the cleaning of the support structure work with a zeal and courage worthy of admiration.
>
> On one side, there are those who "dress up to go"; on the other, those who stay to help and monitor.
>
> In the dressing room, the latter fuss over the former, turning a clasp that was pointing in the wrong direction, adjusting a wayward buckle on one of many tubes, checking everything scrupulously. It's a moving moment. Through the masks and the cowls, one can detect a certain apprehension, fleeting but nonetheless real and quite understandable.
>
> The operation itself begins. A lapse of time that seems very long goes by before a sound link, then a television link is established.
>
> This is where the essence of the operation lies:
>
> On the one hand, the main actor, looking like an astronaut, who has just played mountain climber to hoist himself onto the support structure and who now crawls as best he can, like a spelunker! On the other hand, those in charge of monitoring, who follow the operation extremely attentively, offering advice and recommendations.
>
> It is difficult to explain what stands out in this spectacle, because it is always difficult to translate how looks, gestures, and words contain sympathy and kindness.
>
> This teamwork, accomplished with so much enthusiasm and great team spirit, can only end in complete success, which everyone hopes will come soon.[142]

The astronaut metaphor evoked the ultimate male pioneer: the man who entered a space not made for men, who crossed a frontier previously thought unattainable, who shone as a symbol for the whole world of what other men could accomplish. Mountain climbers and spelunkers were also respectably male heroes. They too performed difficult physical feats under extreme conditions, and they did so with "courage." Equating the nuclear workers with symbols of heroic masculinity simultaneously reasserted and constructed the pioneering nature of their work.

The event itself was construed as a "spectacle," an enthralling performance that captivated performers and spectators alike. The "main actor" stood at the center of the show. His actions propelled the plot forward, and his predicament generated the emotional tension. The supporting characters fussed over him and sustained him in his trial. The emotion conveyed by the performance was subtle and elusive, contained in "gestures" and "looks," but the message of community and solidarity was clear enough. The participants were bound to one another by "sympathy," "kindness," and "team spirit." They formed a team, and belonging to a community involved in a common project filled them with "enthusiasm" and "zeal." The enormity of their task might cause "fleeting apprehension," but solidarity made them fearless. These images and meanings were repeated in many accounts of the cleanup, both before and after the mezzanine intervention.[143] The solidarity evoked by the process was such that not even the CFDT, the labor union most concerned with workplace health and safety, raised the slightest protest over the methods.[144]

Clearly, cleaning up Saint-Laurent 1 was not a purely technological event. It involved transforming physical motions into culturally and politically meaningful acts. The CEA strikes brought engineers and workers together to construct alternative technological scenarios and to contest the techno-economic practices of light-water's proponents. The Saint-Laurent cleanup brought (a different group of) engineers and workers together to construct alternative meanings for the termination of the gas-graphite program and for their role in the national order.

Twenty years later, the ways in which workers talked about the cleanup show how extraordinary the experience was for them and how deeply it etched some of these meanings in their minds. The associations between the accident and the abandonment of the gas-graphite program remained clear for all of them. As one worker put it succinctly, the accident came at a "politically unfortunate" time.[145] Another man mentioned a rumor, which had circulated right after the accident, that Boiteux's announcement had indirectly caused the accident by making workers too jittery to concentrate properly.[146] In retrospect, the men involved experienced the cleanup as the last hurrah of the gas-graphite program, the last time they felt special. It marked the moment when everything changed.

The Battle Fizzles Out

After the CEA strikes, the *guerre des filières* faded quickly from public view. In January 1970 Marcel Boiteux shared the latest PEON report with EDF's board of directors. This report essentially reiterated the points outlined in the "plan of action" sketched the previous year and specified that the

Fessenheim site would house light-water, not gas-graphite, plants. More as a matter of form than anything else, the union members of the board objected that the termination of the gas-graphite program had not been finalized. Their arguments were futile, however, and soon union members turned their energies to struggles they thought they could win. For the rest of 1970, debates continued to rage on the board over the contracting and organization of the Fessenheim projects. Confrontations over the design itself, however, had ceased.[147] The gas-graphite reactors in service or under construction would continue to function, but no more would be built.

*

I moved to France for the first time in 1975. One of my most vivid memories of the cultural landscape from that period was an advertising slogan that seemed to be everywhere—on suburban billboards, in newspapers and magazines, on the radio, and on television: "En France, on n'a pas de pétrole, mais on a des idées" ("In France, we may not have oil, but we have ideas"). My parents and I found this a wonderful phrase. Repeating it and adapting it to different circumstances became a favorite game. At the time, of course, I neither knew nor cared that the phrase was part of EDF's advertising campaign for its light-water reactor program. The other slogan for this campaign was "Tout électrique, tout nucléaire" ("All electric, all nuclear").

Just five years earlier, the *guerre des filières* had ended with a decision to build light-water reactors with an American license. Between 1970 and 1973, EDF broke ground for four Westinghouse-licensed reactors—a "modest" number, as prescribed by the 1970 PEON report. But any impulse to remain modest disappeared during the 1973 oil crisis. In March 1974, Prime Minister Pierre Messmer announced a new energy plan calling for the launch of thirteen 1000-MW light-water reactors within two years. By 1989, when I began my research, France was obtaining more than 70 percent of its electricity from pressurized-water reactors, and engineers were eager to tell me how the light-water system had become *francisé*—Frenchified.[148]

Terminating the gas-graphite program and buying a license from Westinghouse involved a profound rearrangement of industrial and institutional relationships. This had deep repercussions for reactor designers, builders, and workers, whose roles and skills had to change to accommodate the licensing agreements and the new technology. The licensing agreement specified work and safety guidelines that sometimes conflicted with existing practices. The new prescriptions for work practices affected not only those operating the new reactors but also those operating the older gas-graphite reactors.[149] And as the new reactors went up in the 1970s, the first wave of anti-nuclear protesters contested their construction.[150]

The triumph of the light-water design marked the ascendance of the men Robert Frost has called "economist-managers" and James Jasper has called "cost-benefiters": men who measured technological success by purely economic criteria. It also marked the successful reformulation of EDF's technopolitical regime into one that privileged selection criteria defined by economists and took "public service" to mean the support of *private* industry efforts to become profitable on international markets. Light-water reactors were the technopolitics through which these ideologies became policy. Building light-water reactors with an American license meant advocating a France that would evaluate itself in terms of comparative economics—a France measured on a scale whose increments were defined by international institutions and conglomerates.

As we have seen, choosing the light-water system over the gas-graphite system was itself far from a purely economic process. Neither was it purely technological or purely political. It was all these things. The process involved not only competing conceptions of France (independent vs. interdependent) but also complex, ongoing redefinitions of which technological trajectory best embodied those conceptions. The light-water system was either the instrument of American imperialism or the path to French radiance through industrial exports. The gas-graphite system was either the eternal guarantee of national independence or a route to technological and economic obsolescence. The uncertainties inherent in the still experimental breeder system filled it with technopolitical possibility: everybody could agree that it represented France's future, even if they could not agree on what that future should be or how to get there.

In outlining these technological trajectories, the participants in the battle pursued three related strategies. The first involved defining the proper context for technological development and the relationship between context and development. Thus, gas-graphite advocates—including Jules Horowitz, Claude Tourgeron, and CEA labor militants—insisted that the relevant context was the nation. The same technological choice would play out differently in the United States than in France. In the United States, light-water development worked because of contextual conditions that did not apply in France. In the United States, inexpensive enriched uranium, large conglomerates, pollution regulations, pricing structures, and the vastness of the nation made for economies of scale in reactor manufacturing. Meanwhile, France had a need for national independence that—pending the development of breeder reactors—only the gas-graphite system could meet. Advocates of the light-water system ignored these definitions of context and created their own: the international market, a context populated by large conglomerates. To succeed in this domain, France had to develop its own conglomerates, and that would happen only with the jump start provided by the purchase of an American license. French companies could thereby form consortia that would benefit from the experience acquired by American companies without having to incur massive technological and financial risks.

The second strategy involved interpreting the significance of uncertainties in the data used to compare the two technological trajectories. These uncertainties included the lack of significant operational data for light-water reactors, the future performance of the CEA's new fuel rods, fluctuations in the source and price of reactor fuel, the reliability and longevity of reactors (which affected their amortization rates), and potential cost overruns. Advocates of each system claimed that the ambiguity generated by these uncertainties favored their system.

This ambiguity, in turn, prompted the third strategy pursued by both factions: the definition of the appropriate selection criteria. Which combination of possible criteria—technological, economic, or political—should guide the final choice? And how should each type of criteria be weighted?

As we have seen throughout this book, developing and operating the gas-graphite system involved continually associating technology and politics. At the most basic symbolic level, this meant that gas-graphite reactors had come to incarnate the French nation: it was thanks to them that France could fuel its nuclear *force de frappe*, and thanks to them that the country could aspire to energy independence. The gas-graphite system thus enabled a radiant and technological France, the only truly French France. Charles de Gaulle, the nation's biggest hero, stood by these associations.

Clearly, attacking a system that continued to incarnate the French nation would lead nowhere. The only way in which light-water advocates could imagine breaking this powerful association was by rhetorically separating technology from politics. Such a separation undermined the links between gas-graphite technology and the nation. Equally important, excluding politics from technological choice privileged economic selection criteria. Admittedly, the data that constituted these criteria were uncertain. But light-water advocates subjected this uncertainty to quantitative analysis in order to claim control over it. Gas-graphite advocates subjected the uncertainty to qualitative—specifically, political— analysis in order to do the same. As long as de Gaulle remained president, this qualitative reasoning held. Once he stepped down, quantitative reasoning took over. The triumph of light-water meant that it came to be defined as the "economic" system, while gas-graphite became the "political" system.

The effort to separate technology and politics continued in the decade that followed the *guerre des filières*. Consider these retrospective accounts: "The termination of the gas-graphite system was not a political decision but a technological decision; it was a mistake to call it a political decision; a political decision would have consisted of maintaining gas-graphite. The end of gas-graphite was justified by two reasons: its operation was unsatisfactory, and export was very difficult."[151] Separating technology and politics required program leaders to disentangle the gas-graphite system from French identity, an effort which they kept up well after the war's end: "We finally decided in favor of the American system after having lost four years. . . . The explanation [for this waste of time] is purely political. The so-called national system was opposed to the so-called American system. . . . What does that mean? Was it forgotten that in conventional oil-fired plants there are also American licenses?"[152] Questioning the nationality of these systems undermined the legitimacy of the gas-graphite system as a symbol of French glory. Witness the response of a former CEA official to Philippe Simmonot's question about the "French system":

> Oh! It's not as French as all that.
> Technologists had convinced politicians of the value of this system, which was in part copied from the English. And these politicians had become even more avid. . . .
> The French system has two serious defects. First, it uses metal uranium, which is an unstable material and less safe than enriched uranium, for example in the case of fire. Look at what happened at Windscale (Great Britain). Then, the use of gas poses difficult problems; you have to install a continuous loading system, while with [light-] water reactors you can open the pressure vessel just once a year.[153]

Rather than portray British gas-graphite reactors as *similar* to French ones, this speaker alleged that the French had *copied* elements of the British system—a far less glamorous picture, and one that subverted the technical value (and therefore the symbolic value) of the gas-graphite system. The reference to Windscale made it appear as though the accident had provided a reason to stop gas-graphite's development; it elided the fact that the accident occurred before EDF1 was even completed, and was well known to gas-graphite engineers at the time. The former CEA official also made continuous loading appear to be a requirement rather than an option. The two seemingly pure technical reasons for the failure of the gas-graphite system blurred history. Thus the move to separate technology and politics was closely tied with efforts to disentangle the gas-graphite system from French identity and to create a technologically determinist explanation and outlook.

The move to separate technology and politics succeeded only at one rhetorical level. The very effort to disentangle the gas-graphite system from French identity meant that the discourse of nationalism continued to matter in the nuclear program. During the *guerre des filières*, French identity was not removed from reactors altogether; instead, it was transferred onto breeder reactors. Eventually the discourse of nationalism crept back into the light-water program as builders and developers began discussing the *francisation* of the system. The emphasis had merely shifted from making a French technology to making a technology French.

Meanwhile, the effort to separate technology from politics was by no means uncontested. Unionized CEA employees, Saint-Laurent workers and engineers, EDF labor leaders, and design engineers in both regimes all challenged the exclusion of politics from technological choice. The effort to separate technology and politics was a strategy to gain dominance over programmatic choices. Resisting that dominance involved resisting the separation. It meant, indeed, rehearsing the conflation of technology and politics—through strikes, comparative analyses of the two systems, and the repair of a damaged reactor. The victors tried to invent a technological determinism by defining a context in which there *was* such a thing as a single best technology and by defining new standards for "best." The losers resisted that determinism by asserting the technopolitical nature of their system and by continuing to treat their technologies as hybrid entities through which men wrestled for control over their lives and their nation.

Conclusion

In 1996 I was invited to speak at a three-day conference celebrating the fiftieth anniversary of Electricité de France. The talks took place in the Louis Armand Hall[1] of the Museum of Science and Industry in La Villette. The venue would have appealed to Armand's esthetic sense. Elegant canoe-shaped fluorescent lamps, each lined with emerald green along one edge, graced the walls. The stylish charcoal gray chairs had their own audio hookups, which piped simultaneous translations to the audience. The museum, with its geodesic dome and its light, airy architecture, was exactly the sort of thing the members of the Groupe 1985 had in mind when they said that modern French technology could "engender its own beauty." La Villette's 1996 advertising campaign suggested that the links between technological prowess and national radiance—between technologies of the present and monuments of the past—are maintained as actively now as they had been three decades earlier. All over the subterranean passages of the Paris subways, tourists and commuters saw posters that juxtaposed images of the museum's dome with images of Notre Dame and the Arc de Triomphe.

Technologists of the 1990s continue to link technology and French radiance. In the closing speeches at the commemorative conference, Edmond Alphandéry, EDF's new president, affirmed that the utility's success was "recognized by the French as well as by the rest of the world." Technological prowess, nationalization, the state, and French grandeur: these were all part and parcel of the same thing, embodied in the "world's leading firm in the electricity sector," in "one of France's largest exporters." Minister of Industry Franck Borotra amplified these themes. "France," he declared, "has become the leader of sustainable development. Today, EDF is the symbol of the reconciliation of ecology and growth." Recalling the language Charles de Gaulle had used to talk about the Plan, Borotra maintained that EDF, in its unflagging mission

The Saint-Laurent site in 1974. In this EDF photograph, the two Saint-Laurent gas-graphite reactors appear in the background; in the foreground, we see a man in an old boat, fishing in the Loire in the "traditional" way. This photo was also probably meant to reassure people that fish taken from the Loire were as safe to eat as they had always been. Source: EDF Photothèque.

of public service, had an "ardent obligation" to the nation. He praised France's handling of nuclear waste and reactor decommissioning and averred its commitment to nuclear power. Responding to widespread concern that President Jacques Chirac's enthusiasm for the European Union would lead to the privatization of EDF, he invoked the utility's history: "EDF identified almost perfectly with the spirit of the Liberation and the Reconstruction. . . . Fifty years after its birth, EDF is more than ever the instrument of the nation." Then and now, the utility would "remain public"; "the government's resolve [would] not waver on this matter."[2]

The historians sitting with me groaned, squirmed, and shook their heads throughout these proclamations. They seemed embarrassed by such unabashed displays of national pride. They also appeared puzzled by my behavior. Why was I frantically taking notes and grinning so happily? Indeed, these speeches delighted me. The issues surrounding the

Windsurfers in front of the new cooling towers at the Saint-Laurent site in 1979.
Source: EDF Photothèque.

nuclear program had evolved over the intervening decades, but the basic images and interpretive framework remained the same. The heir to the technopolitics whose development I had traced during eight years of research was being staged right before my eyes!

Further evidence of the persistence—and transformation—of this technopolitics could be found upstairs, in the museum itself. The Commissariat à l'Energie Atomique had celebrated its fiftieth anniversary the previous year, and its commemorative exhibit was still on display, arranged along a curved walkway at one end of the museum. At the bottom of the curve, visitors could learn about the CEA's early history. De Gaulle, Joliot-Curie, and Dautry were all there, along with the standard foundational narrative, recounted in the present tense:

The CEA is born of a political demand: the independence of France in the domain of energy supply. . . . Despite the difficult context—the country must be reconstructed—the CEA receives considerable funds right from the beginning, as well as considerable autonomy of action. Means and skills unite around several great names of nuclear research, leading to a research institution capable of making up for the lag experienced during the war in just a few years.

The G2 reactor also took its place in this official history, accompanied by a now familiar description: "Located at Marcoule in a building large enough to hold three Arcs de Triomphe, G2 goes critical in June 1958. The first reactor hooked up to EDF's network, G2 marks the encounter between nuclear research and industry." Was it simply reflex that made the commentator gloss over G2's plutonium production nearly forty years after France officially embarked on a nuclear weapons program? It certainly could not have been secrecy, since at least a quarter of the exhibit displayed French military nuclear achievements.

In any case, the gas-graphite program received little attention. A single panel told the story of its demise: When EDF and the state embarked on a full-scale nuclear power program, they decided that the French system was not competitive, and it was abandoned. In the 1970s, Framatome, a corporate affiliate of the CEA, began building Westinghouse-licensed reactors. The license contract expired in 1984, "after the builder completely Frenchified the new plants." The rest of the exhibit covered the CEA's recent research and offered ample assurances about the safety of nuclear plants and the benign nature of radioactivity.

EDF also sponsored an anniversary exhibit at La Villette, entitled "An Electric Life." In contrast to the conference, this display elided the institution's history altogether. A few turn-of-the-century electrical appliances occupied one corner of the hall. Otherwise, the exhibit focused on contemporary electricity. Modern appliances were suspended in midair. Captions made statements like "electricity: it brings daily comfort; it changes lives." A map displayed France's entire distribution network, giving visitors a chance to apprehend their nation through electricity. A young man standing in front of a scale model of a light-water plant asked visitors whether they would like him to explain how it worked. Another model represented EDF's latest nuclear plant, N4; its caption made the gas-graphite system disappear altogether, alleging that N4 went beyond the Frenchification of a Westinghouse license, representing "the *first* stage of entirely French design."

The nation's nuclear industry has undergone dramatic transformations since the period covered in this book. Proportionally, France is now the world's largest producer of nuclear energy. It derives 75–80 percent of its electricity from nuclear power, and even exports electricity to neighboring European countries. A reprocessing facility in La Hague treats nuclear waste from France, Japan, Switzerland, Germany, Belgium, and the Netherlands. There can be no question that France has attained the goal articulated by the technologists of the 1950s and the 1960s: it has become the world leader in nuclear power. True, the rest of the world no longer views nuclear power in quite the same light—but one could easily forget this while sitting in a high-speed train powered by nuclear-generated electricity, zooming past the nuclear plants that dot the banks of the Loire and the Rhône.

Ironically, France's nuclear triumph came at the expense of the "French system." Yet, in a sense, this too has been forgotten. Only a few years after establishing the licensing agreement with Westinghouse, French engineers proceeded to "Frenchify" the light-water design. The lure of American technology did not last long; ultimately the French technologists only redoubled their efforts to make their large-scale systems French. This "Frenchification" has entailed the rhetorical erasure of the original French system—so much so that in 1996 an employee of EDF's own archives insisted to me that there had been no nuclear program before 1970!

It may be in part because of this erasure that the engineers and workers who built the gas-graphite program look back on it with such fondness. Nostalgia has preserved, probably even amplified, their memories of the "pioneering spirit" that pervaded much of the program forty years ago. They were on a national mission, the success of which ended up entailing the failure of their program. Perhaps this is why their recollections sometimes conveyed the sense that they had made sacrifices for their country.

In one sense, though, they did not fail. Technological prowess has staked a firm claim as a basic element of French national identity. At least rhetorically, the builders of the high-speed train (the TGV), the Minitel communication system, the Concorde airliner, the Ariane rocket, and numerous other technological systems continue to cultivate the association between technology and French radiance—even when these systems are developed in cooperation with other European nations. Only more research can determine whether and how the design and operation of the systems themselves articulated such associations in a manner analogous to the nuclear program of the 1950s and the 1960s.

Imagining a Technological Nation

Clearly, however, the nuclear program was a site for articulating and nego-tiating the meaning of a technological France. The image of a radiant and glorious France appeared repeatedly in the discourse of engineers, administrators, labor militants, journalists, and local elected officials. These men actively cultivated the notion that national radiance would emanate from technological prowess.

Linking technological prowess and national identity was a complex, multidirectional process. Technologists, labor militants, and elected offi-cials invoked apparently eternal characteristics of the nation, which at the most general level were qualities they could all agree made France French: radiance, glory, and grandeur. They simultaneously suggested that France had lost these things through wartime defeat, and/or postwar decolonization, and/or general economic and industrial backwardness. This, in turn, implied that France was no longer fully, truly French. In the scenarios these men envisaged, technological development would restore Frenchness to the nation in a way that made them—as men of action, as heroic male workers and militants, as representatives of their regions— central players. At the same time, they repeatedly invoked the nation in efforts to arbitrate disputes and to legitimate their scenarios. Thus the nation (and/or the national interest) justified particular forms of tech-nological development, while technological prowess defined the nation. This circularity bound conceptions of the nation and of technology more tightly together. Furthermore, the fact that these links were so widely artic-ulated gave them strength and flexibility.

Indeed, the general principle of a technological France drew strength from its multiple manifestations. In some respects, these manifestations supported one another. In both the Gard and the Touraine, for example, local elites and technologists *together* represented nuclear development as a glorious spectacle. Each group had different ideas about the meanings of technological France. Local elites focused on how nuclear sites would bind their region to the nation both economically and culturally, whereas nuclear technologists focused on how reactor development would enact French independence and place them in a position of political and/or industrial leadership. Although different, these visions were compatible; they did not undermine or even compete with each other. In the specta-cle these men co-produced, regional history, national destiny, and tech-nological development all worked together on several levels. Other images of technological France interacted or intersected in parallel ways—for example, those of CGT and CFTC/CFDT labor militants and EDF engineers (especially up until the mid 1960s), or those of CEA and EDF engineers during the *guerre des filières.*

At the same time, though, the very multiplicity of "technological France" made that notion into contested terrain. Ideas about the nation could divide as well as unite. So, for example, while technologists at the CEA and EDF both cultivated ideologies of public service to a technological nation, from the mid 1950s to the mid 1960s they articulated different ideas about what that nation should be, and how best to serve it. They did work together to establish the nuclear program as an arena for defining France's future and identity. But they had competing definitions of the public interest and of the nation's future, which they translated into two distinct technopolitical regimes. The CEA's *nationalist* regime found form in its Marcoule reactors and its "policy of champions." EDF's *nationalized* regime found form in its Chinon and Saint-Laurent reactors and in its early efforts to control the development of private industry through its contracting practices. Each technopolitical regime developed distinct ideas about nuclear and industrial policy, which were simultaneously distinct prescriptions for the nation's future.

Technologists thus sought to define the nation through the example and action of their regime. At the same time, they invoked the nation in discussing, formulating, and implementing their technopolitical projects. So, for example, the national interest justified manufacturing weapons-grade plutonium before the government had decided to build a bomb. After that decision, the national interest warranted extracting plutonium from EDF reactors. National pride justified using prestressed concrete for reactor pressure vessels, as well as designing EDF3 to run at 500 megawatts. French radiance—especially the notion that France had to export technology in order to maintain its status as a world power—played a major role in conflicts over industrial contracting and the overall structure of private industry.

Labor militancy and reactor work also engendered both conflict and accommodation over conceptions of the technological nation. Conflict appeared in the realm of labor union politics. Militants in the three major unions produced distinct visions of a technological France. The CGT dreamed of the glorious technological France that would follow a socialist revolution. Force Ouvrière situated France's technological future in a non-communist international community. The CFTC/CFDT saw technological change as a potential conduit to a better and more just society. None of these visions stood alone; all were produced in counterpoint with the others and in the context of union rivalries. In this sense, technological France was one of several contested terrains in union politics. Viewed alongside the future France imagined by leading state technologists, however, the three unions' scenarios had at least one point in common: each imagined a sociopolitical order that gave workers a more central and better-recognized role in shaping the nation's future. Yet, from the perspective of the Catholics or the Poujadistes, labor militants of all stripes also shared something with the state technologists: despite the differences they imagined in the sociopolitical order, they all contemplated a technological future for France. And indeed, the fact that all three labor unions sought to enroll the technical elite in their programs indicates that militants did think that their vision of technological France was potentially compatible with that of the technical elite—perhaps not the very top layer (at the level of Pierre Massé or Louis Armand or Marcel Boiteux), but conceivably up to the middle level (such as rank-and-file engineers and scientists at the CEA, like those who went on strike during the *guerre des filières*).

The dialectic of conflict and accommodation found yet another set of manifestations at the nuclear sites of EDF and the CEA. In neither case were the labor unions at odds in a significant way. Instead, the dialectic must be considered not so much across technopolitical regimes as within them. Hence the technological France prescribed by the CEA's regime was a source of conflict for Marcoule workers, who could not find a place for themselves in that vision. The nationalist military hierarchy at Marcoule privileged experts and ignored workers. In contrast, the technological France prescribed by EDF's nationalized regime formally made room for workers, according them a significant ideological and technical role in nuclear development. In the 1960s, most of the utility workers at Chinon accommodated fairly well to this vision of the technological nation. While CEA workers cast themselves in an adversarial role with respect to their regime's prescriptions, EDF workers cast themselves as pioneers on a par with their hierarchical superiors.

In 1969 these roles were replayed under somewhat different circumstances as the dialectic between conflict and accommodation acquired yet another configuration. Toward the end of the *guerre des filières*, unionized Marcoule workers joined engineers, scientists, and technicians throughout the CEA in protesting the termination of the gas-graphite system. Inasmuch as they directed this protest against the regime's top administrators (as well as against EDF and the government), Marcoule workers reenacted their familiar adversarial role. Yet joining with others at the

CEA symbolized an accommodation of sorts: though Marcoule workers felt little loyalty to the technopolitical regime instantiated in the gas-graphite system, they were nonetheless willing to defend that system because this also meant defending their jobs. Meanwhile, EDF workers at Saint-Laurent, who had also cast themselves as pioneers, came to interpret the cleanup of the accident there as a reenactment and an affirmation of those pioneering roles.

The year 1969 also provided an occasion for Gardois leaders and Marcoulins to reconcile and to once again declare a common set of interests. The Gardois had been promised a spectacular technological France, a drama of regional salvation through modernization. Their experiences, however, did not reflect this dramatic new nation. Instead, technological France seemed invasive and suffocating. Even the local leaders who had helped to produce the initial spectacle expressed dismay. But when the termination of the gas-graphite program threatened to remove the Marcoulins from the region, Gardois leaders realized that, for better or worse, their region's infrastructure had become dependent on the CEA. At the same time, they recognized that some cultural cross-fertilization had occurred. Though their place in it remained uncertain and conflicted, the technological nation had definitively arrived in the Gard. The events of 1969 made little difference to the Tourangeaux, whose experience, on the whole, tended to match their expectations. And though they did not yet know it, their region stood on the verge of even greater nuclear development.

Meanwhile, 1969 and the *guerre des filières* reconfigured disputes among nuclear technologists over the meaning of the technological nation. The *guerre des filières* showed just how slippery and malleable the concept of the national interest could be. Technopolitical regimes and visions of technological France were rearranged during that conflict. Top administrators at EDF and the CEA began to define the national interest in terms of economics, corporate development, and international markets. Engineers, technicians, and workers at both institutions continued to frame the national interest in terms of technical distinctiveness and energy independence. Once again, "the nation" legitimated competing technological trajectories, just as those trajectories articulated conceptions of the nation.

In 1969, rearranging the meanings of technological France also meant reconfiguring claims about the relationship between technology and politics. During the nearly two decades of gas-graphite development, enacting scenarios for a technological France had meant the deliberate, conscious interweaving of technology and politics. In the process, the gas-graphite system had become an incarnation of the French nation. The only way to unseat the system was to attack the conflation of technology and politics—at least on a rhetorical level.

Technology and Politics

The relationship between technology and politics has interested scholars for a long time. In the past decade or so, most research has proceeded on two related fronts: (1) examining how politics shape technological design and development in particular historical or sociological contexts and (2) identifying the ontological relationship between technology and politics in those contexts. In this book I have explored these avenues, but I have also pursued two other questions: How do technological artifacts and practices, both in the process of being designed and after the completion of their design, function as forms of politics—as political negotiation, action, iconography, and rhetoric? And how do the actors we study 1conceptualize the relationship between technology and politics?

I have argued that technologists—defined broadly to include engineers as well as top administrators of industrial state enterprises, regardless of technical training—created distinct technopolitical regimes in the pursuit of nuclear development. These regimes consisted of linked sets of people, engineering and industrial practices, technological artifacts, political programs, and institutional ideologies, which acted together to govern technological development and pursue technopolitics. Time and again, a key component of technopolitics was the manipulation of flexibility and uncertainty. Flexibility in the basic principle of gas-graphite reactors meant that they could produce both plutonium and electricity. How well they did one or the other depended on the specific design. But the fact that they could do both made possible the production of weapons-grade plutonium in Marcoule's reactors before the government officially decided to build an atomic bomb. This flexibility also made it possible for the CEA to demand plutonium from EDF's reactors: thus technologies could not only enact political agendas but also make possible new political goals.

The manipulation of uncertainty also played a key role in technopolitics, in instances such as the definition of the competitive nuclear kilowatt-hour. Perhaps the most striking use of uncertainty, though, occurred in the *guerre des filières*. There, uncertainties included the lack of significant operational data for light-water reactors, the future performance of the CEA's new fuel rods, fluctuations in the source and the price of reactor fuel, the reliability and longevity of reactors, and potential cost overruns. Advocates of the light-water system claimed that some of these uncertainties—the most relevant ones, in their opinion—could be quantified. Quantification would remove all ambiguity and would make possible a clear choice (in favor of the light-water system). Gas-graphite advocates did not think that the ambiguity generated by these uncertainties could be so easily erased. They argued that this ambiguity militated in favor of qualitative judgments.

In developing the gas-graphite system, technologists in both regimes deliberately conflated technology and politics. This conflation was itself a strategy, and it operated outside the nuclear program as well as within it. Recall the elaboration of the multi-year nation plans or the discourse of labor militants—both instances in which the conceptual conflation of technology and politics defined a way for planners or unions to shape the nation's future. Within the nuclear program, technologists who effected this conflation gave themselves permission to shape policy not just in the nuclear arena but also in the broader arenas of military and industrial development. This is not to say that technologists were the only policy makers in these arenas—clearly there were others—but rather that conflating technology and politics served technologists as a strategy for acquiring legitimacy as policy makers. In addition, politics and policy making gave the reactor projects significance, both within each regime and in the interactions each regime had with its surroundings. For example, EDF1 was important not because it itself would produce economically viable electricity but rather because it constituted the first step in a nationalized nuclear program that would enact and strengthen the utility's ideology and industrial contracting practices. In this instance as in many others, EDF1's technical characteristics were inseparable from its political dimensions. Had EDF1 failed to function properly, or had engineers and workers been unable to garner adequate operational experience from the reactor, the plant would have failed both technically and politically.

Conflating technology and politics created a major resource for engineers. In the debates over industrial organization, for example, EDF engineers reshaped the political meanings of their contracting policy in order to make it fit the priorities of the Fifth Republic. Under de Gaulle's regime, the conflation of technology and politics ultimately provided the gas-graphite system with its most powerful defense. As long as the identification of the gas-graphite system with national independence and identity held, the French system remained unassailable. In sum, conflating technology and politics delineated an arena of action for gas-graphite technologists and created a defense for the system they produced.

In arguing for quantitative selection criteria, light-water advocates simultaneously sought a rhetorical separation of technology and politics. This separation was every bit as much a strategy as the conflation effected by gas-graphite's developers and advocates. This separation entailed aligning quantitative measurement with technology and economics, and qualitative judgment with politics. It also entailed some redefinitions: the redefinition of "politics" as irrational and backward-looking (which was the sense of "politics" that technologists had used to situate themselves as better qualified to make decisions than politicians) and the redefinition of "public service" as the support of the national economy through the support of private industry. Separating technology and politics made it possible to attack the identification of the gas-graphite system with the nation, and thus made it possible to attack the gas-graphite system. This meant inventing a technological determinism by defining a context in which there *was* such a thing as a single best technology and defining new standards for "best."

In sum, light-water's proponents used the separation of technology and politics in exactly the same way that gas-graphite's developers used their conflation: to delineate an arena of action and defend the system they advocated. In separating technology and politics, light-water advocates adopted what Ken Alder has called a "technocratic pose": a stance that rhetorically places technological activity above and beyond the sphere of politics and the reach of politicians.[3]

This "technocratic pose" is far more common in technological development than the deliberate and proud conflation of technology and politics espoused by gas-graphite advocates. It is particularly common in the United States. For example, Paul Edwards and Donald MacKenzie have observed that Cold Warriors in the United States spent a great deal of energy constructing discursive separations between science and technology (on one side) and politics (on the other). The successful prosecution of the Cold War and the concomitant pursuit of big science and complex technology depended on making this separation appear natural. Cold Warriors located momentum for change within science and technology. Conceptualizing science and technology as apolitical was crucial in justifying the vast resources poured into military and industrial development, as well as in legitimating specific technological choices.[4] Science and technology did take on political meanings, as scholars who have studied the politics of display in Cold War America have shown.[5] But if atomic weapons, nerve gas, the moon landing, or any number of other achievements functioned as credible evidence of American superiority, it was precisely because technology was thought to provide an objective, natural, and *inherently apolitical* measure of strength.

Nonetheless, this separation of technology and politics was itself a political strategy. It worked only at the rhetorical level. As Edwards and MacKenzie have argued, computer and missile-guidance systems were not only shaped by political goals but also used as political tools. They were, in effect, forms of technopolitics. Unlike the developers of gas-graphite, however, American engineers would not— perhaps could not—admit that they engaged in political activity through their technological work.

What made the effacement of politics in American technological development an effective strategy? Part of the reason may lie in the McCarthyite construction of "politics"—in the sense of ideologies that competed with democracy— as un-American. In the black-and-white world of the Cold War, "politics" meant what the communists did. A striking instance appears in post-1947 American commentaries on industrial nationalizations in France. Popular publications such *Business Week* as well as trade journals such as *Electrical World* portrayed nationalized French companies (particularly EDF) as dangerous communist strongholds in which politics tainted the pursuit of technological development.[6]

I made this observation in my talk at EDF's fiftieth-anniversary conference, stressing that French technologists, by and large, did not seem to want or need to separate technology and politics. I meant this point to be provocative—after all, the triumph of light-water at EDF had resulted precisely from a separation of technology and politics. But my attempt at controversy failed. Numerous EDF engineers and administrators (the primary audience for this conference) told me afterward that I had been "absolutely right" in my assessment. Indeed, as efforts to "Frenchify" the light-water design in the 1970s also indicate, the rhetorical separation of technology and politics in the French nuclear program does not appear to have lasted very long.

Of course, this is not to say that everyone in France advocated the conflation of technology and politics. As we saw in chapter 1, in the 1950s and the 1960s many French intellectuals argued strongly for a separation of the two and viewed their conflation as a threat to democracy. This struggle between social scientists and engineers over the proper relationship between technology and politics has a contemporary equivalent, crystallized in attitudes toward the work of Bruno Latour and his colleagues at the Centre de Sociologie de l'Innovation in Paris. Latour has argued that the work involved in keeping nature and culture (and technology and society) separate requires enormous intellectual and social energy, without correspondingly significant returns. It would be better, he believes, to think in hybrid categories.[7] Perhaps in part because it threatens the edifice of their theories, many (though by no means all) French social scientists dismiss this suggestion. Technologists, however, seem to find it eminently congenial. The Centre de Sociologie de l'Innovation (itself housed in the Ecole des Mines) regularly receives contracts from institutions such as the CEA, EDF, and the RATP (Paris's public transportation company) to study their scientific and technological histories, methods, and prospects.

Of course, the Cold War critics of technocracy were not entirely wrong. Certainly, the elaboration of French nuclear military policy was anything but democratic. Yet surely the road to technologies that better serve society lies along a different path from those that require a rigid and radical split between technology and politics. If for no other reason, such a separation proves impossible in practice, however attractive it may seem in rhetoric or theory. As historians and sociologists have demonstrated time and again, technologies are produced by institutions and people with stakes and interests—political, social, historical, and cultural. This is neither inherently good nor inherently bad; it simply is. Arguing that technology and politics are or should be separate serves only to obscure these interests and the struggles among stakeholders, which are part and parcel of the processes of technological development. It does not serve to produce better or more democratic technologies.

Although the stakeholders in the gas-graphite program rarely if ever resorted to an American-style separation of technology and politics, I am not suggesting that French nuclear development represents some kind of ideal. Clearly, recognizing the links between technology and politics does not *suffice*. But such recognition is a necessary first step to a deeper, broader, and more useful consideration of the social and political dimensions of technological change. There is nothing wrong or shameful about technopolitics. Technopolitics does not necessarily produce bad or inferior technology. But engineers must work within a framework that openly acknowledges the fact of technopolitics. This need not lessen their technical expertise in any way. They will remain, after all, better qualified than anyone else to build technological systems that work, and to judge which solutions can work and which cannot. Obviously, not all engineering choices are meaningfully political; nor are all technologies equally political. But many fundamental technical choices—such as choices about system design and programmatic development—have significant and inseparable political dimensions. Recognizing this is important not just for social scientists and humanists but also for engineers.

Acknowledging the political dimensions of technological change does not imply that anyone and everyone should be able to influence decisions about technological development; this would be neither feasible nor appropriate. It can, however, breathe fresh air into decision making. Acknowledging and (especially) respecting political arguments in the process of technological decision making would, at the bare minimum, create a more honest process. Developing such respect for the full range of stakeholders in technical decisions is incumbent not simply upon engineers but upon all of us, as human beings who live in a technological world.

Notes

Introduction

1. Quoted in Gildea 1994 (p. 112).

2. Quoted in "Le ministre atomique," *Normandie*, 22 October 1945. (This and all subsequent translations are mine, unless an English-language source is cited. —GH)

3. Frank 1994.

4. *L'Aube*, 17 December 1949, quoted in Weart 1979 (p. 248).

5. Quoted in Renou n.d. (p. 34).

6. Anecdotal evidence suggests that, although the notion of French radiance existed in the late nineteenth and early twentieth centuries (usually in the context of rhetoric about the French empire), "radiance" did not gain widespread currency until after World War II. "Grandeur" has a much longer history, discussed in Gildea 1994.

7. For a recent English-language summary of this crisis of grandeur see Gildea 1996. As Kuisel (1995) notes, American scholars in the early postwar years also worried about France's losing its status as a great nation.

8. On the symbolic meanings of nuclear technology see Boyer 1985 and Weart 1988.

9. Weart (1979) provides an account of French nuclear science in the first half of the twentieth century.

10. For a sampling of arenas in which the notion of Frenchness was debated and contested see Nora 1996. See also Nora 1992.

11. Eric Fassin (1995), who has labeled this type of explanation "culturalism," notes that it is "not so much a set of intellectual rules as a spontaneous practice of interpretation, which is why academics tend to ascribe it to nonacademics: culturalism as 'popular knowledge'" (p. 453). Fassin (ibid., p. 455) describes the

problem with "culturalism" succinctly: "It provides an interpretation without an explanation: the French are x or y because they are French, and the French have always been like that." He argues that, although professional historians try hard to avoid this practice by looking at how historical actors themselves conceptualize Frenchness, they can nonetheless fall into its trap by equating those conceptualizations with the actual construction of a national identity. Though I think Fassin exaggerates the extent to which the historians he cites pursue a "last-minute reluctant reconstruction" of French national identity, and I disagree that comparative national histories provide the only way out of this dilemma, his caution is important to bear in mind.

12. My use of the notion of "hybrid" has been heavily influenced by discussions with scholars at the Centre de Sociologie de l'Innovation at the Ecole des Mines in Paris. One of the more representative elaborations of these CSI ideas is Latour 1993.

13. Hughes 1983.

14. See e.g. Dunlavy 1994 or Kranakis 1997.

15. The literature is vast. For an introduction see Hunt 1989 and Berlanstein 1993.

16. On intersections of technological and cultural history see Scranton 1994. For a discussion of the reciprocity of "technology" and "culture" as analytic constructs see Lerman et al. 1997a, Lerman et al. 1997b, and Lerman 1997. I am particularly indebted to Nina Lerman for many conversations on this issue.

17. For a thorough overview of the history of technology see Staudenmaier 1985. More recent historiographical works include Hounshell 1995, Smith 1994, and Staudenmaier 1990.

18. Examples of this work include Hughes 1983, MacKenzie 1990, Noble 1984, Pinch and Law 1992, and Bijker et al. 1987

19. Some of the best and most explicit critiques of technological determinism can be found in McGaw 1987 and in many of the essays in Smith and Marx 1994.

20. See e.g. Law's essays in Bijker et al. 1987 and in Pinch and Law 1992.

21. Wiebe Bijker has proposed the "sociotechnical ensemble," a unit of analysis in which "all relations are both social and technical." We must not decide ahead of time, he argues, what is social and what is technical—instead, these characteristics derive from the process of constructing ensembles. Bruno Latour and Michel Callon propose analyzing technological systems as "actor-networks" composed of human and non-human "actants" that shape each other. All actants receive the same analytic status; electrons, for example, can act to shape a network in the same way as people. This premise can lead to an implicit or explicit claim that non-human actants have agency. The stability and success of an actor-network cannot be predicted a priori; they depend on the strength of the relationships that bind the actants together. See Bijker et al. 1987; Callon et al. 1986; Latour 1987, 1996.

22. This is a central tenet of the recent work of Bruno Latour, Michel Callon, and their colleagues at the Centre de Sociologie de l'Innovation. A significant subset of this work examines technological "users" and consumers. Some particularly interesting work includes that of Madeleine Akrich, who has used the notion of "scripts" to describe how engineers "write" social programs into their designs; when users follow scripts closely, she argues, technologies can help to define social categories such as citizenship. See also Akrich and Rabeharisoa n.d. and Rabeharisoa 1990. Bryan Pfaffenberger has proposed the metaphor of "techno-logical drama" to describe the endlessly recursive dialogue between a technology's makers and its users. He argues that the social contexts of these dramas are cre-ated alongside the technologies, and that discursively produced myths and ritu-als legitimate and reinforce the pursuit of "politics constructed by technological means." Pfaffenberger 1990 is more satisfying than Pfaffenberger 1992.

23. MacKenzie 1990.

24. Edwards 1996; Alder 1997. There has also been some good scholarship on the spread of technological images and values in American culture. This includes the work of Joseph Corn, Michael Smith, David Nye, Jeffrey Meikle, and Leo Marx. Still, focusing on culture tends to lead these scholars away from the inner workings of technology.

25. For example: Birnbaum 1982; Baumgartner and Wilsford 1994; Cohen 1992; Kosciusko-Morizet 1973; Kuisel 1981; Papon 1978; Thoenig 1987; Suleiman 1974, 1978; Jasper 1990.

26. Alder 1997 is not about the twentieth century.

27. Sewell 1993. For an overview of these debates see the other essays in Berlanstein 1993. Two excellent and more recent works that adopt this "post-materialist" stance are Downs 1995 and Biernacki 1995.

28. The interest of historians of technology in the history of labor has grown sig-nificantly in recent years. Most of this literature, however, treats the nineteenth and early twentieth centuries. Historians of the large-scale technological systems of the late twentieth century have yet to take a serious interest in workplace technologies.

29. Nora 1996a, p. xxiv.

30. Maza (1996) provides a nice overview of how narrative has become a histori-ographical category in "Stories in History: Cultural Narratives in Recent Works in European History."

31. Lebovics (1992) argues that "searches for France embody beliefs, values, pro-jets, in short, ideologies of what France should be." In a different type of scholarly endeavor (Hall and du Gay 1996, p. 4), Stuart Hall makes a similar point: "Though they seem to invoke an origin in a historical past with which they con-tinue to correspond, actually identities are about questions of using the resources of history, language and culture in the process of becoming rather than being: not 'who we are' or 'where we came from,' so much as what we might become, how

we have been represented and how that bears on how we might represent our-selves. Identities . . . relate to the invention of tradition as much as to tradition itself. . . ." On mythologizing the past see Nora 1996a and Gildea 1994.

32. Gildea (1994) discusses the ways in which the notion of grandeur was con-structed as a quintessential characteristic of the French nation through reference to past historical greatness. The related metaphor of radiance underwent simi-lar processes after World War II.

33. Anderson 1991, p. 144.

34. Sahlins 1989.

35. Ford (1993) makes this argument.

36. See e.g. Kuisel 1993; Ross 1995; Noiriel 1996a,b; Lebovics 1992. For an overview of debates about how American scholars have treated the concept of French national identity see Kuisel 1995; Lamont 1995; Fassin 1995.

37. Here I am paraphrasing Stuart Hall's introduction to Hall and du Gay 1996.

38. See especially the scholarship of Bijker and Latour.

39. Others have used this term, but without developing it as an analytical tool—see e.g. Pfaffenberger 1992.

40. On French arms exports see Kolodziej 1997 and Péan 1982.

41. Hughes 1983.

Chapter 1

1. *Le Monde*, 17 December 1957.

2. In writing about Loire Valley châteaux and historical memory, Jean-Pierre Babelon (1992, p. 431) says: "Thanks to its easily memorized silhouette, Chambord has become the paradigm of the Loire château, of which it has all the principal characteristics."

3. The history of modern French engineering has an extensive and diverse his-toriography. Among the works that have informed this overview are Kranakis 1997, Alder 1997, Belhoste et al. 1994, Grelon 1986, Kranakis 1989, Picon 1992, Porter 1995, Shinn 1980a, Shinn 1980b, Smith 1990, Thépot 1986, Weiss 1982a, and Weiss 1982b.

4. Weiss (1982a) and Kranakis (1989, 1997) offer particularly good insights into the difference between state and civil engineers and the conflicts between the two groups.

5. For example, Kranakis (1989, pp. 32–34) argues that the fact that state engi-neers controlled vast areas of technological development, together with their

social background, made their values and technical styles the standard even for civil engineers. See also Picon 1992, pp. 604–612.

6. Crawford 1989, pp. 131–132; Shinn 1980a, p. 207.

7. Quoted in Boltanski 1982 (pp. 129–130) (translation from p. 132 of Crawford 1989).

8. Guigeno 1994. See also Thépot 1986.

9. Quoted in Guigeno 1994 (p. 411).

10. Shinn 1980a, p. 190.

11. Smith 1990.

12. Porter 1995, chapter 6, passim.

13. Porter 1995, p. 121.

14. Smith 1990, p. 681.

15. Crawford, Porter, and others.

16. In some ways, this debate parallels nineteenth- and twentieth-century professionalization and identity conflicts among the strata of French engineering, in which civil engineers accused state engineers of technological backwardness and state engineers denied civil engineers access to the *grands corps* on the ground that their training lacked sufficient breadth.

17. Shinn 1980a, p. 207. Another strand of the debate comes through in Picon's (1994) contribution to a volume commemorating the bicentennial of the Ecole Polytechnique. The essay—entitled "The Ecole polytechnique, an engineering school?"—concludes that the school was and is and engineering school, but one that, because of its status and the breadth of its curriculum, has special responsibilities.

18. Kranakis 1989, pp. 22–23.

19. For a wide-ranging introduction to this scholarship see Hughes 1983, Bijker et al. 1987, and Pinch and Law 1992.

20. Picon 1992, p. 617.

21. Porter 1995, 115–118.

22. Quoted in Porter 1995 (p. 137).

23. Smith 1990, passim.

24. For an analysis of the interwar period and Vichy see Brun 1985 and Kuisel 1981.

25. Kuisel (1981) argues that what differentiated France from other nations after World War II was this sense of national decline and economic backwardness.

Indeed, fear of decline was a recurrent theme in French history; see e.g. Frank 1994.

26. For a quick overview of other forms of cultural reconstruction see "Crises of Modernization" in Forbes and Kelly 1995. On French conceptualizations of the state see Guéry 1992.

27. The politics of the immediate postwar period have been well covered by historians. In addition to the excellent account in Kuisel 1981, see Rioux 1980 and Larkin 1988.

28. Picard 1990.

29. The first head of this Commissariat Général du Plan was Jean Monnet, and the first five-year plan became known as the Monnet Plan. For more on this see Kuisel 1981, Rioux 1980, Bonin 1987, Rousso 1986a, and Pierre Massé, *Le plan ou l'anti-hasard* (Gallimard, 1965).

30. "Republics" in France refer to the instauration of new constitutions. The First Republic was the one created by the revolution of 1789. World War II and the rise of the Vichy regime marked the end of the Third Republic. When de Gaulle returned to power in 1958, he pushed through a new constitution the following year, thereby marking the beginning of the current Fifth Republic.

31. Part of the reason for this had to do with the "synarchie" affair during Vichy. On this and other aspects of technocratic movements before and during World War II see Kuisel 1981, Brun 1985, and Philippe Bauchard, *Les Technocrates au pouvoir* (Arthaud, 1966).

32. Louis Trotabas, in his introduction to the proceedings of this conference in Gaston Berger et al., *Politique et technique* (Presses Universitaires de France, 1958).

33. André Siegfried, "Le Problème de l'Etat au XXe siècle en fonction des transformations de la Production," in Berger et al., *Politique et technique*.

34. Marcel Merle, "L'Influence de la technique sur les institutions politiques," in Berger et al., *Politique et technique*.

35. Marcel Waline, "Les Résistances techniques de l'administration au pouvoir politique," in Berger et al., *Politique et technique*.

36. Cited in Hoffman 1956 (p. 256).

37. P. Le Brun, "Le point du vue d'un syndicaliste," in Berger et al., *Politique et technique*.

38. See the introduction to Berger et al., *Politique et technique*.

39. Anonymous, "Les vrais immortels," *Carrefour*, 31 December 1958: 5.

40. For Massé's version of events see Pierre Massé, *Aléas et Progrès* (Economica, 1984), p. 161.

41. Jean Meynaud, *Technocratie et politique* (Etudes de Science Politique, 1960), p. 7.

42. Ibid., p. 39.

43. Ibid., p. 64.

44. Meynaud wrote several other critiques of technocracy, including the following: "Les Mathématiciens et le Pouvoir," *Revue Française de Science Politique* 9, no. 2 (1959): 340–367 (where he actually came out in favor of more moderate approaches to cybernetics, in which cybernetic techniques might be used as thinking tools, but still argued strongly against mathematicians having any sort of decision-making power); "A propos de la Technocratie," *Revue Française de Science Politique* 11, no. 4 (1961): 671–683; "A Propos des Spéculations sur l'Avenir," *Revue française de la science politique* 13 (1963): 666–688; *Technocracy (Technocratie, Mythe ou Réalité?)* (Free Press, 1964, 1968). Since these all make similar arguments to the book discussed here, I do not analyze them in detail.

45. The sketch of the popular image of the technocrat is taken from a survey of the popular media done by Jean Touchard and Jacques Solé; see their article "Planification et Technocratie," in *La Planification comme processus de décision* (Armand Colin, 1965). For other social science treatments of technocracy see the following works of the sociologist Nora Mitrani: "Reflexions sur l'opération technique, les techniciens et les technocrates," *Cahiers internationaux de sociologie* 19 (1955); "Les mythes de l'énergie nucléaire et la bureaucratie internationale," *Cahiers internationaux de sociologie* 21: 138–148; "Attitudes et symboles techno-bureaucratiques: refléxions sur une enquête," *Cahiers internationaux de sociologie* 24: 148–166. See also Bernard Chenot, *Les enterprises nationalisées (Que sais-je?* no. 695, Presses Universitaires de France, 1956); Georges Lescuyer, *Le contrôle de l'Etat sur les entreprises nationalisées* (1959); Henri Migeon, *Le Monde après 150 ans de technique* (1958); Gabriel Veraldi, *L'humanisme technique* (1958).

46. *L'Express* has received considerable attention as a site for the production of *cadre* discourse; see e.g. Boltanski 1982, Kuisel 1993, and Ross 1995.

47. Alfred Sauvy, "Lobbys et Groupes de Pression," in Berger et al., *Politique et technique*, p. 325.

48. Ibid., p. 326.

49. Louis Armand and Michel Drancourt, *Plaidoyer pour l'avenir* (Calmann-Lévy, 1961), p. 230.

50. Dominique Dubarle, preface to Jean-Louis Cottier, *La Technocratie, Nouveau Pouvoir* (Cerf, 1959), pp. 26–27.

51. Maurice Roy, "Progrès et Tradition," *La Jaune et la Rouge* 140 (1960): 34–45.

52. Cottier, *La Technocratie, Nouveau Pouvoir*, p. 36. (Cottier was the pen name of Jean-Louis Kahn, a civil servant.)

53. Louis Armand, "Technocrates et Techniciens," *La Jaune et La Rouge* 216 (1967): 4–8.

54. Ibid., p. 8.

55. André Léauté, "Les Vertus Cardinales de l'Ingénieur de Grande Classe," *La Jaune et la Rouge* 120 (1958), p. 41.

56. Ibid., p. 42.

57. Armand and Drancourt, *Plaidoyer pour l'Avenir*, p. 107.

58. Ibid., p. 17.

59. Pierre Massé, "Propos incertains," *Revue française de la recherche opérationelle*, 2e trim., no. 11 (1959): 60.

60. Jean Barets, *La fin des politiques* (Calmann-Lévy, 1962), p. 113.

61. Péan 1995, p. 18.

62. Massé, *Aléas et Progrès*. Throughout these memoirs, Massé uses this expression to confer high praise on the reasonableness of certain interlocutors.

63. Jean Barets, "Précisions sur l'objectivisme," *Revue de Défense Nationale*, May 1963, p. 886.

64. Ross (1995) argues that the 1950s and the 1960s witnessed the creation of a new male subjectivity, which she locates in three discursive and spatial sites: the Resistance *maquis*, the university, and the corporation. The "man of action" archetype suggests that this division is a bit too neat, however. Indeed, the "man of action"—while clearly conceptualized as a member of a quite exclusive elite— occupies all three discursive spaces (though the first two more than the latter): he is a fighter and a revolutionary (hence the *maquis*), he is thoughtful and calculating (the university), and he is ultimately highly concerned with questions of economy (the corporation).

65. This is the successor of the stance that Alder (1997), speaking of engineers in the French Revolution, calls the "technocratic pose."

66. Georges Villiers, "Industrie, Technique et Culture," *Prospective* 4 (1959): 21–32. Such statements pervaded the discourse of technologists. For example, the Groupe 1985 argued as follows: "le niveau Scientifique et Technique d'un pays sera dans l'avenir, plus encore peut-être qu'actuellement, un élément essentiel de sa position politique, économique, industrielle, militaire, culturelle et que, pour tenir la place qui lui revient, la France devra consentir de grands efforts, aussi bien sur le plan financier que sur celui—peut être plus difficile, mais certainement tout aussi important—des structures" (*Refléxions pour 1985* (La Documentation Française, 1964), p. 123).

67. Quoted on pp. 4–5 of Wieviorka 1990. For a more sustained discussion of Gaullist discourse on technology and power see the beginning of my chapter 3.

68. Cottier, *La Technocratie, Nouveau Pouvoir*, p. 41.

69. Soutou and Beltran 1995; Renou n.d.

70. See Kuisel 1993 and Ross 1995.

71. See Kuisel 1993.

72. Groupe 1985, *Refléxions pour 1985*, pp. 13–14.

73. Louis Cambournac, "Dix Années de Réalisations Techniques Françaises," *Mémoires de la Société des Ingénieurs Civils de France* 112, no. 1 (1959): 192–207.

74. Albert Caquot, "Le Rayonnement de la France aux points de vue scientifique et économique du constructeur," *La Jaune et La Rouge* 112 (1958): 23–28; Ingénieur général de l'Air Dumanois, "L'Aéronautique Française, sujet de fierté et d'espoir," *La Jaune et La Rouge* 106 (1957): 27–28; Jacques Lecarme, "Un triomphe nationale: 'la Caravelle'," *La Jaune et La Rouge* 118 (1958): 25–42.

75. Pierre Couture, "Explosion de la première bombe A Française," *La Jaune et La Rouge* 136 (1960): 32–35.

76. Pierre Cot, "La Nouvelle Aérogare de l'Aéroport d'Orly," *La Jaune et La Rouge* 158 (1962): 10.

77. The French obsession with the "genius" of the French language is notorious. For a historical treatment see Fumaroli 1992.

78. Georges Combet, "Défense de la langue française: Le Langage Technique," *La Jaune et La Rouge* 100 (1956): 24.

79. Anonymous, "Défense de la langue française: Le Langage Technique," *La Jaune et La Rouge* 107 (1957): 29.

80. "Promotion des Techniques Françaises à l'Etranger," collection "Aux Portes du Monde," radio program aired 13 December 1962. Maison de la Radio archives, KB27571.

81. Groupe 1985, *Refléxions pour 1985*, p. 88.

82. For another example of such assertions see Armand and Drancourt, *Plaidoyer pour l'avenir*, p. 153.

83. Groupe 1985, *Refléxions pour 1985*, p. 85.

84. Ibid., p. 87.

85. Ibid., p. 88.

86. Armand and Drancourt, *Plaidoyer pour l'avenir*, p. 238.

87. "Le Grand Oeuvre: Panorama de l'Industrie Française," film commissioned by Ministère des Affaires Etrangeres in 1957. The first phrase in this title translates literally as "the great work," but the phrase also refers to "the philosopher's stone."

Two copies of the film were ordered the following year by the Ministère de l'Equipement. Today the film can be viewed at the vidéothèque of the Ministère de l'Equipement.

88. Quoted in Fourquet 1980 (p. 237).

89. Armand and Drancourt, *Plaidoyer pour l'avenir*, p. 79.

90. In particular, Gaston Berger, the founder and president of the Centre Internationale de Prospective, was a philosopher—though a rather special kind of philosopher, since he also sat on the scientific advisory board of the CEA. But the seven vice-presidents of the Centre (Louis Armand, François Bloch-Lainé, Pierre Chouard, Jacques Parisot, Georges Villiers, and Arnaud de Voguë) all fit under the label "technicien" in its broadest sense.

91. See the statutes of the organization published at the end of the first issue of *Prospective*, the center's official publication.

92. Gaston Berger, "L'attitude prospective," *Prospective* 1 (1958), p. 4.

93. Ibid., p. 6.

94. Louis Armand, "Vues prospectives sur les transports," *Prospective* 1 (1958): 37–44. He did not, however, specify how operations research might be applied on such a scale or to such problems.

95. Gaston Berger, "Culture, qualité, liberté," *Prospective* 4 (1959): 5.

96. The second issue of *Prospective*, on "Conséquences générales des grandes techniques nouvelles," focused exclusively on these two technologies.

97. See e.g. André Gors, "Avant Propos," *Prospective* 2 (1959): 2.

98. Berger, "Culture, qualité, liberté," p. 6.

99. Gors, "Avant Propos," p. 7.

100. Berger, "Culture, qualité, liberté," p. 7.

101. Ibid., p. 5.

102. The spiritual dimension of the "attitude prospective" was mainly propounded by Berger, a devotee of the Catholic thinker Pierre Teilhard de Chardin. I cannot elaborate on this aspect of his thought here; see Gaston Berger, "L'idée d'avenir et la pensée de Teilhard de Chardin," *Prospective* 7 (1961): 131–153.

103. Berger, "L'idée d'avenir et la pensée de Teilhard de Chardin," p. 151.

104. Berger, "Culture, qualité, liberté," p. 7.

105. Armand and Drancourt, *Plaidoyer*, p. 153.

106. Ibid., p. 157.

107. Villiers, "Industrie, Technique et Culture," p. 21.

108. Jean Meynaud, "A Propos des Spéculations sur l'Avenir. Esquisse bibliographique," *Revue française de la science politique* 13 (1963): 666–688, quoting an interview of Armand in *Nouveau Candide* (3 July 1963, p. 9).

109. Unless otherwise noted, my description of the plans and of the planning process is based on the following sources (in no particular order): "Quatrième Plan de Développement Economique et Social (1962–1965)," Commissariat Général au Plan, 1961; Pierre Bauchet, *La Planification Française: du premier au sixième plan* (Seuil, 1966); François Fourquet, *Les Comptes de la Puissance: histoire de la comptabilité nationale et du plan* (Recherches, 1980); Fondation Nationale des Sciences Politiques et Institut d'Etudes Politiques de l'Université de Grenoble, *La Planification comme processus de décision* (Armand Colin, 1965); INSEE, "Méthodes de Programmation dans le Ve Plan," *Etudes et Conjonctures* 21, 12 (1966); Etienne Hirsch, "Les méthodes françaises de planification," *Mémoires de la Société des Ingénieurs Civils de France* 112, no. 2 (March-April 1959): 81–94; Pierre Massé, *Aléas et Progrès* (Economica, 1984); François Perroux, *Le IVe Plan Français (1962–1965) (2e édition; 1e éd. 1962), Que Sais-Je?* no. 1021 (Presses Universitaires de France, 1963); Bernard Cazes, "Un Demi-Siècle de Planification Indicative," in *Entre l'Etat et le marché,* ed. M. Lévy-Leboyer and J.-C. Casanova (Gallimard, 1991); Kuisel 1981; Lucas 1979; McArthur and Scott 1969; Rousso 1986a.

110. Perroux, *Le IVe Plan Français,* p. 126.

111. "Quatrième Plan de Développement Economique et Social (1962–1965)," pp. 4–5.

112. Armand and Drancourt, *Plaidoyer pour l'avenir,* p. 209.

113. Pierre Massé himself was an ardent supporter of *la prospective.* For an example see his article "Prévision et Prospective," *Prospective* 4 (1959): 91–120.

114. In passing over the details of these models and the processes by which they were built, I in no way mean to imply that these were unproblematic or uncontested. Planners and experts from the INSEE and the Ministry of Finance negotiated over nearly every step. At no point was the planning process seen, either by participants or expert observers, as a "black box." For instance, François Perroux, an applied economist who strongly supported the fourth plan but was not involved in its elaboration, wrote a *Que Sais-je?* volume aimed at both describing the planning process to an educated lay public and critiquing the methods used by planners: *Le IVe Plan Français (1962–1965), Que Sais-Je?* no. 1021 (1963). His critiques argued that planners needed to seek even more intensive quantitative methods in order to produce one final model that would describe the national economy as a single coherent system.

115. "Quatrième Plan de Développement Economique et Social (1962–1965)," p. 153.

116. Ibid., p. 274.

117. Ibid., p. 265.

118. INSEE, "Méthodes de Programmation dans le Ve Plan," *Etudes et Conjonctures* 21, no. 12 (1966), p. 11.

119. Ibid., p. 12.

120. Massé, *Aléas et Progrès*, p. 183.

Chapter 2

1. MacKenzie 1990.

2. For a relatively recent example see Baumgartner and Wilsford 1994.

3. This parallel applies mostly only in the superficial sense that both schools are considered the most prestigious places to study engineering in their respective nations. It was precisely for this reason, of course, that I made this parallel.

4. Nora (1996b, p. 207) has put the ideological "kinship" between Gaullism and communism most eloquently: it was born, he says, "of sharing the most deeply rooted traits of French political culture and tradition: a Jacobin patriotism, a haughty nationalism, a heroic and sacrificial voluntarism, a sense of the state, an understanding of the tragic in history, and a shared hostility to American modernity and the world of capitalism and cash." Both currents "drew their appeal and power to mobilize" from "the historical legitimacy they claimed to embody" (ibid., p. 208), which in turn was "the inexhaustible foundational legitimacy that both derived from Free France and the Resistance" (p. 217).

5. Ordonnance no. 45-2563, 18 October 1945, *Journal Officiel* (31 October 1945), quoted in Scheinman 1965 (p. 8).

6. Ibid.

7. For more on this incident see Weart 1979.

8. Gaillard was Secrétaire d'Etat à la Présidence du Conseil under René Pleven (1951), Edgar Faure (1952), Antoine Pinay (1952), and René Mayer (1953). He had authority over the Haut Commissariat à l'Energie Atomique thanks to the following decrees: 14-8-51; 23-1-52 (no. 52-105); 22-3-52 (no. 52-328); 10-1-53 (no. 53-10). See Lamiral 1988.

9. These are "anciens francs."

10. Cited on p. 82 of Renou n.d.

11. This was exemplified a few months later in a letter from Perrin to Gaillard dated 17 November 1951, accompanied by a report titled "Programme pour assurer la realisation du plan quinquennal au point du vue de personnel." Letter from President du Conseil des Ministres to Ministre de l'Education nationale, 24.11.51; Memo, Francis Perrin, HC to Monsieur Felix Gaillard, Secretaire d'Etat a la Presidence du Conseil, 24.11.51. CEA archives, T5-01-65.

12. Renou n.d.

13. The terms "primary" and "secondary," in use in the early 1950s for reactors, are no longer current.

14. Soutou 1991.

15. Soutou 1995, p. 98.

16. Renou n.d., pp. 79–80.

17. Scheinman 1965, p. 68.

18. Goldschmidt (1987, 1980) also asserts that the military goal of plutonium production was at least tacitly understood by everyone. This assertion was widely confirmed by the many CEA engineers I interviewed in 1989 and 1990.

19. For more on the CEA's experimental reactor program see Weart 1979, Goldschmidt 1962, and Goldschmidt 1967.

20. Interview.

21. Scheinman (1965), citing *New York Times* of 13 November 1951.

22. JO, AN, no. 65, 4 July 1952, p. 3455, quoted by Scheinman (1965, p. 76).

23. JO, AN, no. 65, 4 July 1952, p. 3455, quoted by Scheinman (1965, p. 77).

24. Scheinman 1965, p. 81.

25. The communist deputies, whose party had recently focused on obtaining signatures for the Stockholm Appeal (a worldwide petition calling for a ban on nuclear weapons), tried to introduce a clause into the plan which would formalize France's commitment to the peaceful atom. The rest of the Assembly interpreted this effort as a piece of communist propaganda, and the clause was shelved. There was thus no Parliamentary discussion of the military implications of the plan. For more on parliamentary (in)action see Scheinman 1965.

26. Letter, President du Conseil des Ministres to Ministre de l'Education nationale, 24.11.51. CEA archives, T5-01-65.

27. Péan 1995, pp. 11–20.

28. Stoffaës 1995, p. xxiv.

29. Ibid., p. xix.

30. M. Pascal, interview with André Finkelstein, 29 November 1985.

31. Ailleret was in fact EDF's first representative at the CEA. In principle, other members of EDF who should also have been involved in CEA committees were the Président du Conseil d'Administration (decree 3-1-51), the Directeur Général or one of his adjuncts (19-4-51), and two Directeurs Général Adjoints (12-12-52 and 18-11-52). See Lamiral 1988.

32. Soutou 1991, p. 354.

33. Picard et al. 1995, p. 187.

34. Jean Toutée, "Le Commissariat à l'Energie Atomique: Ses Aspects Juridiques," *Droit Social* 6 (June 1953), p. 319, cited on p. 13 of Scheinman 1965.

35. Interviews; see also Lamiral 1988.

36. M. Pascal, interview with A. Finkelstein, 29 November 1985.

37. Ibid.

38. Virtually all of the CEA personnel that I interviewed stressed that solutions had to be chosen and implemented quickly because they were in a hurry to get sufficient quantities of weapons-grade plutonium to make one or more bombs.

39. Lamiral, vol. 1, p. 26.

40. Interviews; see also Lamiral vol. 1, p. 27.

41. For example, in a letter to the CEA's Direction Industrielle dated 4 March 1958 accompanying a report on the "Installations de Récupération d'Energie G2-G3" (Personal Papers of Claude Bienvenu), Georges Lamiral wrote: "We attract your attention to the following fact: as a result of the latest modifications effected on the fuel load, the Société Rateau put together a study in order to determine the new characteristics of the CO_2 in the [energy] recuperators. The results of this study have not been communicated to us. . . ."

42. These terms, "nuclear" and "classical," were used by CEA and EDF engineers throughout the 1950s and the 1960s.

43. Interviews.

44. Power is a measure of energy per unit time; in the case of a nuclear reactor, it measures how many decays occur in a given time period. G2 operated at a power of around 200 thermal megawatts, and contained 100 tons of uranium. The optimal irradiation (measured in power multiplied by the number of days in the reactor, divided by the tonnage of uranium) was set at 500 MWday/t. To find the number of days, engineers made the following calculation: 500 MWday/t = [(200 MW)/(100 t)] × N days. So N days = [500 MWday/t] × [100t/200 MW] = 250 days, the amount of time a slug should remain in the reactor. (Interview.)

45. From the standpoint of today's reactor technology, one might imagine that EDF would also be interested in loading the core while the reactor was on line in order to avoid losing money in plant downtime. However, the technical and economic considerations that EDF engineers privileged in the mid to late 1950s led them to prefer loading reactors that were stopped over those that were operating.

46. Interviews by G. Hecht; see also Pierre Guillaumat, interview with A. Finkelstein, 11 December 1985; A. Ertaud and G. Derome, "Chargement et Déchargement," *Bulletin d'Informations Scientifiques et Technique* 20 (1958): 69–88.

47. Ibid.

48. Ibid.

49. Yves Chelet, interview with G. Hecht.

50. P. Passérieux and R. Scalliet, "Installations de récupération d'énergie," *Bulletin d'Informations Scientifiques et Technique* 20 (1958): 99–114; J. Kieffer, "La centrale de Marcoule: expérience, résultats et enseignements dans le domaine de la production d'électricté," *Energie nucléaire* 5 (June 1963).

51. Passérieux and Scalliet, "Installations de récupération d'énergie."

52. "Extrait de la 141ème réunion du Comité de l'Energie Atomique, 3 March 1955 (EU archives, JG 53/3); Vallet 1986; Soutou 1995, p. 101.

53. According to the CEA personnel interviewed, they did not gain access to the knowledge that the British had accumulated while building gas-graphite reactors until well into the Marcoule projects. Rumors exist that some scientific and technical knowledge was secretly imparted to the French by the British, but even if these are true the amount of knowledge transferred in the mid 1950s is still likely to have been rather small.

54. Interview.

55. "Réunion sur l'information du public, 12 May 1958," JR/mc-58/1336, 13 May 1958. CEA archives, M7-18-34.

56. "Compte-rendu de la conférence de presse faite par le directeur du centre de Marcoule, le 16 mai 1958," 23 May 1958. CEA archives, M7-18-34.

57. What is distinctive is the frequency with which articles specified that the reactors were French, or that the site was. See R. Papault, "Le Centre de Production de Plutonium et d'Energie Electrique d'Origine Nucléaire de Marcoule (Gard)," *Le Génie Civil*, 134 (1 October 1957): 389–398; M. de Rouville, "Le Centre de production de plutonium de Marcoule: sa place dans la chaine industrielle de l'énergie nucléaire," *Revue de l'Industrie Nucléaire* 40 (June 1958): 483–489; "Nous avons visité pour vous . . . le Centre Français de production de Plutonium à Marcoule," *Energie nucléaire* 1, no. 3 (July-September 1957): 141–144.

58. F. Perrin, "Avant-propos," *Bulletin d'Informations Scientifiques et Techniques* 20 (1958).

59. de Rouville, ""Le Centre de production de plutonium de Marcoule," p. 486.

60. Papault, "Le Centre de Production de Plutonium et d'Energie Electrique d'Origine Nucléaire de Marcoule (Gard)," pp. 389 and 398 (emphases mine). Measuring the worth of a large technological program in terms of its general value to the nation rather than in terms of direct economic returns on investment was an argument with a long tradition among French public engineers. See e.g. Smith 1990, p. 683.

61. Aline Coutrot, "La politique atomique sous le gouvernement de Mendès-France," in Bédarida and Rioux 1985. See also Simmonot 1978 and Scheinman 1965.

62. Quoted in Simmonot 1978 (pp. 228–229). Interestingly, there is still conflict over what Mendès-France meant to do at this meeting. Some think that he did give the basic go-ahead to pursue a bomb project; others, including Guillaumat, maintained that he didn't. See e.g. Stoffaës 1995 and Duval 1995.

63. See Larkin 1988, Rioux 1983, and Bédarida and Rioux 1985.

64. Pierre Guillaumat, interview with A. Finkelstein, 11 December 1985.

65. For example, in April 1955 Gaston Palewski, the minister for atomic affairs, attended a meeting of the CEA steering committee to discuss the progress of the institution's development plan and talk about the new government's view of atomic energy. He specified that Marcoule's plutonium would be used in an "explicitly civilian framework. The Government feels that the time has not yet arrived to take a decision about which sectors France will seek to apply atomic energy to. The versatility of the program adopted by the Government is conceived in such a manner that, should unfortunate circumstances so demand, [the program] could quickly be reconverted toward national defense purposes." In the meantime, he exhorted the steering committee to o help him "objectively shape [public] opinion" in favor of a more congenial view of atomic energy. See "Extrait du Procés-Verbal de la 23ème réunion plénière et 143ème réunion du Comité de l'Energie Atomique," 5 May 1955 (EU archives, JG 53/3). The main argument made in Scheinman 1965 is that French military atomic policy in the 1950s was essentially made covertly, by CEA leaders with the help of a small group of people from a few ministries and the military. Scheinman does not, however, examine how the technologies of the nuclear program enacted that policy.

66. These proposals included calculations for the length of time that fuel rods would have to remain in G2 in order to produce reactor-grade plutonium fuel, as well as estimates of the amount of plutonium required for a breeder program. See "Etude d'un développement possible de l'industrie nucléaire française," DPP 55/316, 12 December 1955; "Note pour le Comité, Programme de développement général 1955–1957," CEA no. 143, question no. 1, 5 May 1955 (EU archives, JG 53/2).

67. Pierre Guillaumat, interview with A. Finkelstein, 11 December 1985.

68. These designs are discussed at greater length in Soutou 1991.

69. For the frankest discussions of the French bomb projects see Mongin 1997 and Scheinman 1965.

70. "Compte-rendu de la réunion tenue à Saclay le 9 novembre 1956 sur le projet de second plan quinquennal," DP 56/1532, 21 November 1956 (EU archives, JG 53/4).

71. Commissariat Général au Plan, "Compte-rendu de la réunion d'information du 2 février 1957 relative au deuxième plan quinquennal de l'énergie atomique," circulated on 9 February 1957 (EU archives, JG 27/1).

72. CEA, "Note sur l'usine de séparation des isotopes," 26 July 1957 (EU archives, JG 49/1); Direction Industrielle, "Plan quinquennal 1957–1961, la séparation des isotopes de l'uranium," DI no. 3469, 5 November 1956 (EU archives JG 27/1); Daviet 1995.

73. For more on the nationalization of EDF see Badel 1996, Frost 1991, and Picard et al. 1985.

74. Frost 1991, p. 70.

75. For more on labor unrest in the postwar period see Chapman 1991, Holter 1992, and Frost 1991.

76. "Extraits de l'allocution de M. Pierre Simon, Réunion des Comités d'Entreprises d'EDF du 26 février 1947," appendix 6 of FNCCR, *Bulletin d'Information* 100 (special, February 1973): 75–77, quoted and translated in Frost 1991 (p. 83).

77. For more on the creation and activities of EDF see Frost 1991, Picard et al. 1985, and Lamiral 1988. In his thorough analysis of worker-management relations at EDF, Frost shows that the utility did not live up to its social potential: by 1968, it had a "symbiotic relationship with [the capitalist economy]" (p. 246). But the utility retained its *symbolic* importance as an icon of improved worker-management relations, and work culture within the utility remained solidly left-wing.

78. Within EDF there was considerable competition between these two technologies. See Picard et al. 1985.

79. Indeed, this struggle persists today, albeit greatly attenuated.

80. One example among many can be found in "Les prévision du IVe plan: l'avenir," *Contacts* 43 (1963): 12–25.

81. For the text of some of these agreements see Lamiral 1988, annexe 2.

82. The first members included Guillaumat, Perrin, and Taranger from the CEA and Gaspard, Ailleret, and Guiguet from EDF. The two institutions began seriously outlining joint development plans in 1955. "Memorandum intérieur de la reunion EDF-CEA du 16 mars 1955." CEA archives, F6-13-18.

83. Nine EDF engineers were among the first enrolled in the CEA's "Génie Atomique" course (Lamiral 1988, volume 1, p. 20).

84. "Memorandum de la réunion EDF-CEA du 18 novembre 1955," 21 November 1955. CEA archives, F 6-13-18.

85. CEA attitudes were manifest, e.g., in "Programme de centrales nucléaires EDF," DPP 55/505, 29 June 1955 (EU archives, JG 27/6).

86. Tension was such that Ailleret insisted on naming the reactor "EDF1" rather than G4—or, as it was later (once EDF had firmly established its place in the nuclear program) known, Chinon A1. (Interview)

87. This role was officially termed "industrial architect." The industrial architect coordinated the overall project, fitting the various components of the reactor into a whole and managing the contracts passed with individual companies.

88. "Etude Préliminaire d'une Installation de Récupération d'Energie sur EDF1," 1956, personal papers of Claude Bienvenu.

89. Ailleret's declaration to the Conseil Economique et Social, 27 June 1963, personal papers of Claude Bienvenu. Ailleret's post was called Directeur Général Adjoint.

90. Claude Bienvenu, quoted in Picard et al. 1985 (p. 191).

91. Jean Cabanius, quoted in Picard et al. 1985 (p. 191). EDF's Direction de l'Equipement, the division in charge of building the reactors, had already espoused this working method while developing the hydroelectric sites.

92. Interviews. Taranger had in fact pushed very hard for private industry to take the role of industrial architect, and was furious when the EDF engineers rejected this course of action. The animosity left over from this meeting did much to increase the tension between the two institutions.

93. Lamiral 1988, volume 1, p. 280; "Memo RETN 1," 25 July 1957, personal papers of Claude Bienvenu.

94. In order to expedite the development of EDF1's design, Ailleret formed a Comité Nucléaire within EDF. This committee met around once a month to discuss technological problems associated with reactor design as well as EDF's overall nuclear policy. It was in the course of these meetings that EDF engineers agreed on the basic characteristics for EDF1.

95. "Etude des Réacteurs Enérgétiques EDF, Projet d'Organisation dans le cas d'un réacteur du type Uranium Naturel - Graphite - CO_2," 8 March 1957, personal papers of Claude Bienvenu.

96. The minutes of the EDF-CEA meeting of 16 March 1955 (quoted in Lamiral 1988, volume 1, p. 29) read: "Mr. Guillaumat feels that it is essential that an original French technology exist. But Mr. Gaspard [president of EDF] does not want to play the role of patron that much. . . ."

97. Interviews; see also Lamiral 1988, volume 1, p. 280.

98. Before hitting on this they had considered various other possibilities, including one that combined steel and prestressed concrete (interviews).

99. Interviews; see also Jean-Pierre Roux, "La Centrale Nucléaire EDF1 de Chinon," *Mémoire de la Société des Ingénieurs Civils de France*, tome 110, fascicule IV (July-August 1957): 294–309.

100. "Etude Préliminaire d'une Installation de Récupération sur EDF1," 1956; Ailleret's declaration to the Conseil Economique et Social, 1963, personal papers of Claude Bienvenu.

101. Starting up any gas-graphite reactor for the first time inevitably requires that some rods be removed before they are fully irradiated, thereby generating what was known as "fatal plutonium." Thus, the CEA did recover some plutonium from spent EDF1 slugs, but only a minimal amount. The point remains that EDF1 was not designed to produce plutonium. Later EDF reactors, however, did have loading devices that functioned while the reactor was on line. This was not solely a matter of the CEA imposing its will; EDF engineers gradually became convinced that such a system also benefited the economics of running a nuclear power plant.

102. Leo, Kaplan, and Segard, "Problems of fuel loading and unloading in reactor EDF1," *Geneva Conference, 1958*, pp. 582–590.

103. Interviews.

104. Interviews made it quite clear that simplicity of design was a major goal for EDF engineers. It should be noted here that while EDF engineers strove for low cost in designing EDF1, they did not attain that goal (in part because of an unforeseen event: the spherical steel containment vessel cracked, and repairing the crack added tremendous costs and delays to the project.). But EDF1 represented a first step towards optimizing reactor costs, and the general goal of minimizing costs held throughout EDF's reactor program.

105. See Picard et al. 1985.

106. Yvan Teste, "Les Installations de Production d'Energie de Marcoule et la Centrale Nucléaire de Chinon," *Mémoires de la Société des Ingénieurs Civils de France*, tome 110, fascicule II (March-April 1957), p. 73.

107. Ibid.

108. Martine Bungener, "L'Electricité et les trois premiers plans: une symbiose réussie," in Rousso 1986.

109. Picard et al. 1985.

110. Roux, "La Centrale Nucléaire EDF1 de Chinon," p. 309.

111. Teste, "Les Installations de Production d'Energie de Marcoule et la Centrale Nucléaire de Chinon," p. 75.

112. Duval 1995, p. 43.

Chapter 3

1. Quoted on p. 447 of Stoffaës 1991.

2. Wieviorka 1990.

3. Quoted in Wieviorka 1990 (pp. 4–5).

4. Wieviorka, p. 5.

5. Quoted in Wieviorka 1990 (p. 9).

6. Quoted in Wieviorka 1990 (p. 4).

7. See Frost 1991, especially chapter 4.

8. Rousso 1986b, pp. 27–40.

9. RETN1, BS/JCr, "Memento pour la réunion du 17.7.57 sur EDF2," 15 juillet 1957, personal papers of Claude Bienvenu.

10. Indeed, the CEA ended up funding and designing a second loading machine for EDF1 in order to extract fuel rods from that reactor more quickly.

11. RETN1, BS/JCr, "Memento pour la réunion du 17.7.57 sur EDF2," 15 juillet 1957, personal papers of C. Bienvenu.

12. Ibid.; interview.

13. Lamiral 1988, volume 1, pp. 38–39.

14. Ibid., p. 120; interview. Though this concern was never voiced in the documents that changed hands between the two institutions in 1957, subsequent events and recent interviews showed that it loomed large. Recall from chapter 2 that weapons-grade plutonium is that element's 239 isotope. Pu239 was produced in a reactor when a U238 atom absorbed a neutron and decayed, first into U239, then into Pu239. If the Pu239 stayed in the reactor long enough, it would absorb the neutrons produced by the fissioning of the uranium and decay into Pu240 and Pu241. These isotopes were undesirable for weapons-grade plutonium, as they could cause the plutonium to undergo fission unpredictably. They were also impossible to separate from the Pu239 in the spent fuel. Therefore the CEA wanted to remove fuel from the reactor as quickly as possible, whereas EDF wanted to leave it in as long as possible in order to extract the maximum amount of energy.

15. Interview.

16. A newly welded piece had not yet undergone the thermal treatment required to relax the internal stress produced as the metal cooled off. The fissure occurred when the metal suddenly released the internal energy it had thus accumulated. Interviews; Lamiral 1988, volume 1, p. 47.

17. A solution was eventually found, but the accident pushed back EDF1's startup date by 3 years. It did not begin operation until September 1963. Interviews; Minutes of meetings of EDF Comité d'Energie Nucléaire, 1959; Lamiral 1988, volume 1, pp. 47–51.

18. Interviews; Picard et al. 1985.

19. See Frost 1991 for other attacks on EDF.

20. This advice did not appear connected to the issue of plutonium production.

21. My use of the term "nuclear engineers" does not imply that these people had received professional degrees in nuclear engineering; it merely refers to the engineers involved in the nuclear program.

22. Minutes of meeting of EDF Comité d'Energie Nucléaire, 12 June 1959, personal papers of C. Bienvenu.

23. Econometrics was founded in France by economists at the Ecole des Mines and the Ecole des Ponts et Chaussées. It involved applying mathematical modeling techniques to economics. See Picard 1990.

24. Systems analysis was initially developed primarily by scientists in the Rand Corporation in the United States as a tool for defense strategy. The heir of wartime operations research, it provided a way of analyzing complex problems through quantification and modeling. See Edwards 1996.

25. In his autobiography, *Aléas et Progrès* (Economica, 1984), Massé claims that he and his colleagues unknowingly developed linear programming independently of its American and Soviet inventors. He does, however, acknowledge contact with systems analysts at Rand beginning in the late 1950s.

26. Massé created this division in 1955. It depended directly on the *Direction Général* (rather than on one of the three big *Directions*: Equipement, Etudes et Recherches, and Production et Transport). For more on Massé and his economic thought see Massé, *Aléas et Progrès* and *Le plan ou l'anti-hasard*; Picard et al. 1985; Frost 1991.

27. Interview. Typically, the division developed consumption forecasts for 5–10 years in the future. These forecasts were then used to justify current construction of power plants. Prediction of France's overall energy demand would be based on the GNP forecasts publicized by the Plan. At first, according to one of Boiteux's employees, the division used classic econometrics models that posited the growth of energy consumption to be a function of the expected rate of growth of the GNP. The elasticity coefficient—an expression of the ratio of energy use to the GNP—was set almost at 1. During the interview, this employee remarked that he had had no idea that this coefficient had been set at an artificially high value in order to promote electricity production: at the time, he said, he had thought that the coefficient expressed a law of nature.

28. Interview.

29. This was by no means confined to France; various forms of economic analysis became popular throughout the western world in the 1950s and the 1960s. On the rise of cost-benefit analysis in the United States, see chapter 7 of Porter 1995; on the role of systems analysis in US defense strategy, see chapter 4 of Edwards 1996.

30. Interview. See also Pierre Ailleret, "Les besoins d'énergie à long terme et l'énergie atomique," *Energie nucléaire* 4 (1962), no. 1.

31. Massé, *Aléas et Progrès*.

32. Picard et al. 1985, p. 190.

33. Minutes of meeting of EDF Comité d'Energie Nucléaire, 12 June 1959, personal papers of C. Bienvenu.

34. Minutes of meeting of EDF Comité d'Energie Nucléaire, 16 November 1959.

35. Minutes of meetings of EDF Comité d'Energie Nucléaire, 1958–1960; interview with Claude Tourgeron by Picard, Beltran, and Bungener, 19 February 1981; Lamiral 1988, volume 1, p. 41; author's interview with Pierre Tanguy, 8 January 1990.

36. Minutes of meeting of EDF Comité d'Energie Nucléaire, 23 December 1960.

37. Interviews. Edwards (1996, chapter 4) cites Joseph Weizenbaum as saying that computers "put muscles on [the] techniques" of systems analysis.

38. Interview.

39. Interviews; Picard et al. 1985; Massé, *Aléas et Progrès*. Modeling, from optimization studies to energy consumption forecasts, became a trademark of EDF. Pioneered in the utility, these techniques became something that the whole institution grew proud of, and which marked the work done there, at least during the 1960s, as unique. Although these methods spread to a few other industries, EDF economists and engineers remained the acknowledged experts in the domain.

40. Minutes of meetings of EDF Comité de l'Energie Nucléaire, 18 January and 23 November 1960; interviews.

41. Although I will not go into detail here, there is plenty of evidence for the use of optimization studies in the making of design decisions. The engineers I interviewed in 1989 and 1990 discussed these practices extensively. They would, for example, seek to optimize the relationship between reactor size and power, or the thermal cycle of the reactor, in such a way as to minimize the cost of a kilowatthour; reference costs were provided by the Service des Etudes Economiques Générales. Their accounts were confirmed and enriched by the minutes of meetings of EDF Comité de l'Energie Nucléaire (especially those of 7 July 1961, 7 February 1962, 7 March 1962, 17 May 1962, and 28 September 1962, in personal papers of C. Bienvenu).

42. See Frost 1991 and chapter 10 of Picard et al. 1985.

43. Interviews. One interviewee added that the Ministry of Finance people would even get irritated at the high quality of EDF's economic studies.

44. Latour 1983.

45. On "matters of fact," see Shapin and Schaffer 1984.

46. On this shift, see chapter 3 of Frost 1991. Dictionaries translate *rentabilité* as "profitability"; however, this is a misleading for EDF's sense of the term, since as a public institution it could not make a profit.

47. Interviews; minutes of meetings of EDF Comité de l'Energie Nucléaire, 21 October 1960, personal papers of C. Bienvenu.

48. Interview.

49. Minutes of meeting of EDF Comité de l'Energie Nucléaire, 4 May 1961, personal papers of C. Bienvenu.

50. Minutes of meeting of EDF Comité de l'Energie Nucléaire, 7 July 1961, personal papers of C. Bienvenu.

51. Pfaffenberger (1992) makes the point that technologies are more than simply instruments for attaining pre-defined political goals—they help to shape those goals, and that's why they provide a different means of conducting politics.

52. Memo, Directeur des Productions to Chef du Département des Programmes, 17 July 1963 (CEA archives, F6-13-18).

53. CEA engineers had been contemplating getting plutonium out of Chinon from the very beginning. See CEA, "Programme de centrales nucléaires EDF," DPP 55/505, 29 June 1955 (EU archives, JG 27/6).

54. Lamiral 1988, volume 1, p. 56.

55. Minutes of meeting of EDF Comité de l'Energie Nucléaire, 15 June and 8 July 1960, personal papers of C. Bienvenu. At the second meeting, the Comité concluded that it had to "study as of now what would be the loss in energy value . . . that would ensue from unloading [rods] after a short irradiation [period] as a function of the average flow of fuel rods that one would use in this fashion."

56. CEA technologists had engaged in rudimentary calculations of the relative value of plutonium since the mid 1950s. See CEA, "Programme de centrales nucléaires EDF," DPP 55/505, 29 June 1955 (EU archives, JG 27/6); "Calcul approché du prix de l'énergie d'origine nucléaire," DPP 56/1581, 4 December 1956 (EU archives, JG 27/2); "Estimation des matières premières et ouvrés nécessaires à un programme de production d'énergie nucléaire," DPP 56/284, 16 March 1956 (EU archives, JG 115).

57. Lamiral 1988, volume 1, p. 57.

58. Minutes of meeting of EDF Comité de l'Energie Nucléaire, 28 September 1962, personal papers of C. Bienvenu (emphasis mine).

59. Minutes of meeting of EDF Comité de l'Energie Nucléaire, 29 March 1963, personal papers of C. Bienvenu.

60. Ibid.

61. "Esquisse d'un programme CEA-EDF de réacteurs de puissance," 5 juin 1963, pour la réunion du 17.9.63, personal papers of C. Bienvenu.

62. Minutes of meeting of EDF Comité de l'Energie Nucléaire, 12 March 1964, personal papers of C. Bienvenu (emphasis mine).

63. Jacques Gaussens, "Faut-il fixer un prix du plutonium?" *Bulletin d'Informations Scientifiques et Techniques* 81 (June 1964), p. 10 (emphasis in original). See also Jean Andriot and Jacques Gaussens, *Economie et Perspectives de l'Energie Atomique* (Dunod, 1964).

64. Memorandum, Taranger to Perrin, "Conclusions de la réunion EDF/CEA du 17.4.62" (CEA archives, F 6-13-18). EDF Conseil d'Administration, meeting # 232, 22 October 1965. "Convention relative aux conditions de contribution d'Electricité de France aux frais d'études engagés par le Commissariat à l'Energie Atomique"; "Convention particulière relative à la contribution d'Electricité de France aux frais d'études du Commissariat à l'Energie Atomique sur EDF 1, 2, et 3 et aux frais d'études d'ordre général sur les réacteurs de la filière uranium naturel, graphite, gaz"; "Convention générale pour la mise à disposition d'Electricité de France de combustibles nucléaires par le Commissariat à l'Energie Atomique." All three "conventions," or agreements, signed 2 July 1965 by R. Hirsch (CEA) and A. Decelle (EDF). Personal papers of C. Bienvenu.

65. On the electronics industry see Botelho 1994 and Mounier-Kuhn 1994. On the aerospace industry see McDougall 1985.

66. Stoffaës 1991.

67. Wieviorka 1990.

68. Cazes 1991.

69. Minutes of meetings of EDF Comité à l'Energie Nucléaire, 6 June 1961.

70. Minutes of meetings of EDF Comité à l'Energie Nucléaire, 7 March 1962.

71. Minutes of meeting of EDF Comité à l'Energie Nucléaire, 28 September 1962; "Elément Combustible EDF 4: Réflexions à la suite de la réunion du 5 Mars 1963."; SEGN, "Les appareils de chargement et de déchargement du combustible," pour la réunion programme CEA-EDF du 17.9.63.; EDF, Direction Production et Transport (Service de la Production Thermique), "Centrale Nucléaire à une tranche de 500 MW, réacteur graphite-gaz, Estimation des Frais d'Exploitation," 27.12.63. Personal papers of C. Bienvenu.

72. "Compte-rendu de la réunion EDF-CEA du 1er avril 1963," DPA/63/317, 26 avril 1963 (CEA archives F6-13-18).

73. BS/MLa, "L'Evolution des Relations CEA-EDF," 23.10.64, personal papers of C. Bienvenu.

74. Ibid.

75. Ibid.

76. Direction des Piles Atomiques, "Fiche pour Monsieur l'Administrateur Général, Remarques générales sur la collaboration EDF-CEA," 25 February 1964, CEA archives, F3-24-25.

77. Interview.

78. Memo, Directeur des Piles Atomiques to Administrateur Général, DPA no. 1.362, 9 November 1964. CEA archives, F6-13-18.

79. Memo from H. de Laboulaye to Hirsch, Perrin, et al., DPg/S64/1137-HL/eg, 3 December 1964 (CEA archives, F6-13-18).

80. Memo from Directeur des Relations Extérieures et des Programmes to Administrateur Général & Haut Commissaire, DREP 64-654, 14 December 1964 (CEA archives, F6-13-18).

81. Robert Hirsch, Administrateur Général, Délégué du Gouvernement, à Monsieur le Directeur Général d'EDF, letter dated 16.12.64, personal papers of C. Bienvenu.

82. "Filière graphite-gaz, Problème de répartition des commandes," 19.10.65, personal papers of C. Bienvenu.

83. Letter from A. Buchalet (Schneider & Cie, le directeur adjoint à la gérance) to Renou, 7 January 1964 (CEA archives, F3-24-25).

84. "Filière graphite-gaz, Problème de répartition des commandes," 19.10.65.

85. EDF, REN2, CB/SPi, "La Politique Industrielle d'EDF," 25.11.65, personal papers of C. Bienvenu.

86. Ibid.

87. Ibid.

88. Ibid.

89. For more examples of how EDF engineers defended their ideas and practices see EDF, REN2, "Complément à la note sur la Politique Industrielle d'EDF," 27.12.65; EDF, REN1, "La 'Chaudière Nucléaire'," 6.12.65; Memo from Yves Cordelle, SENA, to SEGN, EDF, re: "Marché d'ensemble 'clé en main'," 20.12.65; EDF, Direction de l'Equipement, "Politique Industrielle EDF pour la filière gaz graphite," 6.1.66, personal papers of C. Bienvenu.

90. See e.g., Cazes 1991.

91. "Politique Industrielle EDF dans les centrales de la Filière Gaz Graphite," accompanied by note from Roux to Bienvenu dated 1.3.66 indicating that this was a draft of a note to be presented on 24.3.66 and "Projet de Rapport Général (Ve Plan: Commission des Industries de Transformation): II: Objectifs et Méthodes de l'Amélioration des Structures." Personal papers of C. Bienvenu.

92. Quotes in Massé, *Aléas et Progrès* (p. 137).

93. For example, in a meeting of the Commissions des Finances in December 1965, Didier Olivier-Martin had to defend EDF against accusations that engineers always wanted to build new things and did not pay much attention to cost. See minutes of 16.12.65 meeting of the Commission des Finances on "Dépenses d'Investissements."

94. See Frost 1991, especially chapter 4.

95. EDF, "Filière Gaz Graphite, Lotissement des Commandes," 28.12.65; EDF, Direction de l'Equipement, "Politique Industrielle EDF pour la filière gaz graphite," 6.1.66; "Extrait du Compte-Rendu de la Réunion de Direction du 17 Février 1966," personal papers of C. Bienvenu.

96. Frost 1991, p, 194.

97. For example, "Note sur Politique Industrielle d'EDF en matière de Centrales Nucléaires," 3.1.66; EDF, REN2, "Répartition des Commandes," 21.1.66, personal papers of C. Bienvenu; and many others.

98. EDF, meeting of Conseil d'Administration no. 238, 25 March 1966.

99. Ibid

Chapter 4

1. Of course, the specific relationship between confederation and federations varies by union and by federation within the union. I do not mean to suggest that federation militants do not participate in formulating union ideology; I mean to suggest only that, in order for such ideology to translate into official, union-wide policy, it must pass through the confederal level at some point.

2. Willard 1995; Chapman 1991; Holter 1992; Frost 1991.

3. Even in the CGT, some militants went through periods when they denounced the nationalizations as a sham. In such cases, it was not the concept of nationalization which they criticized, but its execution. In particular, they accused nationalized companies of submitting to the will of large capitalist corporations in their contracting procedures. By and large, however, the CGT upheld a vision of a society in which all the major industrial sectors were nationalized.

4. Dreyfus 1995.

5. Bergounioux 1984.

6. On the relationship between the nationalism of Gaullists and communists, and its origins in the myths of the Resistance and the liberation, see Nora 1996b.

7. Bergounioux 1984.

8. Bergounioux 1975.

9. For a masterly treatment of the transformative years of the CFTC/CFDT's history see Georgi 1995. For a broader temporal sweep see Branciard 1990. For a brief English-language treatment see Mouriaux 1984.

10. An overcomplex statistical assessment of union constituencies in 1970 can be found in Adam et al. 1970.

11. Figures for EDF and the CEA come from Papin 1996 (pp. 213–214). See also *Rayonnement,* June 1960 and February 1966.

12. I do not treat union discourse on automation in the body of the chapter; it occupied a relatively small place in union discourse during the 1950s and the 1960s, and it is not immediately germane to my argument. I do wish to provide a discussion in these notes, however, for readers interested in this topic. Much interesting research remains to be done on the automation debate in France; a good introduction to the debate can be found in du Tertre and Santilli 1992. Some of my discussion is based on this source. The debate was by no means confined to labor unions; indeed, it was spawned by technologists and sociologists. For most participants in the debate, automation was one of the two major technological advances—for better or worse—of the postwar period (the other being nuclear power). It represented a modern ideal, which many argued France ought to attain. The subject of automation prompted unions into deeper reflection on the role of technological change in workers' experience of work. But automation—in the strict sense of technologies that replaced human workers with machines controlled by automatic feedback loops—had not yet penetrated French industry very deeply in the 1950s or even the 1960s. Much of the discussion, therefore, used examples drawn from the United States and Britain. Automation in this period appeared less as reality than as potential. Writing in *L'Usine nouvelle* and in similar publications, industrial leaders extolled the virtues of the (still theoretical) automated factory. Sociological analyses ranged from models in which—with technology driving society—automation would create a more highly skilled working class and free up leisure time for all, to Marxist models in which automation would lead to deskilling and unemployment unless social structures and attitudes changed first. (See works by the following authors in the primary source bibliography: Georges Friedmann, Alain Touraine, Pierre Naville, Claude Durand et al., and Serge Mallet.)

Labor unions, struggling to make sense of this hypothetical technological change, were torn between these projections. Should France, as a nation, pursue automation? Most apparently agreed that the technology was critical to France's future. The Conseil Economique issued a formal statement on the subject: "France can occupy a prominent role [in the development of automation] if it sets precise goals focused primarily on quality. . . . Instead of letting itself be outdistanced by foreign countries, . . . France should take a serious interest in this new technology which can contribute to the improvement of its economic situation." The question of how to appropriate the technologies of automation for the nation's benefit extended to linguistic considerations as well. Should the French word be *automation,* or *automatisation?* The translator of one American book on automation argued against the first option, evoking linguistic purism and adding:

". . . if the word automation designates an automatic operation, why not translate it as *automatisme*, and if it defines the process that makes [something] automatic why not translate it as *automatisation?*" (Maurice Rustant, *L'automation* (Editions ouvrières, 1959); see also p. 26 of du Tertre and Santilli 1992). CGT militants writing in the 1950s were pessimistic about the possibilities of automation. While this technology might herald a "new industrial revolution," historical odds dictated that the result would not benefit workers. Only large capitalist corporations could afford substantial automation. These companies would only automate in order to maximize their profit. The resulting automation would encourage the development of more capitalist monopolies and lead to unemployment. "Theorists" (meaning, presumably, sociologists) who argued otherwise served capitalism and did not care about the best interests of the worker. CGT economist André Barjonet, who issued some of these stern warnings, nonetheless also defended the use of "automatisation," accusing those who used the word "automation" of being "Anglomaniacs [*anglomanes*]." Indeed, for him the entire craze over automation appeared to be a capitalist-driven "Anglomania" (A. Barjonet, "L'automatisation et ses problèmes," *Le Peuple* 495 (15 November 1955), p. 3.

Though it lay in the future, automation loomed large because apparent precursor technologies already existed in French factories. CGT militants discussed the psychological costs of demands for a faster work pace. The fatigue that accompanied work as a switchboard operator, for example, was notorious. Rather than seeing automation as a potential solution to such difficulties, CGT militants darkly predicted that automated factories would induce even greater fatigue and bring more mind-numbing jobs. At a 1967 conference on automation, one CGT militant asserted his opposition to any argument that held that automation, and technological change more generally, would lead to social progress. Social progress could only occur in a different political system. At the bare minimum, the sectors most likely to develop automation (insurance, steel, oil, electronics, atomic energy, and more) had to be nationalized. Then, perhaps, automation might occur in a manner beneficial to workers. Those who predicted a golden age labored under a dangerous illusion by "imagining automation by itself, outside of the social and political context in which it is necessarily inscribed." [R. Duchet, "Au Centre Inter de Lille: Femmes ou Robots?" *La Vie Ouvrière* 913 (28 February 1962), p. 10; P. Vernoux, "A propos de l'utilisation des techniques modernes de production, du recours à l'automation," *Le Peuple* 777 (1–15 June 1967), p. 8]

Force Ouvrière, in contrast, saw automation as inevitable. "One cannot stop technological progress," wrote one militant. Another wrote that union militants could not afford indifference to automation, or to technological progress more generally. These would occur no matter what, and those who (like the CGT) opposed progress simply wasted their time. Instead of opposing change, militants had to remember that automation was a means, not an end. The ends included improving standard of living, increasing leisure time, and reducing fatigue. (Here *Force Ouvrière* writers echoed, though they did not directly cite, the conclusions of some industrial sociologists.) Like nuclear energy, automation could lead workers to the "the threshold of Eden or the edge of chaos." In order to improve the odds in favor of paradise, unions had to participate in training the men who would operate the new machines. Indeed, education in general had to undergo reform in order to adapt to technological progress. Further, the distinction between office and factory workers would no longer make sense in an automated plant. Such a change would require defining new skill hierarchies and developing new (unspecified) means of negotiating labor contracts. For Force Ouvrière, therefore, automation promised much, provided its effects were correctly predicted and handled. The union preserved this optimism through the end of the 1960s ["Rapports présentés au 5e congrès confédéral," *Force Ouvrière Informations* 53 (September 1956); "Au seuil de l'eden ou au bord du chaos par l'automation et l'énergie nucléaire," *Force Ouvrière Hebdomadaire* 581 (11 April 1957), pp. 6–7; Max Rolland, "Après les journées d'information de l'automatisme, réformes de nos méthodes d'enseignement . . . ," *Force Ouvrière Hebdomadaire* 523 (23 February 1956), pp. 6–7; "Automation et énergie nucléaire: révolution permanente des techniques," *Force Ouvrière Hebdomadaire* 582 (18 April 1957), pp. 6–7; Janine Goret, "Le colloque de Grenoble," *Force Ouvrière Hebdomadaire* 1090 (numéro spécial, 1967), pp. I–VIII].

Some CFTC/CFDT writers evinced a similar, albeit more cautious, optimism toward the potential of automation. The Reconstruction intellectuals justified their optimism by citing sociological studies such as Alain Touraine's investigation of a Renault factory—especially his prediction that the machine tender of the future would be selected not for his manual abilities, but for his intelligence, seriousness, and sense of responsibility. Other militants expressed more reserve, arguing that automation would only improve working lives if it were conducted according to democratic planning. Planning could ensure the introduction of automation for the right reasons (for example, to help workers handle toxic substances from a distance and thereby improve job safety) rather than the wrong ones (such as helping large corporations gain greater profits at the expense of workers). Planning would also help the nation manage the change in social structure that automation would bring (fewer blue-collar workers, more technicians and engineers). While Force Ouvrière militants focused on how their union should deal with the social effects of automation, CFTC/CFDT militants argued that workers had to participate in decisions about when and how to introduce automation (either directly, or indirectly, through democratic planning). [P. Cournil, "Automation et mouvement ouvrier," *Cahiers des groupes reconstruction* 29–30 (February-March 1956); "Les travailleurs seront-ils victimes du progrès?" *Syndicalisme* 666 (March 1958).] The approach of CFTC/CFDT militants to automation, and more generally to technological change in the workplace, grew more nuanced over the course of the 1960s. They continued to draw upon sociological studies that pointed to the erosion of distinctions between skilled workers and technicians and to the changing structure of the working class. Such changes, they insisted, demanded greater worker participation in management: "changes in the organization of the firm and in the nature of work—which increasingly place technology at the root of decisions about the pace of work, salary, hiring, and promotion—reinforce trade union demands for the close involvement of workers in these decisions, an involvement which should be provided for in labor settlements." Some argued that the effects of technology varied by sector and refused to countenance broad generalizations on the subject. In order to control technological development and its consequences, unions had to get involved at all stages of the process, from the decision to automate to the negotiation of an appropriate labor contract. [Charles Savouillan, "Progrès technique et conventions collectives," *Cahiers Recontruction* V (September–November 1961); Jean Berthon, "Le progrès technique et ses conséquences sur l'emploi," *Syndicalisme* 1096 (30 July 1966): 6]

The analyses of automation presented by militants in each union were thus consistent with their confederations' overall positions on matters of technological change. For CGT militants, nothing good could happen unless the entire social system changed. At a minimum, industries had to nationalize before they could automate in a beneficial manner, and more detailed discussion was pointless until this occurred. Force Ouvrière militants ignored questions pertaining to the design and deployment of automation, and focused instead on how to handle its effects. CFTC/CFDT militants saw design, deployment, and effects as interrelated, and argued for labor involvement at each stage of the process.

13. Pierre Le Brun, "Le point du vue d'un syndicaliste," in Gaston Berger et al., *Politique et technique* (Presses Universitaires de France, 1958), p. 340.

14. Henri Beaumont, "Le débat sur l'électronique," *Le Peuple* (16–31 May 1966): 5–7.

15. Le Brun, "Le point du vue d'un syndicaliste," p. 340.

16. Beaumont, "Le débat sur l'électronique."

17. *Le Peuple* was the weekly for active militants and *La Vie Ouvrière* the weekly aimed at rank-and-file members.

18. Daniel Deschamps, "A Tancarville j'ai vu jaillir du sol LE PLUS GRAND PONT SUSPENDU D'EUROPE," *La Vie Ouvrière* 674 (31 July 1957), pp. 18–19.

19. "Bonnes clés pour le futur: Les Nationalisations," *La Vie Ouvrière* 1173 (22 February 1967), p. 9.

20. Ibid., p. 11.

21. Ibid., p. 12.

22. "Nouvelles Techniques Françaises: les Voitures à turbines," *La Vie Ouvrière* 618 (3–9 July 1956), p. 18.

23. Robert Sautereau, "Et ceux qui ont fabriqué le "Mille-pattes"?" *La Vie Ouvrière* 653 (5–11 March 1957), p. 4.

24. "L'équipement atomique française et la bombe A," *Le Peuple* 571 (1 January 1959), p. 5.

25. Ibid.

26. Georges Léon, "Une Grande Famille: Les Joliot-Curie," *La Vie Ouvrière* 604 (27 March–2 April 1956), p. 18 (continued in *La Vie Ouvrière* 605, 606, and 614).

27. Georges Léon, "Les Curie," *La Vie Ouvrière*, 606 (10–16 April 1956), p. 18.

28. "L'équipement atomique française et la bombe A," p. 5.

29. "Ce qu'il faut savoir de la bombe atomique et de retombées radio-actives (1)," *Le Peuple* 593 (1959).

30. Jean Schaefer, "La Défense de la paix question politique interdite aux syndicats?" *Le Peuple* 699 (1964), p. 15.

31. R. Telliez, "Bombe 'A' Gaulliste: Y a-t-il de quoi crier hourrah?" *La Vie Ouvrière* 808 (24 February 1960), p. 9.

32. Roger Guibert, "Pourquoi pas nous?" *La Vie Ouvrière* 1010 (8 January 1964), p. 2.

33. Gérard Carrère, "Peut-on recolter la science en cultivant la 'force de frappe'?" *La Vie Ouvrière* 1024 (15 April 1964): 5–10.

34. André Barjonet, "l'Euratom contre la classe ouvrière," *Le Peuple* 502 (1956), p. 3.

35. Jean Serdias, "Après la CECA et avec le 'Marché Commun,' l'Euratom c'est la relance d'une CED armée atomiquement et la mise de l'avance atomique française à la disposition d'une Allemagne réarmée," *Le Peuple* 530 (1957).

36. Georges Beaulieu "La famille Euratom," *La Vie Ouvrière* 644 (1–7 January 1957), p. 9.

37. "Les incidences du progrès technique sur l'orientation, la formation professionnelle, et le travail de la jeunesse," *Force Ouvrière* 861 (3 October 1962): 6–7.

38. T. Ottavy, "Progrès technique et progrès sociale," *Force Ouvrière* 489 (23 June 1955), p. 9.

39. "Les incidences du progrès technique sur l'orientation, la formation professionnelle, et le travail de la jeunesse."

40. A few examples: "La Science électronique n'a pas fini de nous étonner. . . ," *Force Ouvrière* 492 (14 July 1955): 6–7; "Du Laboratoire à l'usine," ibid. 494 (28 July 1955): 6–7; "Le Bureau , lui aussi, sera boulversé par les techniques industrielles," ibid. 495 (4 August 1955): 6–7; "Des réalisations révolutionnaires . . . ," ibid. 830 (21 February 1962): 5; "L'électricité en France," *Force Ouvrière Informations* 146 (January-February 1965): 5–22; "Les Télécommunications en France," ibid. 149 (May 1965): 149–158; "Le temps perdu de la recherche," ibid. 155 (December 1965): 485–506.

41. "Pour le meilleur et pour le pire, notre sort est lié à l'atome," *Force Ouvrière* 484 (19 May 1955): 6–7; "Euratom," *Force Ouvrière Informations* 60 (April 1957).

42. Consider, for example, this quote: "The cobalt 'bomb' that just arrived in France is the latest scientific discovery in the fight against cancer. The science of the atom, after having caused the most serious worries, can tomorrow open for us unknown horizons in the most varied domains."*Force Ouvrière* 472 (24 February 1955), p. 7.

43. "Euratom," *Force Ouvrière Informations* 60 (April 1957).

44. "Realisation internationale, Euratom est une oeuvre constructive," *Force Ouvrière* 545 (26 July 1956), p. 1.

45. Conversation with Jean-Pierre Alliot, editor of *Force Ouvrière*, 1996.

46. "La protection des utilisateurs de substances radio-actives," *Force Ouvrière* 584 (2 May 1957): 6–7.

47. "L'Atome, enfant gâté à la croissance rapide, est en âge de recevoir une bonne éducation syndicaliste," *Force Ouvrière* 507 (3 November 1955): 6–7; "Les méthodes de protection des travailleurs contre les radiations," ibid. 619 (9 January 1958): 3; Charles Veillon, "Perspectives atomiques," ibid. 684 (18 September 1958): 2.

48. "L'Atome, enfant gâté à la croissance rapide, est en âge de recevoir une bonne éducation syndicaliste." See also "La protection des utilisateurs de substances radio-actives," *Force Ouvrière* 584 (2 May 1957): 6–7.

49. Max Rolland, "Nos techniciens bâtissent les cathédrales du XXe siècle. Grâce à une technique éprouvée de l'alliance du béton et du fer, ils changent la face du monde," *Force Ouvrière* 522 (16 February 1956): 6–7. Emphasis mine.

50. René Gruissan, "L'Electricité de France, Une entreprise nationale en prise directe sur l'Europe," *Force Ouvrière* 881 (20 February 1963): 12.

51. Yet another related reason can be found in the availability of sources. Scholars can access the CFDT's confederation archives easily, while neither the CGT nor Force Ouvrière have catalogued confederation archives open to researchers. Access to internal confederation documents would doubtless reveal more complicated understandings of technology in both the CGT and Force Ouvrière than does a reading of their official publications. The point, however, is that such complexities remain hidden from public view.

52. Jean Berthon, "Quelques aspects du progrès technique et de ses conséquences sur l'emploi," *Formation* 51 (January-February 1963).

53. "Une merveille de la technique et des hommes," *Syndicalisme* 735 (18 July 1959): 1.

54. Pierre Papon, "La Recherche scientifique: Nouveau problème de politique syndicale," *Cahier reconstruction* (October 1964): 101–136.

55. "Atome, espoir de l'avenir?" *Syndicalisme* 507 (February 1955): 9.

56. "A Marcoule, G1 (première installation atomique industrielle de notre pays) met la France à l'heure atomique," *Syndicalisme* 557 (19–25 January 1956).

57. "Marcoule: Cité de l'énergie atomique industrielle est aussi, pour les militants syndicalistes, Le Chantier de La Peur!" *Syndicalisme* 559 (February 1956); "Marcoule (ses licenciements sans motif, ses licenciements en cours de maladie,

ses refus d'embauche) est toujours le chantier de la peur!" ibid. 590 (September 1956).

58. "La Protection des travailleurs condition indispensable de l'utilisation des radiations," *Syndicalisme* 726 (9 May 1959).

59. "Devant le Conseil Economique: La radioactivité et la protection des travailleurs," *Syndicalisme* 690 (30 August 1958), pp. 1–2.

60. Alfred Williame, "Euratom, une nécessité moderne"; *Syndicalisme* 609 (26 January 1957), p. 3; "L'Euratom première étape vers les Etats-Unis d'Europe?" ibid. 559 (February 1956); Alfred Williame, "L'Euratom devra nous permettre de faire avec d'autres ce que nous ne pouvons pas faire seule," ibid. 583 (19–25 July 1956): 1–2; "La France est Bien Placée . . . face à la Révolution Nucléaire!" ibid. 607 (12 January 1957): 3; Alfred Williame, "Dans le cadre d'Euratom, voici un problème actuel: L'Usine de Séparation Isotopique," ibid. 619 (April 1957), special edition; "Ressources atomique de l'Europe des six" ibid. 696 (11 October 1958): 3; "Un rapport de l'Euratom détermine . . . Les perspectives de l'énergie nucléaire jusqu'en 1980," ibid. 793 (3 September 1960); Francois Picard, "Euratom prépare l'ère de l'électricité atomique" ibid. 812 (March 1961),

61. CFTC, Secteur Politique, "Note sur la force de frappe" 26 April 1963 (CFDT archives, 7H287).

62. "Syndicalisme et Force de Frappe (II)," *Cahiers Reconstruction* 63-II (August 1963): II-37–II-44.

63. Syndicat National du Personnel du CEA, section de Saclay, "Réponse à une note sur la force de frappe du secteur politique de la CFTC adressée le 6 mai 1963 aux Fédérations—Union Départementales et Permanents." CFDT archives, 7H287.

64. Socialisme et Démocratie, "Force Stratégique Nationale, Défense et Politique Etrangère," *Cahier Reconstruction*, March 1965, no. 6, p. 28.

65. The one quasi-exception to this came when the Socialist Jules Moch invited the CFTC to join his Ligue nationale contre la force de frappe in 1963. Eugène Descamps, the CFTC's secretary-general, declined the invitation on the grounds that this was a political issue, and that demonstrations and other actions should therefore be led by political, not union leaders. Part of his refusal undoubtedly stemmed from the fact the union prepared for its transformational 1964 congress. With ongoing debates about the appropriate doctrines and policies for a secularized union, joining an organization led by a prominent Socialist probably seemed likely to cause more discussion than it was worth. My point here still stands, though, because in contrast to Force Ouvrière members, CFTC/CFDT militants continued to speak freely on the *force de frappe* and other national political issues that did not relate directly to working class interests. Moch-Descamps correspondence: CFDT archives, 4P40; Letter from Jules Moch, 21 June 1963, to Eugène Descamps; Letter from Descamps to Moch, 2 October 1963; "Ce que pourrait être une position CFTC," (no date); Descamps to Moch, December 1963. See also Box

7H287: CFTC Secteur Politique, note no. 2. "Notre Position sur les problèmes d'action contre la force de frappe," conseil confédéral, 12–14 December 1963.

66. Quoted on p. 88 of Georgi 1995. "Consumer culture": *la civilisation du confort.* The other three questions were: What size should the union seek to attain, and how should it structure itself? How should relationships with the other major unions be conceptualized? Is a doctrinal basis for action necessary, and if so how should this be formulated?

67. Georgi 1995.

68. "Face à l'évolution des hommes, des techniques, et du Monde. Eléments de réflexion sur les responsabilités de la CFTC de demain," Rapport présenté par le Conseil Confédéral sur proposition de la Commission pour l'étude des problèmes d'orientation. suppl. *Formation* 48 (September-October 1962).

69. Ibid., p. 24.

70. Ibid., p. 27.

71. Ibid., questionnaire attached. Again, this was one of several questions posed by the questionnaire. We must keep in mind that the main function of the survey was not to determine what union members thought about technological change, but to give confederation leaders a sense of how many members supported secularization. But secularization was not an isolated issue; for if the union did abandon Christianity as the basis for its doctrine, it had to find a new way of defining itself ideologically. Qualitatively, then, the responses served to outline the directions that members thought the unions should take and to provide a sense that all members could participate in shaping this direction.

72. Response 240, Section syndicale CFTC du Crédit Lyonnais de Paris (CFDT archives, 7H24).

73. These words were in both response 256 (Section syndicale du CNEP) and response 308 (Syndicat du Personnel des Banques de la Region Parisienne). CFDT archives, 7H24.

74. Response 281 (Livre-papier-carton, Roubaix-Tourcoing) CFDT archives, 7H24.

75. Response 70 (Fédération des syndicats des personnels civils de la Défense Nationale, Section du Bas-Rhin), CFDT archives, 7H24.

76. Response 269 (Syndicat du Personnel de la Securité Sociale et des Organismes Sociaux, Bourses du Travail), CFDT archives, 7H24.

77. Response 291 (Section fédérale des LGD), CFDT archives, 7H24.

78. Response 319 (Section de Saclay) CFDT archives, 7H24.

79. See especially Ross 1995, Boltanski 1982, and Kuisel 1993.

80. Daniel Mothé, quoted on p. 171 of Ross 1995. A *cadre* did not necessarily have a technical job, but most rank-and-file engineers counted as *cadres*.

81. This chorus is discussed at length in Gilpin 1968.

82. Maryvonne Besson, "La France a besoin de 10.000 d'ingénieurs nouveaux par an . . . et nous en formons seulement 4000," *Force Ouvrière* 629 (20 March 1958): 6–7; Jacques Laurin, "Un aspect d'une promotion sociale défaillante. Grave pénurie d'ingénieurs pour l'économie française," ibid. 678 (5 March 1959).

83. "Orientation et formation des ingénieurs et cadres supérieurs," *Le Peuple* 527 (1957): 7.

84. Besson, "La France a besoin de 10.000 d'ingénieurs," pp. 6–7.

85. Sylvain Morgan, "Enseignement technique et scientifique inadapté; Promotion ouvrière à peu pres inexistante: Nous manquons d'ingénieurs et de techniciens," *Force Ouvrière* 580 (4 April 1957): 6–7.

86. Pierre Papon, "La Recherche scientifique: Nouveau problème de politique syndicale," *Cahier reconstruction* (October 1964): 101–136

87. For a better history of the UGIC and other manifestations of *cadre* unionism, see Descostes and Robert 1984.

88. "Une préoccupation pour tous nos syndicats: les ingénieurs, cadres et techniciens," *Le Peuple* 605 (15 May 1960): 16–17.

89. René le Guen, "En vue du 34e congrès: la préparation de la conférence nationale des ingénieurs, cadres, et techniciens," *Le Peuple* 671 (1–10 March 1963): 13–16.

90. "Après la conférence nationale des ingénieurs, cadres et techniciens— Documents adoptés," *Le Peuple* 681 (1963): 17–18.

91. "Les ingénieurs, cadres et techniciens dans les organismes où siègent le élus des salariés," *Le Peuple* 724 (15–30 April 1965): 9–15.

92. R. Guibert and G. Carrère, "Ingénieurs, cadres et techniciens: 2 millions aujourd'hui, 4 millions demain," *La Vie Ouvrière* 1006 (11 December 1963): 16–20.

93. Bergounioux 1975, pp. 229–231.

94. "Le Rôle et la Place des Ingénieurs et Cadres dans le mouvement syndical," *Force Ouvrière Informations* 110 (November 1961): 976–980.

95. Ibid.

96. Claude Perrignon, "Les cadres: des privilegiés ou des nouveaux prolétaires?" *Syndicalisme* 1163 (December 1967): 13.

97. Ibid.

98. Response 261 (DGI Gard); similar response given by response 256 (Sect. Synd. du CNEP). CFDT archives, 7H24.

99. Response 291 (section fédérale des LGD); response 299 (syndicat général de la sécurité sociale et des organismes sociaux de la région lyonnaise). CFDT archives, 7H24.

100. Response 319 (Section de Saclay). CFDT archives, 7H24.

Chapter 5

1. Frost 1991, p. 53.

2. Ibid., p. 70.

3. Papin 1996, p. 213.

4. In his thorough analysis of worker-management relations at EDF, Frost (1991) shows that the utility did not live up to its social potential: by 1968, it had a "symbiotic relationship" with the capitalist economy (p. 246). But the utility retained its symbolic importance as an icon of improved worker-management relations, and work culture within the utility remained solidly left-wing.

5. See Mazuzan and Walker 1984 and Walker 1992.

6. This is not to say that nuclear countries drew no benefit from each other's experience. But communication about such matters was limited, and the CEA and EDF did not merely pattern their approach on that of Britain or the United States.

7. The basic premise of this scholarship is that, in a world filled with mortal dangers, people choose what to worry about not on the basis of some absolute, decontextualized sense of safety or danger, but for cultural reasons. Much of this work has taken a structural approach. For example, Rayner (1986) examines how employees in American hospitals behave in the presence of radiation hazards and found that people respond to risk according to how much autonomy, power, and trust they have in their jobs. He develops a matrix model to try to predict whether individuals will tolerate, deny, take, or avert risks according to their position in the institutional structure of a hospital. This approach builds on the work of Douglas and Wildavsky (1982), who analyze American society in terms of three forms of social organization—hierarchy, market, and sects—and argue that people's place in this scheme explains their behavior in the face of danger. Although I draw on some of the basic premises of such scholarship, my analysis also demonstrates the limits of this kind of structural approach. Describing an institutional structure and using that structure as an explanation for behavior is not enough. If, instead of taking the organization of social relations (and, more specifically in this case, the organization of work) for granted, we examine its provenance, then we begin to see how broader social and political ideologies shape the cultural meanings of work and risk in particular settings and at particular historical moments. The cultural construction of risk in the workplace must

be understood in the larger context of the cultural construction of that workplace.

8. An excellent example of this kind of study is Downey 1986.

9. Maurice Pascal, 1956, quoted on p. 9 of the first volume of Lamiral 1988. See also p. II.14 of *Marcoule et sa vocation dans le Languedoc-Rhodanien: document établi en commun entre la Direction et les Organisations Syndicales du Centre de Marcoule: CFDT, CGC, CGT, CGT-FO, SPAEN*, October 1969 (henceforth referred to as *Marcoule et sa vocation*).

10. And they were men: the reactor had to run 24 hours a day, and French legislation at this time prohibited women from working night shifts. One suspects, of course, that other barriers also contributed to the lack of female nuclear workers.

11. *Marcoule et sa vocation* gives a figure of 1241 overall at Marcoule in 1958. No figure given for G2 for 1958, but a total of 421 people, of whom 326 were CEA agents, worked on G2 and G3 together in 1963 (p. II.50). The figure of 200 is thus my rough estimate. (See also p. II.61.)

12. Everyone interviewed confirmed that such tests and checks were conducted, but the content of both the tests and the security checks remain inaccessible still today, years after the reactors were shut down for good: in 1994, I was denied access to the security investigations, located in the files of the Renseignements Généraux at the Archives départementales du Gard in Nîmes.

13. Interviews. Owing to the difficulties I experienced in obtaining sources, I could not answer some obvious questions with much precision. I do not know the precise nature of the training course; nor can I be any more precise about the educational background of those hired at Marcoule.

14. Interview.

15. Interview.

16. The four areas of specialization were the loading machine and the internal mechanics of the core, the electronic devices, the electricity distribution system, and the cooling circuits and turbines.

17. Interview.

18. The *salle de commande* was what is known today as a *salle de contrôle*. I have kept the term "command" because it was the one used by the CEA in the 1950s and the 1960s and significantly evokes the military atmosphere that reigned at Marcoule.

19. The term *chef de quart*, meaning "officer on watch," came from the military.

20. "Pile" was a term used at the time to designate the core of a reactor. A "pile" consisted, quite literally, of a pile of fuel elements; although it did in the case of the Marcoule reactors, a "pile" did not necessarily include the other accoutrements of a reactor. The term used for G2 and G3 in the 1950s and the 1960s was *pilotage*, which again carried military connotations—in this case, the "pilotage" of a Navy vessel. I have translated *conducteur* as "pilot" rather than as "operator" to

reflect more accurately the terminology used in the 1950s and the 1960s; today, such employees are known as *opérateurs*.

21. *Marcoule et sa vocation*, p. II.23.

22. Technologies were classified as either "nuclear" or "conventional."

23. A. Ertaud and A. Pagès, "Le contrôle des réacteurs G2 et G3," *Bulletin d'Informations Scientifiques et Techniques du CEA* (henceforth *BIST*) 20 (1958): 126–139.

24. Ibid. See also M. Lignières, J. Bobigeat, and A. Darré, "Installations électromécaniques," *BIST* 20 (1958): 141–155.

25. Ertaud and Pagès, "Le contrôle des réacteurs G2 et G3."

26. In addition to the men who made up these shifts, a host of other workers, technicians, and engineers were associated, to greater or lesser extents, with the reactors. Personnel were needed for maintenance and repair of the instruments and machines in the reactors, to conduct tests and experiments either within the reactor or on materials or instruments that had or would sojourn in the reactor, to clean the floors and bathrooms, to process administrative paperwork and other such secretarial work, etc. It is neither possible nor (for the purposes of this chapter) instructive to examine these jobs.

27. A. Ertaud and G. Derome, "Chargement et déchargement," *BIST* 20, p. 86.

28. Interviews.

29. Interviews.

30. *Cahier de bord* was yet another term derived from the Navy, where it was used to designate a ship's log book.

31. In rare cases, the four engineers consulted CEA researchers outside Marcoule.

32. Interviews.

33. *Exploitant* translates directly as "operator." To avoid confusion with reactor operators, I will keep the French term.

34. Several engineers who had designed parts of the Marcoule reactors and participated in their startup phases refused to stay on as *exploitants*, disdaining such jobs because they did not think they would require or even allow for much ingenuity or creativity.

35. Interview.

36. Interview.

37. J. Rodier, "Evolution de la protection dans le Centre de Marcoule," *BIST* 72–72 (1963): 5–10; interviews.

38. The main accomplishment of the SPR agents during G2's startup phase was to provide breathing equipment to men who worked in an environment where CO_2 might leak. Carbon dioxide gas presented two potential dangers. First, if present in sufficient quantities, it could asphyxiate or poison anyone who inhaled too great a quantity. Second, even a minute amount of renegade CO_2 would contain radioactive particles: if inhaled, it would cause internal radioactive contamination, which presented even more danger to humans that external skin exposure. The metal and concrete "shields" provided by reactor designers did not, in the opinion of SPR agents, adequately protect workers against this and other risks of internal contamination. Source: J. Chassany, "La radioprotection des piles G1, G2, G3," *BIST* 72–72 (1963): 11–17.

39. Experimenting with various fabrics led SPR agents to conclude that all outfits should be made out of cotton, which could be decontaminated more easily than other materials. An agent exiting a regulated zone would scan his body with a radiation detector. If the detector recorded dosages above the accepted norms, he would undergo decontamination and discard his clothes into the designated bin, from which they would travel to the SPR's new laundry facility. See J. Rodier et al., "Les problèmes vestimentaires posés par l'exploitation des installation actives et leurs solutions," *énergie nucléaire* 4, no. 1 (1962); J. Rodier et al., "Le travail en milieu radioactif et ses problèmes," *énergie nucléaire* 5, no. 4 (1963): 291–301.

40. France had adopted international radiation exposure standards set by the early 1960s. For people directly engaged in work involving exposure to radiation, this meant that their overall organism could not be exposed to more than $5(N-18)$ rems per year, where N was the age of the workers (legally, no one under 18 could work in a radioactive workplace). In addition, such people should not be exposed to more than 3 rems in any 13 week period, and women should never be exposed to more than 1.3 rems. External exposure to radiation varied for different parts of the body: the limits for the skin and bones were 30 rems/year, for most organs were 15 rems/year, and for the hands, forearms, feet, and ankles were 60 rems/year. Source: R. Vial, "Protection des travailleurs dans l'industrie nucléaire," *énergie nucléaire* 6, no. 5 (1964): 305–312.

41. Rodier et al., "Le travail en milieu radioactif et ses problèmes"; interviews.

42. Rodier et al., "Le travail en milieu radioactif et ses problèmes"; Chassany, "La radioprotection des piles G1, G2, G3"; Rodier et al., "Les problèmes vestimentaires posés par l'exploitation des installation actives et leurs solutions"; interviews.

43. Chassany, "La radioprotection des piles G1, G2, G3." Implementing such measures meant that the number of SPR agents had to increase dramatically. By 1963 the SPR employed 320 agents and contained several subdivisions. The two main ones were devoted to radiation protection in the reactors and in the plutonium factory. An additional section took charge of decontamination problems. Furthermore, a whole host of labs acted as support services for these three subsections; these specialized in chemistry, biology, physics, photometry, masks and aerosols, effluents, electronics, and statistics. See Rodier, "Evolution de la protection dans le Centre de Marcoule."

44. J. Rodier, J. Castain, and C. Guerin, "Information et éducation en matière de radioprotection," *BIST* 72–73 (1963).

45. Interview.

46. The CEA was the only institution in French society, apart from hospitals, that dealt with radiation on a regular basis (and it had much more exposure, so to speak, than did medical institutions). Thus, it had considerable say over what kind of information the public got about radiation, and thus a big hand in shaping how the wider society perceived nuclear risks throughout the 1960s. Of course, this changed somewhat as protests against nuclear power gained momentum throughout Europe in the 1970s.

47. Interviews.

48. I could not find any evidence to indicate that violators of safety measures were subject to formal disciplinary measures. Interview evidence clearly points to the heroism of engineers such as those described here. I suspect that if workers had been disciplined, this would have been mentioned either in the articles that discussed the violations, or in the labor union literature.

49. The anthropologist Françoise Zonabend analyzes the differing meanings of exposure and contamination and cleanliness and uncleanness in *La presqu'île au nucléaire* (1989), her study of present-day work in France's plutonium reprocessing plant in La Hague.

50. Such behavior in the workplace is not unique to the French nuclear industry. The heroes of Marcoule have their counterparts in the nineteenth-century mining and railroad industries; indeed, as Zonabend (1989) points out, some present-day nuclear workers in the plutonium reprocessing plant in La Hague consider themselves the heirs of their nineteenth-century miner forefathers. And the tensions between the social meanings of exceptional and mundane risks exist on the battlefields of just about any war.

51. This newsletter, published by union leaders at the Saclay research center, included articles written by militants at all the CEA's sites.

52. *Rayonnement*, November-December 1959; also March 1963 and April-May 1963.

53. *Rayonnement*, January–March 1960.

54. *Rayonnement*, July-August 1961

55. *Rayonnement*, September-October 1961

56. *Rayonnement*, December 1962-January 1963.

57. *Rayonnement*, February-March 1963.

58. Ibid.

59. *Rayonnement*, March-April 1961.

60. Interview. Indeed, I myself was lodged there during my first visit to Marcoule.

61. Interviews. The interviews also confirm that the worlds of the CEN and those of the Centres de Production were very different.

62. *Rayonnement*, November-December 1959, March-April 1961, December 1962-January 1963, and March-April 1964. The situation began to improve in 1966, apparently as a result of union activism.

63. *Rayonnement*, March-April 1964.

64. The CFTC argued that three conditions should suffice for an injured worker to receive compensation: the victim should be regularly engaged in activities presenting a risk of radiation exposure, the illness suffered should be on the list of illnesses possibly related to radiation exposure, and when filing for compensation the victim should present a medical certificate within a defined period of time after exposure. Source: *Rayonnement*, February-March 1963.

65. Ibid.

66. Ibid. Early in 1963 the CFTC proposed a series of measures to improve job safety at the site. The union asked that restrictions on the release of information imposed by military secrecy not apply to information relevant to the health of the workers. It asked for a better system of accounting for employee exposure to radiation, claiming that Marcoule's medical service did not keep proper track of the level of contamination of the personnel. (See *Rayonnement*, December 1962-January 1963 and February-March 1963.) Other demands included requests that the personnel delegates to the commission on health and safety have equal status and rights as other personnel delegates and that these delegates receive the technical and legal training necessary to discuss issues of health and safety with the various heads of division on the site. Marcoule, said the union, was "reticent" to apply international safety norms, and the dressing rooms in G2 and G3 did not even meet French legislative standards. In 1966, the management agreed to refurbish the dressing rooms; it is not clear whether it ever acceded to the rest of the CFTC's demands (*Rayonnement*, February 1966).

67. "Rapport établi en vue de la séance du 19.7.56. Projet d'Arrêté portant déclaration d'utilité publique des travaux de construction d'une centrale thermique nucléaire à Avoine (Indre-et-Loire)," Affaire no. 167. Archives of the Ministry of Industry.

68. Interviews.

69. Jacques Veuillet, "La Leçon d'électricité dans un parc," *Contacts électriques* 5 (March-April 1957): 9.

70. See e.g., René Dupuy and Michel Brigaud, "Au rendez-vous de Nanterre," *Contacts électriques* 17 (May 1959): 10–18.

71. "Ouvriers de la lumière," *Contacts électriques* 29 (May 1961): 6.

72. Chinon's organizational structure paralleled that of EDF's thermal power plants. Jobs at the site fell into three main divisions; here I concentrate primarily on the *Service Exploitation* (production division). This organization was first put into effect at the thermal plant at Nantes-Cheviré. Whereas before there were as many services as there were technological subunits in the plant (machine rooms, boiler rooms, etc.), this new method greatly centralized the organization. The personnel were divided into three main "functional" categories, each of which was placed under the authority of a single engineer. See R. Fays and J. Pupponi, "La conduite des centrales nucléaires de Chinon," *énergie nucléaire* 5, no. 8 (1963): 562–572.

73. Description taken from Fays and Pupponi, "La conduite des centrales nucléaires de Chinon," and *Rapport de Sécurité EDF2* (1965), Chapitre XII: "Organisation du personnel et procédures d'exploitation."

74. This comparison is difficult to make, because there is more accessible documentation on the Chinon training programs than on the Marcoule ones. Even the Chinon training documentation is sketchy and incomplete. As far as I could determine, Chinon engineers either had university engineering degrees or followed EDF's engineering training program. Most had also taken the CEA's course in *Génie Atomique* (atomic engineering; this course preceded university programs on the subject, offering an in-depth introduction to nuclear physics, chemistry, and engineering) and followed EDF training programs in radiation protection. The *chefs de quart* possessed either a standard high school diploma or one from an Ecole Nationale Professionelle (ENP), thereby holding the more general title of *agents de maîtrise principale*. They completed elementary courses in "nuclear science" either at Chinon or at Marcoule, took radiation protection courses, and attended EDF sessions on general factory safety. This training generally lasted from 3 to 6 months, depending on an individual's previous experience. In addition, most also took full-time 6-month courses in electronics. The *chefs de bloc/ techniciens de radioprotection* were known as *agents de maîtrise de conduite et d'exécution*. A few also had ENP diplomas. They took the same courses in "nuclear science" as the *chefs de quart*, as well as similar courses in electronics. They did radiation protection internships at the CEA's Saclay site, and engaged in more practical training sessions at Chinon. Many had previously been *assistants conducteurs* or *rondiers* at one of EDF's conventional power plants. As first EDF2 and then EDF3 neared completion, the *rondiers* of EDF1 became the *chefs de blocs* of EDF2 or EDF3. *Rondiers* and *assistants conducteurs* fell under the lowest category in Chinon's overall hierarchy: *ouvrier professionel*. The minimum hiring requirements for this category were literacy and basic arithmetic skills. These workers received elementary courses in "nuclear energy" upon arrival at the site. Some also took Chinon's radiation protection courses; according to EDF2's safety report, however, many had difficulty with the abstract material in these courses, and the *techniciens de radioprotection* often resorted to "demonstrative methods" to teach radiation protection. After a month or two of courses, these men went straight to the reactors, where the *chefs de bloc* trained them for specific tasks. *Rapport de Sécurité EDF2* (1965),

Chapitre XII: "Organisation du personnel et procédures d'exploitation," interviews. A description of training programs at EDF more generally can be found in Pascal-Michel Pipon, "La Gestion des Hommes au S.P.T (1946–1992)," Mémoire de Maîtrise, Université Paris IV-Sorbonne, 1993–1994.

75. *Rapport de Sécurité EDF2* (1965), Chapitre XII: "Organisation du personnel et procédures d'exploitation."

76. Frost 1991, p. 216.

77. This sense is confirmed in the interviews conducted by Pascal-Michel Pipon for "La Gestion des Hommes au S.P.T (1946–1992)," Mémoire de Maîtrise, Université Paris IV-Sorbonne, 1993–1994.

78. That EDF2's design was largely completed by 1957 shows that this design did not build upon the operating experiences of G2.

79. Interview.

80. Interviews.

81. EDF started using *salle de contrôle* rather than *salle de commande* in the late 1960s.

82. EDF used the term "manual" in this context to mean that a human being had to operate instruments, turn switches, etc., in order to operate the reactor; "automatic" meant that once an operator initiated an automatic sequence the reactor would be run by computer or numerical calculator.

83. Some governed the same components as the primary section instruments (such as the cooling system) but performed different functions. For example, during startup the cooling pipes were filled using instruments on the secondary block, whereas during routine operation the carbon dioxide was regulated from instruments on the primary section.

84. *Rapport de Sécurité EDF2* (1965), Chapitre IX: "Contrôle et Instrumentation." Readers attentive to my notes will notice that much of the information in this section is based on this safety report and other written sources, and may wonder about the extent to which actual practices conformed to these written descriptions. Evidence from interviews and reactor tours suggests that, even though practice did not always conform to these written words, the general outlines of work organization and practice described in these documents and related in this essay did correspond to practices in the 1960s.

85. Running EDF2 was more complicated than running G2, in part because EDF had defined two possible "modes" of functioning. The first based the reactor's operation on the fission reaction, the second on the demands of France's electrical network. As I showed earlier in this chapter, the operators of G2 completely ignored network demands; plutonium production in the core determined G2's power, and any excess heat generated by that power was, in essence, discarded. But EDF had no intention of wasting energy. Safety concerns meant that EDF2

operators had to have the option of operating the plant based on the state of the fission reaction. But under "normal" conditions, the *chef de bloc* and his assistants tried to base their operation of EDF2 on the demands of France's electrical network. In this "normal" mode, network demands led the *chef de bloc* to regulate the turbines to accommodate a specific flow of steam. The steam flow defined the pressure of the heat exchangers, and constant pressure was maintained by regulating the flow of the cooling gas. The flow of cooling gas determined the temperature of the reactor core. This temperature, in turn, was maintained by guiding the control rods—in other words, regulating the intensity of the fission reaction. See Dardare and Lyon, "Contrôle et commande de la centrale de Chinon," *énergie nucléaire* 4, no. 7 (1962): 582–588; *Rapport de Sécurité EDF2* (1965), Chapitre IX: "Contrôle et Instrumentation."

86. The fuel loading schedule emerged from politically charged negotiations between the CEA and EDF. The CEA did in fact succeed in obtaining some plutonium from the Chinon reactors. This forced concession only strengthened EDF's resolve to orient reactor operation as much as possible toward electricity production.

87. *Rapport de Sécurité EDF2* (1965), Chapitre VII: "Dispositif Principal de Manutention."

88. Frost 1991; Papin 1996.

89. Chinon Sous-CMP, "Réunion du 21 mai 1965"; "Réunion du 21 octobre 1965"; "Réunion du 25 avril 1967"; "Réunion du 7 juillet 1967"; "Réunion extraordinaire du 17 juillet 1967"; "Réunion du 17 octobre 1967"; "Compte-rendu des réunions des 25 et 28 février 1969 sur la nouvelle organisation de la centrale"; Memorandum from GRPT 5 to M. Bigeard, "Modifications d'organigramme année 1966," 10 May 1965. All in CFDT-Chinon archives.

90. Chinon Sous-CMP, "Réunion du 21 octobre 1965," CFDT-Chinon archives.

91. This service was created with approval from the Ministry of Public Health and the Ministry of Industry. *Journal Officiel*, 24.6.55, letter from Dr. Aujaleu (directeur de l'Hygiène Sociale) to the Direction du Gaz et de l'Electricité (27 February 1956), letter from the secrétaire d'Etat à la Santé Publique et à la Population to the secrétaire d'Etat à l'Industrie et au Commerce (Direction du Gaz et de l'Electricité, 20 May 1957) [Ministry of Industry archives]. Lamiral, *Chronique*, p. 289. Interview.

92. Commission de Coordination des Comités d'Hygiène et de Sécurité, "Réunion du 30 mai 1963," CFDT Fédération de l'Energie archives.

93. Commission de Coordination des Comités d'Hygiène et de Sécurité, "Projet de Procès Verbal, Réunion 19 février 1964," CFDT Fédération de l'Energie archives.

94. Indeed, I was struck by how little discussion there was of radiation protection in the Commission de Coordination des Comités d'Hygiène et de Sécurité.

Between the records kept by the energy federations of both FO and the CFDT, I saw about 90 percent of the CCCHS minutes for the 1950s and the 1960s (the EDF archives did not grant access to these papers, but the minutes at the unions were the same ones I would have seen at EDF). In all these meetings, only two included substantive discussion of radiation protection.

95. Responsibility for overall site safety lay with Chinon's three organizational subdivisions.

96. Interviews. See also P. Beau, A. Douillard, and J. J. Martin, "La radioprotection des travailleurs dans les centrales nucléaires d'Electricité de France à la lumière de huit années d'exploitation," *énergie nucléaire* 13, no. 5 (1971): 350–359.

97. *Rapport de Sécurité EDF2* (1965), Chapitre XII: "Organisation du personnel et procédures d'exploitation."

98. Beau et al., "La radioprotection des travailleurs dans les centrales nucléaires d'Electricité de France à la lumière de huit années d'exploitation"; interviews.

99. Comité d'Hygiène et de Sécurité Chinon, "Réunion du 16 décembre 1966," CFDT-Chinon archives.

100. Chinon Sous-CMP, "Réunion du 15 janvier 1964," CFDT-Chinon archives.

101. Interviews.

102. Philippe Curval, "A Chinon," *Contacts électriques* 59 (Mai 1966): 12–17.

103. Ibid., p. 16.

104. Ibid.

105. Ibid.

106. These observations also hold true for work at Saint-Laurent, EDF's second nuclear site, also located in the Loire Valley and begun in the late 1960s.

107. Curval, "A Chinon," p. 14.

108. Ibid.

109. Interview.

110. Chinon Sous-CMP, "Réunion du 20 novembre 1964," CFDT-Chinon archives.

111. Every canal had a *poubelle* (waste bin) at the bottom to catch anything than might happen to fall in or come loose within the canal. Needless to say, these waste bins were highly radioactive.

112. Interview.

113. Ibid.

114. Comité d'Hygiène et de Sécurité Chinon, "Réunion du 15 décembre 1964," "Réunion no. 1 du 17 mars 1967," "Réunion du 23 août 1967," "Réunion no. 3 du 13 novembre 1967," CFDT-Chinon archives.

115. Comité d'Hygiène et de Sécurité Chinon, "Réunion du 23 août 1967," CFDT-Chinon archives.

116. Comité d'Hygiène et de Sécurité Chinon, "Réunion no. 4 du 10 janvier 1969," "Réunion no. 2 du 17 juin 1969," "Réunion no. 3 du 14 octobre 1969," CFDT-Chinon archives.

117. Comité d'Hygiène et de Sécurité Chinon, "Réunion no. 4 du 20 janvier 1970," CFDT-Chinon archives.

118. Comité d'Hygiène et de Sécurité Chinon, "Réunion du 15 décembre 1964," "Réunion no. 1 du 25 mars 1969," CFDT-Chinon archives.

119. Ministry of Industry archives, questions écrites 15461 (24 July 1965), 13982 (20 April 1965). Further exploration of this type of subcontracting activity would doubtless be highly revealing.

120. Commission de Coordination des Comités d'Hygiène et de Sécurité, "Réunion du 30 mai 1963," "Projet de Procès Verbal, Réunion 19 février 1964," CFDT Fédération de l'Energie archives.

121. Comité d'Hygiène et de Sécurité Chinon, "Réunion n ° 3 du 25 octobre 1968," CFDT-Chinon archives.

122. Comité d'Hygiène et de Sécurité Chinon, "Réunion du 23 mars 1971," CFDT-Chinon archives.

123. This is clear from a survey of CFTC/CFDT newsletters throughout the 1960s. Beginning in 1965–66, moreover, the three unions began to work together (while maintaining separate memberships and offices, of course). For a more general analysis of unions at EDF see Frost 1991 and Papin 1996.

124. For example, the war in Algeria, or (in the case of the CFTC) democratic planning. See "Informations CFTC" no. 2 (n.d., ca. May 1962), no. 3 (n.d., ca. July-August 1962), CFDT-Chinon archives.

125. "Informations CFTC" no. 3 (n.d., ca. July-August 1962), CFDT-Chinon archives; Claude Tourgeron, "La Centrale de Chinon doit revenir à sa destination pacifique," *Bulletin GNC* 1 (1963): 22–25.

126. Chinon Sous-CMP, "Réunion du 25 juin 1968," "Réunion du 7 octobre 1968," CFDT-Chinon archives.

127. In March 1969, impatient with the slow response of management, the Chinon union locals joined their federations in protesting that the co-management commissions (the SCMP in the case of Chinon, the CMPs at the national level) were a farce, and temporarily withdrew their participation. This protest could hardly have shocked the utility's national managers, but Chinon's director

was somewhat surprised and wished that the militants had opened discussions before withdrawing. Indeed, the union locals had previously expressed considerable satisfaction with the SCMP, and we might suspect that they were doing little more than following their federations on this issue. Chinon Sous-CMP, "Réunion du 26 mars 1969," "Réunion des 17 et 24 septembre 1969," "Réunion du 11 mai 1970"; "Informations CFDT," no. 7 (October 1965). CFDT-Chinon archives.

128. For sophisticated discussions of relationships between skill and social hierarchies, see Lerman 1997 and Downs 1995.

Chapter 6

1. The relationship betwen representation and experience has been debated at length by cultural historians, especially those concerned with gender and/or labor. For a sample of these debates, see Scott 1988, 1991, 1993; Sewell 1990, 1993; Reid 1993; Berlanstein 1993; Downs 1993, 1995.

2. Popular representations of nuclear technology have been examined by Boyer (1985) and Weart (1988). Except in their analyses of protest movements, however, scholars have not paid much attention to the reception of such representations.

3. Numerous scholars have argued that the transformation of material culture into a spectacle has become an integral part of modern capitalist societies. On the spectacle of commodity culture see Debord 1977, Clark 1984, and Richards 1990. On the politics and culture of technological displays and spectacles in the United States see Smith 1983, 1994; Corn 1983, 1986; Nye 1990, 1994. On Britain see Agar, forthcoming.

4. This practice was strongly encouraged by the advertising executive Marcel Bleustein-Blanchet. (See the epigraph to my chapter 1.) He deplored the apparent indifference of the French public to national industrial achievements, arguing that such indifference stemmed not from lack of interest but from ignorance. He encouraged industrialists to engage in a new, more "psychological" style of publicity, which he called "industrial information." This would include not only regular print ads but also articles in the mass press, organized debates on radio and TV, public opinion polls, specially designed brochures, and so on. "The great industrialists," he wrote in *Le Monde*, "must lucidly accept the goal, which consists of providing objective, documented information to the public and to state institutions, in order to shore up the foundations of their companies." ("L'industrie française doit sortir de la clandestinité: l'information industrielle," 17 December 1957) The page on which Bleustein-Blanchet's article appeared offered an exemplar: there were several sidebars covering a variety of Publicis's commercial activities.

5. From a survey that appeared in the journal *Sondages* as "La Presse, le public, et l'opinion" (1955 (3)). Not surprisingly, white-collar employees and those in the *professions libérales* tended to read Paris papers more, while agricultural workers tended to read them much less (70 percent read only a regional paper). Blue-

collar workers followed the overall averages. Those who lived in towns of less than 5000 had a greater tendency to read only regional papers, as did those with lower levels of education. Overall, people tended to be most interested in local news, *faits divers*, cartoons, and politics. One-third of the people surveyed (42 percent of men and 25 percent of women) claimed to read science columns regularly, another third on occasion, and another third never. According to Gildea (1996, p. 161), the regional press continued to dominate the "national" press in circulation figures throughout the postwar period.

6. On the Gard see Hecht 1997b; on the Touraine see Hecht 1992.

7. Here I paraphrase Peter Sahlins (1989).

8. Sahlins 1989.

9. Le Goaziou 1988.

10. For a detailed account of Joliot's scientific work see Weart 1979.

11. Available for viewing at the Vidéothèque de Paris.

12. For example, "La science 'dirigée' . . . M. Joliot-Curie haut commissaire à l'énergie atomique fait un cours de morale communiste," *Le Parisien Libéré*, 26 July 1947; Raymond Aron, "La Cité déchirée: l'état et les communistes," *Le Figaro*, 11 April 1948.

13. "Joliot-Curie et la science française remparts de l'indépendance nationale," *l'Humanité*, 19 May 1949.

14. Another, somewhat different conflation of drama and history took place in the 1968 television miniseries "Les atomistes," which ran for 5 weeks. *Le Figaro* described the series (which I was not able to view) as follows ("Le feuilleton T.V. de l'univers atomique français," 12 February 1968): "With this new miniseries, we enter resolutely into the modern world. The heroes of our evenings will be scientists in white coats. The research centers of Cadarache, Saclay, and Fontenay-aux-Roses will provide their field of research. . . . The filming took place in this French atomic universe. . . . The CEA opened its doors wide to the actors and the team directed by Leonard Kriegel. Real atomists helped their doubles, and even participated in the programs themselves. The experiments underway in the laboratories were in no way interrupted during the filming. Here the imaginary is etched onto daily life." The plot centered around a young researcher who makes a discovery that could make "science take a considerable leap forward of ten years." He puts together a team. The initial results are good, but then "the human element takes over from the disinterestedness of science" and "an accident jeopardizes the research." The article left the rest for viewers to discover, noting only that there was a "woman atomist" in the show (a character made possible, doubtless, by Marie Curie and her daughter Irène). Two years later the TV series "Au théâtre ce soir" presented *Frédéric*, a play about a maintenance worker in an atomic plant in the French Pacific who inadvertently learns about some sensitive scientific secrets (Inathèque, CPF86604273).

15. "La Session d'information sur l'énergie nucléaire—'La France a besoin très vite d'ingénieurs et de techniciens,' déclare notamment le premier conférencier M. Fraudeau," *La Nouvelle République du Centre-Ouest*, 7 and 8 May 1957.

16. A typical headline in *l'Humanité* (5 May 1959): "It is not enough to talk of 'French grandeur'! The crisis in scientific research threatens the future of our country."

17. "Techniques modernes . . . carrières d'avenir," advertisement in *Science et Vie* 530 (November 1961), p. 15.

18. Henri de Turenne, "Ainsi naquit la 'Caravelle,' l'enfant prodige de l'aviation française," *France-soir*, 30 July 1957.

19. Ibid.

20. This includes *l'Humanité*, although the heroes of its week-long series on the Caravelle were not individual, elite engineers but ordinary, nameless engineers, workers, and technicians. The villains of *l'Humanité*'s story (Jean Mérot, "Aéronautique française 57: Classe internationale oui, mais . . . ," 9–17 May 1957) were American companies hungrily seeking to snatch the Caravelle away from its nationalized French manufacturer. In another example, from the opposite end of the political spectrum (Pierre Voisin, "'Caravelle,' ce monstre rugissant, parfaitement domestiqué, est un avion de rêve," *Le Figaro*, 22 April 1957), the Caravelle was called "the second French revolution . . . indescribable, unimaginable, unreal." As Rosemary Wakeman (1992, 1997) shows, the city of Toulouse, which housed the French aviation industry, went into even more intense raptures of joy over the salvational powers of most aviation accomplishments. Aviation provided Toulouse's special spectacle.

21. A few examples: Armand Macé, "L'industrie française devant l'Europe," *Le Figaro* 7–8 July 1962; Nicolas Vichney, "Le programme spatiale français," *Le Monde*, 14 September 1965; "Concorde: 'l'avion était fait pour ça' constate son commandant de bord après avoir, pour la première fois, percé le mur de son," *l'Humanité*, 2 October 1969.

22. Thierry Maulnier, "L'Armée défend en Algérie la chance africaine de la France: les réalités commandent," *Le Figaro*, 2 May 1957. Similar articles include two by Max Clos in *Le Figaro*: "Le départ massif des techniciens, des industriels et de colons français serait une catastrophe pour la Tunisie" (10 April 1957) and "Pour que la France s'assure une place de choix dans les relations transpacifiques, l'aérodrome de Tahiti doit être construit" (15 May 1957). For a history of technological and architectural efforts to "save" northern Africa see Wright 1991 and Rabinow 1989.

23. G. Verpralt and P. Levaillant, "Inaugurant une puissante centrale sur le grand canal d'Alsace, M. René Coty: 'Fessenheim, témoignage des progrès incessants de la technique française," *Le Figaro*, 9 July 1957.

24. The 1960 promotional film *La Chance Nord-Sud* provides an example of a variation on this theme. Commissioned by the Ministère de l'Equipement, this film

was aimed at persuading local officials to support a canal project that would link the Rhône to the Rhine, thereby restoring France to its rightful role as "the natural meeting place of Europe" and indeed saving French civilization after its decline. Another promotional film that celebrated French accomplishments was *Orly 1964–66: la construction de la Tour de Contrôle* (1966), commissioned by Aéroport de Paris (Vidéothèque de Paris).

25. Jacques Chapus and Serge Maffert, "De Gaulle a visité les usines de 'Caravelle' où 2000 ouvriers ont rompu les barrages pour lui tendre les mains," *France-Soir*, 15–16 February 1959.

26. Including inaugurations of nuclear reactors. Prime Minister Georges Pompidou visited EDF's nuclear site at Chinon shortly after its first reactor went on line in 1963. ("Pompidou en Touraine," *Journal Télévisé*, 25 July 1963; Inathèque CAF94058300). A few years later, de Gaulle was reportedly furious when technical problems prevented him from inaugurating EDF's third reactor at the same site.

27. "Voici les premières photos et le premier récit de l'explosion atomique française," *Journal du Dimanche*, 14 February 1960.

28. The paper listed Pierre Guillaumat, Francis Perrin, Pierre Couture, Pierre Taranger, Bertrand Goldschmidt, General Albert Buchalet, and Yves Rocard.

29. "Le film de l'explosion," *Journal du Dimanche*, 14 February 1960.

30. *Le Figaro* offered coverage similar to that of *France-soir*: see for example "Première bombe A française—explosion réussie," and "Reggane 13 février 7h04: La France est devenue puissance atomique," *Le Figaro*, 15 February 1960.

31. "Prix de l'Urbanisme 1959—Bagnols-sur-Cèze" (Ministère de Construction, 1960; no page numbers), Archives municipals, Bagnols-sur-Cèze.

32. Ibid.

33. Ibid.

34. Ibid.

35. *Le Midi Libre*, 26 February 1957.

36. This toy can be seen in the reading room of the CEA's archives in Fontenay-aux-Roses.

37. For a discussion of these tropes see Corbin 1996.

38. Alfred Chabaud, "Les Perspectives Nouvelles de la Région Bagnolaise: L'Electro-métallurgie et l'Energie atomique dans la basse-vallée du Rhône," *Revue d'Economie Méridionale* (premier trimestre 1957): 16. (The page numbers given here are for an extracted copy of this article located in the Archives Départementales du Gard; the actual page numbers in the journal are 61–81.)

39. This too is an old theme in French history. See Lebovics 1992, Adas 1989, and Ross 1995.

40. Chabaud, "Perspectives Nouvelles," p. 16.

41. Ibid., p. 17.

42. "Visite au Grand Centre Atomique Français," *Le Provençal*, July 1957 (n.d.), reproduced in Cazals, "La Résistance au Nucléaire dans la Région du Gard Rhôdanien." In both French and English, the word "pile" was used at the time to designate a reactor core.

43. Etienne Anthérieu, "Ce que j'ai vu au grand centre atomique français," *Le Figaro*, 17 July 1957; Jean Boiset, "Le kilowatt 'A' trop cher . . . ," *Science et Vie*, 494, November 1958. *Science et Vie* used these comparisons as a way of heightening its critique of the economics of France's nuclear program, discussed later in this chapter.

44. Though most prominent there, this kind of epiphanic language was not confined to nuclear achievements. For example, a promotional film about the Donzère-Mondragon hydroelectric plant (*Donzère-Mondragon*, Ministère de l'Equipement, 1953) offered the following description in a voice-over: "Around and above the machine at Donzère, walls rise like the nave of a modern cathedral." The camera showed the building from the inside, panning over enormous windows, supported by a complex concrete lattice structure, with rays of sunshine pouring inside, indeed reminding the viewer of light streaming through the stained-glass windows of a cathedral.

45. Departmental administrators also engaged in analyses of the region which reinforced the interpretive framework discussed here. See e.g. Jean Driol, "La Notion de Croissance des Agglomérations: Etude du coût & du financement de la croissance de Bagnols-sur-Cèze" (Section d'Administration Economique et Financière, Nîmes, Décembre 1958), p. 1 (Archives départmentales du Gard).

46. Chabaud, "Perspectives Nouvelles," p. 13.

47. Pierre Gamel, "Marcoule donnera au Gard de demain un rayonnement national que n'avait pu lui apporter le charbon," *Le Provençal*, 1957 (n.d.), reproduced in Cazals 1982–83.

48. Two examples among many include a series in *Le Provençal* titled "Bagnols-sur-Cèze: Cité Mediévale et ville Champignon" (11–13 July 1957) and "Une insidieuse offensive vauclusienne," *Le Midi Libre*, 21 January 1956.

49. "Technique et esthétique," *Le Midi Libre*, 19 February 1958, reproduced in Cazals 1982–83.

50. See for example articles in *Le Midi Libre*, 20 March 1957 and 8 February 1958; Département du Gard, *Livre d'Or régional: économique et touristique*, 15 July 1964; Bagnols-sur-Cèze, *Revue Municipale*, 1960 (Archives municipales, Bagnols-sur-Cèze).

51. Département du Gard, *Livre d'Or régional: économique et touristique*, 15 July 1964 (Archives municipales, Bagnols-sur-Cèze).

52. Interviews.

53. Procès-Verbaux du Conseil Municipal de Bagnols-sur-Cèze, 9 March 1962 (Archives municipales, Bagnols-sur-Cèze).

54. Another example can be found in Procès-Verbaux du Conseil Municipal de Bagnols-sur-Cèze, 29 December 1966 (Archives municipales, Bagnols-sur-Cèze).

55. Préfecture d'Indre-et-Loire, Conseil Général, *Procès-Verbaux des Séances et Délibérations du Conseil Général*, 10 May and 17 October 1955 (Archives départmentales, Indre-et-Loire).

56. "Une exemple de l'exode rural: le Véron, une des plus riches terres du Chinonais, a perdu 1500 habitants en 70 ans," *La Nouvelle République du Centre-Ouest*, 26 January 1955. This kind of depiction of the traditional, prolific French woman has a complex history in the pro-natalist discourse of the interwar period. See notably Downs 1995 and Roberts 1994.

57. Préfecture d'Indre-et-Loire, Conseil Général, *Procès-Verbaux des Séances et Délibérations du Conseil Général*, 10 May 1955.

58. Préfecture d'Indre-et-Loire, Conseil Général. *Procès-Verbaux des Séances et Délibérations du Conseil Général*, 30 January 1956, p. 47.

59. Ibid., p. 48.

60. Préfecture d'Indre-et-Loire, Conseil Général. *Procès-Verbaux des Séances et Délibérations du Conseil Général*, 1949–1970.

61. Lettre du 14.2.56 de J.P. Roux au Ministère de l'Industrie et du Commerce [Archives du Ministère de l'Industrie, 9000960, no. 870342/27]; Préfecture de l'Indre-et-Loire, Conseil Général. Procès-Verbaux des Séances et Délibérations du Conseil Général. 30.1.56; *La Nouvelle République du Centre-Ouest*, 20, 24, and 25 December 1955; interview.

62. "La Centrale Nucléaire du Véron ne nuira pas à l'esthétique du site célèbre de Candes-Monstsoreau, affirme M. Teste, contrôleur général à l'EDF," *La Nouvelle République du Centre-Ouest*, 28 and 29 January 1956.

63. Préfecture de l'Indre-et-Loire, Conseil Général. *Procès-Verbaux des Séances et Délibérations du Conseil Général*, 17 May 1956 (Archives départmentales, Indre-et-Loire).

64. Lettre de M. Chalumeau à M. Lamouroux, 22.12.55. Archives du Ministère de l'Industrie, 9000960, no. 870342/27.

65. Préfecture de l'Indre-et-Loire, Conseil Général. *Procès-Verbaux des Séances et Délibérations du Conseil Général.* 30.1.56; La Centrale Nucléaire du Véron ne nuira pas à l'esthétique du site célèbre de Candes-Monstsoreau, affirme M. Teste, contrôleur général à l'EDF."

66. Raymond Dreulle, "Sur 80 hectares, un morceau de Touraine est passé . . . ," *La Nouvelle République du Centre-Ouest*, 23 and 24 May 1959.

67. "Marcoule: une cathedrale de béton et d'acier élevée en pleine campagne," *La Nouvelle République du Centre-Ouest*, 21 June 1955.

68. For a discussion of how Loire Valley châteaux have become "realms of memory" see Babelon 1992.

69. "Gardée comme une place-forte, la centrale nucléaire entr'ouvre ses grilles aux touristes du dimanche," *La Nouvelle République du Centre-Ouest*, 3 February 1959.

70. The journalist, who evidently had not mastered the information imparted to him during the tour, mistakenly gave these figures in kilowatt-hours, not kilowatts. I have inserted the correct units to avoid confusion.

71. *La Nouvelle République du Centre-Ouest*, 9 May 1958.

72. Préfet de l'Indre-et-Loire avec le concours des services d'EDF, "L'Indre-et-Loire entre dans l'ère Atomique, brochure édité à l'occasion de la Grande Semaine de Tours 1957" (Bibliothèque municipale de Tours, F2139).

73. R. Mille, P. Decker, and EDF, "La Centrale Chinon-Avoine—sous le sigle EDF—trace la voie de l'expansion énergétique du Centre-Ouest," *La Nouvelle République du Centre-Ouest*, 16 and 17 February 1963.

74. See *La Nouvelle République du Centre-Ouest*, 23 and 24 May 1959, 26 July 1960, 16 and 17 February 1963, and 30 and 31 July 1966. Many other articles on this subject appeared in between. "Hexagon" is a nickname for France that refers to the nation's shape.

75. Boyer 1985; Weart 1988.

76. "Cette petite bombe," *Ouest-France*, 18–19 September 1945, p. 1, cited on p. 40 of Pace 1991.

77. Pace 1991, p. 40.

78. Cited on p. 45 of Pace 1991: Léopold Vigneron and Raymond Chastel, *L'Energie atomique ou calamité?* (Paris, 1949); Daniel Declos, *Bombe atomique: Vers la destruction totale ou le paradis terrestre* (Paris, 1945); Pierre Rousseau, *La Fin du monde; ou l'âge d'or? Histoire documentée de l'énergie atomique* (Lyon, 1946). Pace gives more examples of such books: Edouard Jacquet, *Vers la fin du monde par la bombe atomique ou vers un avenir merveilleux?* (Lyon, 1945); Maurice Déribé, *L'Energie atomique: La Bombe atomique annonce-t-elle notre fin prochaine ou le début d'une ère nouvelle?* (Paris, 1945).

79. André Fontain, "Les radiations atomiques préparent-elles notre suicide collectif?" *France-soir*, 21–29 May 1957.

80. Although I cannot pursue this point here, it is interesting to note that the heroes were members of the elites and the victims were not.

81. Fontain, *France-soir*, 27 May 1957.

82. Charles-Noël Martin, "100 Mégatonnes, qu'est-ce que cela veut dire?" *Science et Vie* 530 (November 1961): 38–44.

83. Jérôme Gauthier, "Eh bien, chiche!" *Le Canard atomique—Le Caneton, supplément au Canard Enchaîné*, 26 June 1957. *Canard* prose is notoriously difficult to translate; the effect in English is never the same. I am thus providing the French version of the text as well and including a bit more of the passage: "On a tari les sources du spirituel. On a déboulonné le Dieu croquemitaine tout juste bon à faire peur aux serins, pendant que les corbeaux se gavaient. On a chassé le Mensonge, mais aucune Vérité n'est venue prendre sa place. La démolition nécessaire accomplie, de la superstition, de l'absurde, du fanatisme, on s'est retrouvé face à face avec l'infini, l'inexplicable, l'incompréhensible. L'homme, robot pensant, guéri de ses visions, s'est redressé aveugle, au milieu d'un fatras de connaissances qui ne lui font pas une miette de certitude. Les saints ont fait faillite et le seul Juste don on nous ait parlé, sert d'enseigne, sur sa croix, aux derniers carambouilleurs d'évangiles. Vers qui, vers quoi se tourner? . . . .La bêtise de l'autre Dieu nous faisait rire: l'intelligence du vôtre nous fait peur, cardinaux de la Haine. Toute science, dans vos pattes, devient l'arme d'un crime. De toute lumière, vous tirez un foudre, de toute énergie une agonie, de toute force un droit. Vous avez ficelé l'humanité sur la chaise atomique, et votre gaminerie de bourreaux joue avec la manette: 'Chiche?'

"Eh bien. CHICHE! Et MERDE! Et que tout saute! . . .

"Qu'on efface tout, mais qu'on ne recommence rien! Et que la planète Terre, froide, exsangue, roule enfin, sans vie, sans pensée, tête coupée du grand Tout, dans le panier de silence."

84. Cazals 1982–83, p. 38. I found remarkably few written traces of Lanza del Vasto anywhere, though everyone knew who he had been. Even his name seemed to be a subject of some confusion: he was also referred to as Lanzo del Vasta, and even Lama del Vasto (the latter in "Succès des rassemblements de Marcoule et Villejuif contre le péril atomique," *l'Humanité*, 26 October 1959).

85. "Nous Sommes Dix-Huit" (flyer, courtesy of Jacques Bonnaud).

86. No one I interviewed was able to provide his name, and I could find no written traces of him.

87. Jean Boiset, "Le kilowatt 'A' trop cher . . . ," *Science et Vie* 494, November 1958. Select other examples of articles that criticized nuclear development or focused on foreign achievements to the expense of national ones can be found in the September 1954, April 1958, September 1958, November 1961, August 1964, and April 1967 issues of *Science et Vie*.

88. Public relations folder, CEA archives, M7-18-34. See my chapter 2.

89. In French, "Faites-en tout un plat," which also means "Make a big deal out of it." *Le Canard enchaîné*, 17 February 1960.

90. The word *éclat* means "flash." The phrase cannot be translated without sacrificing its punch, but the overall effect is to make fun of de Gaulle's equation of

himself, the French nation-state, and the atomic bomb. Other puns on that page included "Ce de Gaulle n'a pas fini de nous détonner" ("That de Gaulle has not done detonating us," where *détonner* is a play on *d'étonner*, "to astonish") and "Fisson accomplie Mongénéral" ("Fission accomplished, general").

91. *Le Canard enchaîné*, 17 February 1960.

92. *Le Canard enchaîné*, 8 June 1966, p. 3. Le Nôtre was the designer of the gardens at Versailles. This cartoon appeared next to a regular column, "La Cour" (The Court), that often referred to de Gaulle as "the king."

93. See e.g. "Hélène Langevin: une bombe atomique française ne servira pas la grandeur de notre pays," *l'Humanité*, 19 May 1959.

94. "Succès des rassemblements de Marcoule et Villejuif contre le péril atomique," *l'Humanité*, 26 October 1959.

95. *L'Echo du Midi*, 20 July 1957 and 24 August 1957.

96. *L'Echo du Midi*, 7 September 1957.

97. *L'Echo du Midi*, 26 October 1957, 16 November 1957, 15 November 1958.

98. *L'Echo du Midi*, 26 October 1957, 25 January 1958, 5 April 1958.

99. "Le propos du Cévenol," *L'Echo du Midi*, 7 September 1957.

100. *L'Echo du Midi*, 31 August 1957. Tavel was a local rosé.

101. Closely following the rest of the Poujadiste movement, the paper briefly indulged in supporting the government when de Gaulle returned to power in 1958. Supporting the government meant supporting Marcoule, and for a few months the paper let up on its criticism of the site. De Gaulle's strong statist agenda soon disillusioned the Poujadistes, and by early 1960 the paper had returned to its opposition stance.

102. M. Gauthier, "Quand Marcoule sera conté" (Pièce jouée à Bagnols les 11, 12, et 18 mai 1957) (typescript, Bibliothèque municipale, Bagnols-sur-Cèze). The script came complete with photographs of the scenes as they were acted, suggesting that only part of the play was written before hand, and the rest worked out by the cast in rehearsals. This was confirmed in an interview with the Menjaud family. They noted that the play's title and form was inspired by the two Sacha Guitry films that had appeared a few years earlier: *Si Paris nous était conté* (1955) and *Si Versailles m'était conté* (1953). Modeling the town play on these films—and thus placing Marcoule on the same narrative plane as Paris and Versailles—may well have been intended to convey a certain irony. The two films can be viewed at the VDP.

103. For a discussion of the technological symbolism of the Eiffel Tower see Levin 1989. Loyrette (1992) talks about how the tower became the quintessential symbol of both Paris and the nation, and about the role of Eiffel Tower memorabilia in narrations of visits to Paris by *provinciaux*.

104. Ross (1995) discusses cultural representations of housekeeping and modern couples in postwar France. Newly crafted associations between romance and modernized housekeeping were widespread in this period. Ross (p. 96) cites a song by Boris Vian, "Complainte du progrès," that evokes many of the same sentiments as this scene in the play:

> Autrefois pour faire sa cour
> On parlait d'amour
> Pour mieux prouver son ardeur
> On offrait son coeur
> Aujourd'hui c'est plus pareil
> Ça change, ça change
> Pour séduire le cher ange
> On lui glisse à l'oreille
> Ah . . . gudule . . . viens m'embrasser
> Et je te donnerai
> Un frigidaire
> Un joli scooter
> Un atomiseur
> Et du Dunlopillo
> Une cuisinière
> Avec un four en verre
> Des tas de couverts
> Et des pelles à gâteaux
> Une tourniquette
> Pour faire la vinaigrette
> Un bel aérateur
> Pour bouffer les odeurs
> Des draps qui chauffent
> Un pistolet à gauffres
> Un avion pour deux
> Et nous serons heureux

Translation (based on Ross's translation, with my corrections): Before when we went courting/We spoke of love/To prove our passion/We offered our heart/Today it's not the same/It's changing, it's changing/ To seduce the dear angel/We murmur in her ear/Oh . . . honey-pie . . . come give me a kiss/And I'll give you/A refrigerator/A pretty scooter/An atomizer/And some Dunlopillo/A stove/With a glass oven/Piles of cutlery/And a cake server/A whirligig/To make salad dressing/A lovely fan/To chase the odors away/Nice warm sheets/A waffle iron/An airplane for two/And we'll be happy.

Chapter 7

1. The validity of polls, the extent to which they reflect "popular opinion," and the problem of their design are extremely complex issues which I cannot address here. For a good treatment of these matters, with relevance well beyond the

French case, see Cowans 1993. For a much briefer and less reflective treatment see Cowans 1991.

2. *Sondages*, 1956 (3): 39–40.

3. In table 7.1 "yes" corresponds to "satisfied" as and "no" to "unsatisfied." It is possible, of course, that people who declared themselves "unsatisfied" thought that the *force de frappe* should be stronger still; however, this seems unlikely, since there were no public critiques of Gaullist policy along those lines.

4. Fourgous et al. 1980, pp. 22–36.

5. In both cases, the polls broke down responses by various standardized census categories. This breakdown confirmed stereotype: in both cases, men appeared better informed than women; adults under 50 better informed than those over 50; white-collar workers and professionals better informed than blue-collar and agricultural workers; residents of larger towns better informed than those of smaller towns.

6. Fourgous et al. 1980, pp. 22–36.

7. *Sondages*, 1957 (3), pp. 17–18. The rest of the sample responses for those who preferred peaceful development broke down as follows (using the *Sondages* categories): **France can use the bombs of its allies (4%)**: "We don't have to build weapons; they will be given to us, maybe even forced on us"; "France isn't likely be alone in a war, it's enough for the allies to have the bomb." **France is already too far behind the USSR and the US (4%)**: "We'll never be as good as the Russians, it's not worth trying"; "We can never compete with the great powers"; "France can't compete with foreign atomic weapons manufacturing; might as well use the little she has to peaceful ends." **Atomic weapons are too monstrous (2%)**: "Atomic weapons make you fear the worst things." **Atomic war will not take place (2%)**: "There's no point in making atomic bombs, since we wouldn't use them"; "I don't believe in an atomic war that would bring the extermination of the world"; "Atomic weapons won't be any good, or it would be the end of the world."

8. *L'Echo du Midi*, 6 October 1957 (reprint of portions of an article by Pierre Accoce that appeared in *Constellation*, September 1956). *L'Echo* was the region's Poujadiste paper and extremely critical of Marcoule. Obviously these attitudes shaped the sorts of stories it printed. Still, all the *Echo* stories I quote from were told to me in a variety of forms by local residents whom I interviewed in 1994, which (in combination with the fact that nobody seemed to remember even the existence of *L'Echo*) suggests that the stories did in fact originate among local residents rather than among the journalists reporting on them.

9. *L'Echo du Midi*, 12 October 1957 (continued reprint of portions of an article in *Constellation*, September 1956).

10. Ibid.

11. Interview with Maurice Fabre.

12. Procès-verbaux, Conseil Municipal de Chusclan, 28 August 1953; Procès-verbaux, Conseil Municipal de Codolet, 6 September 1953.

13. *Le Provençal*, 2–5 October 1958; interviews; Procès-verbaux, Conseil Municipal de Codolet, 1958–1960.

14. Interview with Louis Anglezan.

15. The only mention I found of this statement was in one set of town meeting minutes: Procès-verbaux, Conseil Municipal de Chusclan, 5 February 1953. Although many interviewees mentioned that they had first heard about Marcoule in this oblique manner, none mentioned Chusclan's initial formal opposition.

16. *Le Provençal*, 26 June 1957.

17. Applicants to Marcoule underwent a security check by the services of the Renseignements Généraux (see chapter 5). Most of the residents I interviewed mentioned this process, even though they themselves had never applied to Marcoule for a job. I was denied access to these records, housed in the Archives départementales du Gard.

18. Souchon 1968.

19. Interviews; *L'Echo du Midi*, 20 July 1957; *Courier Français du Témoignage Chrétien*, 20 August 1964.

20. Procès-verbaux, Conseil Municipal de Chusclan, 20 December 1956.

21. *Le Midi Libre*, 1 December 1956.

22. Procès-verbaux, Conseil Municipal de Codolet, 13 August 1959; interview with Louis Anglezan.

23. Procès-verbaux, Conseil Municipal de Chusclan, 1956–1966; Procès-verbaux, Conseil Municipal de Codolet, 1956–1965.

24. Interview with Francis Lemesle.

25. Procès-verbaux, Conseil Municipal de Bagnols, 12 December 1963.

26. Henri Duprat, "Etude Economique du Site de Marcoule," *Rapport CEA* 2541 (August 1964): 15. According to this study, the new inhabitants of Bagnols made 1.5 times what the established residents made and nearly twice what the average villager made. Most of the people interviewed discussed this economic disparity.

27. Procès-verbaux, Conseil Municipal de Bagnols, 23 December 1963 and 20 February 1964.

28. Suzanne Frère, *Bagnols-sur-Cèze: Enquête Sociologique* (J. & M. Pailhé, 1968), p. 142. Archives départmentales du Gard.

29. In addition to Souchon, Duprat, and Frère (all cited above), see the notes to the section on the Gard in chapter 6.

30. This was the first paragraph in the column "On se Montre," Y1, *Midi Libre* (Bagnols), 26 July 1994. Subsequent paragraphs commented on the appearance of 4-meter-high sunflowers, the effects of a hailstorm on local farms, tourists napping in their cars in a town parking lot, the complaint of the new president of the Conseil Général that the press gave lopsided information by only reporting the technical problems in local nuclear sites, and the death of a Brazilian spelunker.

31. Nelkin and Pollak 1981.

32. Interview with André Voisin.

33. Interview. Another resident told a similar story about the Bretons.

34. Interview.

35. "La Rabelaisie en Touraine, capitale Chinon" (undated tourist brochure from Office de Tourisme, Chinon).

36. Interview.

37. Interviews with M. Raffault and G. Joubert.

38. Dossier Déclaration d'Utilité Publique EDF1, EDF2 (Archives du Ministère de l'Industrie, 9000960, no. 870342/27).

39. Actes, Conseil Municipal d'Avoine, 31 March and 8 May 1956.

40. Préfecture de l'Indre-et-Loire, Conseil Général, Procès-Verbaux des Séances et Délibérations du Conseil Général, 30 October 1956.

41. Procès-verbaux, Conseil Municipal, Bourgueil, 12 November 1965; interview.

42. EDF, Service National, Recueil des Conférences (Stage de Nainville-les-Roches du 12 au 15 juin 1961) (Bibliothèque Municipale, Tours).

43. Procès-verbaux, Conseil Municipal, Bourgueil, 22 August 1961.

44. Interview with Jean Chamboissier.

45. "La conférence d'information: la conquête de l'énergie atomique ouvre une ère nouvelle," *La Nouvelle République du Centre-Ouest*, 9 May 1957.

46. *La Nouvelle République du Centre-Ouest*, 10 May 1957.

47. Interview.

48. Interview.

49. Interview with Jean Chamboissier.

50. Actes, Conseil Municipal d'Avoine, 23 February 1957.

51. Interview.

52. *La Nouvelle République du Centre-Ouest*, 23 and 24 May 1959.

53. Interview.

54. Procès-Verbaux, Conseil Municipal de Bourgueil, 1961.

55. Actes, Conseil Municipal d'Avoine, 1958–1965.

56. Actes, Conseil Municipal d'Avoine, 1958–1965; interviews.

57. Actes, Conseil Municipal d'Avoine, 1957; Cahiers, Conseil Municipal de Chinon, 1957.

58. Cahiers, Conseil Municipal, Chinon, 15 December 1961.

59. Procès-verbaux des délibérations, Conseil Municipal, Bourgueil, 15 June 1965.

60. Préfecture de l'Indre-et-Loire, Conseil Général, Procès-Verbaux des Séances et Délibérations du Conseil Général (1960–65); *La Nouvelle République du Centre-Ouest* (1960–69); interviews.

61. Interview.

62. Cahiers, Conseil Municipal, Chinon, 8 March 1957.

63. The same year, Avoine's population was around 970; Beaumont's was around 1800; Savigny's was around 1000.

64. Procès-Verbaux, Conseil Municipal de Bourgueil, 1961.

65. Procès-Verbaux, Conseil Municipal de Bourgueil, 1965.

66. Interview.

67. Another example, from an interview: "Then there were the people from the Midi, very nice people. The whole family would come, the kids would go to school."

68. Interview with Jean Chamboissier.

69. *La Nouvelle République du Centre-Ouest*, 17 August 1964.

70. Ibid.

71. Interview.

72. Interview. Other residents made similar points.

73. *La Nouvelle République du Centre-Ouest*, 6 June 1963.

74. Interview.

75. Interviews.

Chapter 8

1. "Pourquoi l'Energie Atomique en Grève" (CFDT, CGT, CGT/FO tract, 6 November 1969) (Laponche papers).

2. Notably, the television news program "Panorama" devoted a special 15-minute segment to the strikes on 27 November 1969. The program can be viewed at the Inathèque in Paris.

3. Especially Frost 1991 and Soutou 1991.

4. For more on these various projects see Soutou 1991, Lamiral 1988, Floquet 1995, and Bupp and Derian 1978.

5. In 1964, the Commission pour la Production d'Electricité d'Origine Nucléaire—a government advisory group composed of officials from EDF, the CEA, and private industry—issued a report speculating that light-water reactors might have lower investment costs than gas-graphite reactors; it emphasized, however that pursuing the light-water option would mean either depending on the United States to supply enriched uranium fuel or building an expensive enrichment plant that would negate the still uncertain cost advantage of the light-water system. Thus, the PEON commission decided, France should stay on the gas-graphite track (Simmonot 1978, pp. 237–245). In forwarding this recommendation to Prime Minister Georges Pompidou, Gaston Palewski (the minister in charge of atomic affairs) urged Pompidou to make a rapid decision to engage in "massive" development of gas-graphite reactors. The consequences of the choice, Palewski said, mattered not only for France but also for the rest of Europe. Were France to give up its own reactors, "the ensuing technical lag would be felt in our economy and in our policy" (G. Palewski to Georges Pompidou, 4 July 1964, CEA archives, box F3-24-25). Pompidou approved the PEON plan.

6. This argument was made by Bupp and Derian (1978, p. 49). Many other writers have supported this analysis. Frost (1991), Jasper (1990), Puiseux and Saumon (1977), Pringle and Spigelman (1981), and several of the people interviewed by Simmonot (1978) assert that light-water reactors ended up costing far more than the initial estimates suggested.

7. EDF managers and engineers had been making unofficial inquiries into this possibility since at least 1965. See letter from C. Bienvenu to Melèse, 30 December 1965 (Bienvenu papers).

8. Letter from A. Decelle to R. Hirsch, 7 March 1966 (Bienvenu papers).

9. Groupe de Travail Commun CEA-EDF sur les Filières à Uranium Enrichi, 4 May 1966; note from A. Decelle to R. Hirsch, 4 May 1966 (Bienvenu papers).

10. EDF Conseil d'Administration 240, 27 May 1966, p. 37.

11. EDF Conseil d'Administration 244, 25 November 1966; 249, 10 March 1967.

12. Nicholas Vichney, "La centrale nucléaire EDF3 de Chinon est arrêtée pour six mois," *Le Monde* (2 December 1966), p. 7; Vichney, "Les incidents survenus à la centrale de Chinon amènent à poser le problème des rapports entre l'Electricité de France et le Commissariat à l'énergie atomique," *Le Monde*, 24 January 1967.

13. "Le crêpage de Chinon," *Le Canard Enchaîné*, 7 December 1966, p. 2.

14. EDF Conseil d'Administration 244, 25 November 1966.

15. In fact, Vichney at least had mentioned all these points, but only in passing; evidently he did not emphasize them in the way that EDF's leaders would have liked.

16. EDF Conseil d'Administration 245, 22 December 1966.

17. In January 1967, in an effort to reassure the personnel, Pierre Massé wrote a letter to André Decelle in which he confirmed that EDF's nuclear teams had carried out their duties with "exemplary devotion" (Bienvenu papers). Building reactors was a difficult task, Massé affirmed, and was getting more difficult all the time. But Massé also warned that the personnel could expect some structural change within the institution, which, he hastily added, was only normal in such a large establishment. The Direction de l'Equipement would soon receive union representatives to discuss these changes.

18. "Communiqué commun CEA et EDF," 25 January 1967 (CEA archives, F3-24-25).

19. "Un communiqué intersyndical à la Presse, après le récent article de M. Nicolas Vichney," memorandum (unsigned) from département de sûreté et de protection du secret, 22 February 68 (CEA archives, F3-24-25).

20. Letter, Robert Hirsch to Hubert Beuve-Méry, 24 April 1968. Beuve-Méry responded in a conciliatory tone on 30 April, but he did not agree to publish a correction (CEA archives, F3-24-25).

21. Transcript of TV program broadcast on 7 May 1968 in series "Demain Commence Aujourd'hui," with Henri Polad and J. C. Bollardot (CEA archives, F3-24-25). According to one of the CEA's public relations officers, Vichney had developed a personal grudge against the CEA. The journalist had called one morning to inquire about a plan that *Le Monde* had, with the CEA, to devote a full page to Phénix, the breeder reactor whose construction had recently been approved. Upon learning that the CEA had decided to abandon the plan, Vichney turned hostile. According to the officer, he replied: "So it's war . . . and that's why I didn't receive an invitation to your next luncheon. Well, we'll see who laughs now!" The public relations officer continued his report as follows: "On this last phrase, his voice became sarcastic. I pointed out to him that the luncheon in question was more amicable than professional, and that an absence of invitation did not mean a rupture in our professional relations. 'I'll say,' Vichney replied in a grinding voice. 'But I wouldn't have come to this lunch anyway, because I found Mr. Goldschmidt's unfriendly remarks to me on the last occasion intolerable.'" Vichney apparently accused the CEA of not following through on its promises to give *Le Monde* special access to its news. According to the CEA's public relations service, Vichney had been demanding exclusive scoops on all of the CEA's major projects. But when the CEA granted those scoops, bitter feelings often resulted because Vichney would do nothing with the information he had received, even

after CEA scientists and engineers had gone to much trouble and taken much time to arrange exclusive meetings and interviews. In sum, "the unceasing attacks of Nicholas Vichney against the CEA, his polemic with the labor unions, the development of his campaign to help the schemes of Mr. Ambroise Roux, and his threats to the CEA have made the collaboration so painstakingly developed difficult to continue." (Correspondence of Goldschmidt, Renou, and Benoit, Service des Relations Publiques, 14 and 22 May 1968, CEA archives, F3-24-25)

22. EDF Conseil d'Administration 249, 10 March 1967, p. 18.

23. Ibid.

24. Ibid., p. 21.

25. EDF Conseil d'Administration 249, 10 March 1967, p. 29.

26. CFDT, Federation EGF, "Direction de l'Equipement: La Situation de l'Equipement Nucléaire," 5 May 67 (Bienvenu papers).

27. Ibid.

28. Jean Cabanius, "Rapport du Groupe de Travail Placé sous la responsabilité de Monsieur Cabanius (EDF)," 25 January 1967, p. 2.

29. J. Horowitz, "Examen des filières électro-nucléaires dans le contexte français actuel" (1 February 1967).

30. Cabanius, "Rapport du Groupe de Travail," pp. 4–5.

31. Horowitz, "Examen des filières électro-nucléaires."

32. Cabanius, "Rapport du Groupe de Travail," p. 9.

33. Ibid., p. 10.

34. Horowitz, "Examen des filières électro-nucléaires," p. 10.

35. They based their calculations of the cost of the pressurized-water kilowatt-hour on a meeting they had with Framatome, the company that had managed the construction of the Chooz reactor under license to Westinghouse. They had also tried to get figures for the cost of building a boiling water reactor, but no French company or consortium had a license with General Electric yet. (Memorandum from Jules Horowitz to Administrator-General, "Conditions de construction de centrales à eau ordinaire par l'industrie française. Réunions du 7 juillet avec Framatome-Westinghouse, du 13 juillet avec Alsthom-GECO," 27 July 1966, CEA archives, F 6-13-20)

36. Only in an appendix to his report did Cabanius discuss (in half a page) the fact that the American estimates were extrapolated from a very limited operational experience. He also used the appendixes to elaborate on the technical uncertainties of the gas-graphite system. See "Annexe III: les données actuelles et les perspectives futures des centrales classiques et nucléaires," in Cabanius, "Rapport du Groupe de Travail."

37. Horowitz, "Examen des filières electro-nucléaires," p. 26.

38. Ibid., p. 25.

39. Cabanius, "Rapport du Groupe de Travail," p. 25.

40. Although in practical terms the Horowitz-Cabanius committee resolved nothing, some nuclear leaders attempted to extract a resolution from the reports. By occasionally referring to them as "the Horowitz-Cabanius report," they erased the differences. Not surprisingly, the nature of this supposed "consensus" varied. For example, Pierre Ailleret of EDF said that, whereas the main arguments in favor of light-water system and against gas-graphite were political, engineers had to continue to search for technological improvements to the gas-graphite design. Other EDF leaders emphasized that the numbers in both reports favored light-water, and that the utility simply had to bide its time until economic rationality could prevail. For the moment, the least politically problematic way to pursue light-water technology was to participate in the Franco-Belgian and Franco-Swiss reactor projects. This would enable French industry to gain much-needed technological experience while escaping the "political vicissitudes of enriched uranium." Clearly, light-water reactors were economically desirable. But, although EDF was "committed, as an industrial and commercial Establishment, to defend its point of view on the economic considerations relevant to its interests and those of French industry," it "still had to bow to the political considerations that oppose this project." Opposition to light-water technology appeared as a politics separate from (and opposed to) technology and economics. See memorandum from P. Ailleret to J. Renou, R. Hirsch, F. Perrin, P. Massé, and A. Decelle, 20 February 1967, "Au Delà du Consensus Cabanius-Horowitz," CEA archives, F3-24-25; Direction de l'Equipement, "Position d'EDF devant les problème nucléaires actuels (mise à jour juillet 1967 du texte du 29 mai)," Bienvenu papers. The Franco-Belgian and Franco-Swiss projects were discussed in EDF Conseil d'Administration 254 (22 September 1967) and 257 (22 December 1967). De Gaulle signed the official authorization for the Franco-Belgian project in December 1967 (Présidence de la République, Secretariat Général, "Conseil restreint du 7 Dec. 67 relatif au programme nucléaire civil. Relevé des Décisions" (CEA archives, F3-24-25).

41. Simmonot 1978, p. 84.

42. Interview with Maurice Schumann, conducted by Alain Beltran, Martine Bungener, and Jean-François Picard.

43. Ibid.

44. Ibid. See also Maurice Schumann, "La politique électronucléaire de la France," speech given to meeting of groupe X nucléaire, 6 March 1968 (CEA archives, DEDR-DIV 219 DPA).

45. Interview with André Decelle, conducted by Jean-François Picard and Alain Beltran, 4 May 1981.

46. EDF Conseil d'Administration 253, 12 September 1967.

47. "M. André Decelle, directeur-général de l'EDF donne sa démission. Partisan de l'exploitation de la filière uranium enrichi, il était en désaccord avec le gouvernement," *Le Figaro,* 12 September 1967.

48. EDF Conseil d'Administration 253 (12 September 1967), p. 9.

49. Soutou (1991) discusses this part of the story at greater length.

50. See interviews in Simmonot 1978; Hecht, interview with Marcel Boiteux.

51. Shortly after Decelle's resignation, for example, Massé told EDF's board of directors that, for the moment, the program's future rested with the PEON commission—without bothering to reiterate that EDF was represented there. PEON thus appeared completely separate from the utility. Its objectivity and legitimacy derived from this separation, since the separation meant that it was above interests of any kind but national. Meanwhile, the ministries—which until the late 1960s had had very little substantive say in the direction of the nuclear program—also appeared eager to promote PEON as the ultimate and objective source of resolution for the *guerre des filières.* The Ministry of Finance, for example, had considered the Horowitz-Cabanius discussions supremely unsatisfactory. It declared that only PEON (which included a Ministry of Finance representative for the first time in 1967) could produce an objective and useful evaluation of the relative merits of the two systems. Sources: EDF Conseil d'Administration 257, 22 December 1967; anonymous memorandum, Ministère des Finances, "Propositions de questions à inscrire à l'ordre du jour de la Commission Consultative pour la Production d'Electricité d'Origine Nucléaire," 15 March 1967 (Bienvenu papers).

52. Commission Consultative pour la Production d'Electricité d'Origine Nucléaire, Groupe de Travail Général, "Rapport de Conjoncture" (n.d.); "Prix des fuels à moyen et long terme" (9 November 1967); "Hypothèses de travail" (27 September 1967); "Hypothèses de travail concernant le développement nucléaire" (8 November 1967); réunion du 15 septembre 1967, "Compte rendu de l'activité du groupe de travail général" (rédigé par Jacques Gaussens) (all in CEA archives, F6-13-20).

53. "Rapport de Conjoncture."

54. Memorandum, Pierre Tanguy to M. le Directeur des Piles Atomiques (Jules Horowitz), "Observations sur le Projet de Rapport soumis à la Commission le 29 février," 28 February 1968 (CEA archives, DEDR DIV 219, DPA).

55. Ibid.

56. Memorandum, Jules Horowitz to Administrateur Général, 23 February 1968 (CEA archives, DEDR DIV 219, DPA).

57. JB/MW, "Remarques concernant le rapport PEON d'avril 1968," 6 November 1968 (CEA archives, DEDR DIV 219, DPA).

58. Pierre Massé, *Aléas et Progrès: Entre Candide et Cassandre* (Economica, 1984), p. 136.

59. Letter from Ambroise Roux to Jean Couture, 29 March 1968 (also signed by Baumgartner, Blancard, de Calan, Gaspard, Glasser, Jouven, Malcor) accompanying "Note pour la Commission PEON," 28 March 1968 (CEA archives, DEDR DIV 219, DPA).

60. Reported in Tanguy, "Observations sur le Projet de Rapport soumis à la Commission le 29 février."

61. Memorandum from Jules Horowitz to Administrateur Général, 23 February 1968 (CEA archives, DEDR DIV 219, DPA).

62. "Observations sur le Projet de Rapport soumis à la Commission le 29 février."

63. Quoted on p. 248 of Simmonot 1978.

64. Apparently, one of the industrialists proposed this last recommendation as a way to ease the pain of moving to an American license, since the Canadians were known favorites of de Gaulle's. See pp. 199–200 of Picard et al. 1985.

65. "Le Rapport Couture" (editorial), *Revue Française de l'Energie* 201 (May 1968), p. 435.

66. Claude Bienvenu/HB, "Les Filières Nucléaires," 4 July 1968, p. 10 (Bienvenu papers).

67. Ibid., p. 12.

68. Claude Tourgeron, "La production d'électricité d'origine nucléaire en France," *économie et politique* January 1969, p. 2.

69. Ibid., p. 13.

70. Marcel Boiteux/CH, "Politique des Réacteurs Nucléaires" (Bienvenu papers), p. 6.

71. Ibid., p. 7.

72. Memo, CEA, EDF, "Programme d'action dans le domaine des centrales électronucléaires," 21 April 1969 (Bienvenu papers). This was approved by both Boiteux and Hirsch with identical letters to each other, dated 24 and 25 April respectively.

73. EDF CA, 274, 25 April 1969, p. 9.

74. Ibid., p. 14.

75. Memoranda: Cabanius to Directeur de la REN1, Chef du SEPTEN, Chef du SEGN, 29 November 1968; Direction de l'Equipement to Directeur de la REN1, Chef du SEPTEN, Chef du SEGN, n.d. (probably late 1968); SETPEN, DG/MGo, "Centrale Nucléaire a Eau Légère, Organisation du travail SEPTEN-REN1," 13 December 1968 (all in Bienvenu papers).

76. Simmonot 1978, pp. 254–259.

77. De Gaulle resigned after losing a referendum vote in April 1969. The loss is widely interpreted as an aftershock of the 1968 strikes. For more on French politics in this period see Gildea 1996, Berstein 1989, and Chapsal 1981.

78. EDF Conseil d'Administration 276, 27 June 1969, p. 17.

79. Ibid., p. 23.

80. "L'avenir du Commissariat à l'Energie Atomique," *Cadres et Profession (Mensuel de l'Union Confédérale des Ingénieurs et Cadres CFDT)* 234 (June 1969), p. 12.

81. Ibid., p. 13.

82. Westinghouse Electric International Company, "Westinghouse et un projet européen pour l'industrie de la construction électrique," 24 January 1969 (courtesy of Jean-Claude Zerbib).

83. "L'avenir du Commissariat à l'Energie Atomique," p. 13.

84. CGC, CFDT, CGT, CGT-FO, SPAEN, "Pour un Politique Française de l'Energie Nucléaire (Déclaration des Organisations Syndicales du Commissariat à l'Energie Atomique)" (flyer, 8 October 1969) (Laponche papers).

85. Ibid.

86. Ibid.

87. "Un passé qui est un exemple," *Le Compagnon d'Energies Nouvelles (Journal des Ingénieurs et Cadres CGT du CEA)* 7, suppl. 143 (1969): 1.

88. Les sections d'entreprise PSU de l'énergie atomique, "un seul adversaire, un seul combat," 1969 (flyer) (Laponche papers).

89. A memorandum from Hirsch to the M. le Ministre du Développement Industriel et Scientifique, dated 15 October 1969 (CEA archives F3-24-25), provides further confirmation that Hirsch and Boiteux worked out their decision in regard to the American light-water plant together. The memo outlines their recommendation for how the decision should be shaped and worded.

90. Nicholas Vichney, "Abandonnant la filière française, l'EDF affirme sa volonté de construire des centrales nucléaires de type américain," *Le Monde*, 18 October 1969, p. 1.

91. Pierre Juin, "Une réussite de la technique française . . . La centrale nucléaire de Saint-Laurent-des-Eaux risque de n'avoir pas de descendance," *L'Humanité*, 18 October 1969, p. 1.

92. Pierre Juin, "En moins de sept mois d'activité la production d'énergie de 'Saint-Laurent-1' a dépassé un milliard de kilowatts-heures," *L'Humanité*, 18 October 1969, p. 6.

93. *Le Canard Enchaîné*, 22 October 1969, p. 3.

94. Ibid., p. 4.

95. EDF Conseil d'Administration 278, 24 October 1969.

96. Ibid.

97. Interview.

98. Laponche papers.

99. Bernard Gonel et al., tract, 27 October 1969 (Laponche papers).

100. Force Ouvrière and CFDT, "Conference de Presse, Hotel de Ville de Massy (Essonne)" (flyer, 30 October 1969) (Laponche papers).

101. Strike days: 10, 22, 27, 30 October, 3, 6, 13–18, 19–24 November. Not all centers went on strike during each of these times. Marcoule was on strike for all the November dates. For a complete list of which centers went on strike when, see "L'Action dans les Centres," *Energies Nouvelles (CGT, FSM, Journal du syndicat national des travailleurs de l'énergie atomique)* December 1969: 3 (Laponche papers).

102. CGT, CFDT, CGT-FO, "Hier à Palaiseau, 1200 Grèvistes manifestaient. . . " (tract, 31 October 1969) and "Pourquoi la grève du 6 novembre" (tract, 5 November 1969) (Laponche papers).

103. CGC CFDT, CGT, CGT-FO, SPAEN, "Appel aux cadres du CEA" (tract, 17 November 1969) (Laponche papers).

104. "Pourquoi l'Energie Atomique en Grève."

105. "Conference de Presse, Hotel de Ville de Massy (Essonne)."

106. Ibid. The bureaucrat in question was Marcel Boiteux.

107. CGT CFDT, CGT-FO, "Pourquoi une grève à l'échelon national au Commissariat à l'Energie Atomique? Pourquoi cinq agents du CEA en sont aujourd'hui à leur onzième jour de grève de la faim?" (flyer, 1969).

108. "Pourquoi l'Energie Atomique en Grève."

109. "L'Action dans les Centres."

110. CGC, CFDT, CGT, CGT-FO, SPAEN, "Comparaisons Economiques et Politique Industrielle dans le Domaine Electronucléaire," 20 November 1969 (copies provided by Jean-Claude Zerbib and Bernard Boudouresques), p. 5.

111. Ibid., p. 6.

112. Ibid., p. 6.

113. Ibid., p. 1.

114. Ibid., p. 10.

115. Procès-verbaux, Conseil Municipal de Bagnols, 7 July 1969.

116. J.-C. Michau and J. Trélin, "Effectifs du CEA: 2600 en moins pour 1971, 5500 pour 1975, 15000 pour ? L'Action plus que jamais nécessaire," *Energies Nouvelles (CGT, Journal du syndicat national des travailleurs de l'énergie atomique)*, December 1969: 1–4.

117. Procès-verbaux, Conseil Municipal de Bagnols, 6 November 1969.

118. Procès-verbaux, Conseil Municipal de Bagnols, 15 November 1969.

119. By this time, unionized workers had been at odds with the directors of the Marcoule site for several years. See Hecht 1996a.

120. *Marcoule et sa vocation dans le Languedoc-Rhodanien: document établi en commun entre la directions et les organisations syndicales du centre de Marcoule: CFDT, CGC, CGT, CGTFO, SPAEN*, October 1969 (courtesy of René Bernard).

121. Suzanne Frère, *Bagnols-sur-Cèze: Enquête Sociologique* (J. & M. Pailhé, 1968).

122. *Marcoule et sa vocation*, p. III.27.

123. Ibid., p. III.30.

124. Ibid., p. III.28.

125. Procès-verbaux, Conseil Municipal de Bagnols, 15 November 1969.

126. For a more complete analysis of this incident see Hecht 1997a.

127. "Filière graphite-gaz, Problème de répartition des commandes," 19 October 1965; EDF, REN2, "La Politique Industrielle d'EDF," 25 November 1965 (both in Bienvenu papers).

128. *Journal de Saint-Laurent* 6 (June 1969), p. 2 (Saint-Laurent papers).

129. Such sentiments were expressed in interviews I conducted with workers who had been working on the Saint-Laurent site since the construction period (1965–1969).

130. Saint-Laurent workers and engineers expressed their hopes in a variety of forums: training sessions, workplace practices, and the site's local newsletter, the *Journal de Saint-Laurent*.

131. Interview.

132. This account of the accident and the extent of the damage is based on the following sources: Centrale de St. Laurent des Eaux (Electricité de France, GRPT C), "Pollution du réacteur, analyse des signaux DRG," Dépannage du réacteur SL1, rapport no. 1 (Saint-Laurent papers); Lamiral 1988, volume 2, pp. 109–112.

133. The word "pollution" occurs in virtually all reports of the accident and its cleanup. (See, e.g., "Pollution du réacteur, analyse des signaux DRG" and "Mesures de pollution.") The anthropologist Françoise Zonabend gives a fascinating analysis of the varying cultural meanings of words such as "pollution" and

"contamination" in the contemporary nuclear industry in her 1989 book *La presqu'île au nucléaire.*

134. *Journal de Saint-Laurent* 10-11 (October-November 1969), p. 2 (Saint-Laurent papers).

135. "Pollution du réacteur, analyse des signaux DRG"; Centrale de St. Laurent des Eaux (Electricité de France, GRPT C), "Etat d'avancement des études et des travaux au 17.12.69," Dépannage du réacteur SL1, Rapport no. 2 (Saint-Laurent papers).

136. Centrale de St. Laurent des Eaux (Electricité de France, GRPT C), "Choix fondamentaux pour la suite des opérations de dépannage," Dépannage du réacteur SL1, Rapport no. 3 (Saint-Laurent papers).

137. "Mesures de pollution."

138. "Choix fondamentaux pour la suite du dépannage, édition revue et corrigée du rapport no. 3"; "Programme du dépannage et de la remise en service de la tranche," Dépannage du réacteur SL1, Rapport no. 7 (Saint-Laurent papers).

139. M. J. Grand and M. J. Hurtiger, "Aspect de radioprotection pendant les interventions de Saint-Laurent-des-Eaux," *Bulletin de l'Association Technique pour la production et l'utilisation de l'Energie Nucléaire,* 91 (1971): 48; Centrale de St. Laurent des Eaux (Electricité de France, GRPT C), "Etat d'avancement des études et travaux, planning au 1er juin '70," Dépannage du réacteur SL1, Rapport no. 13 (Saint-Laurent papers).

140. Grand and Hurtiger, "Aspect de radioprotection. . . "; Centrale de St. Laurent des Eaux, "Etat d'avancement des études. . . ."

141. Ibid.

142. *Journal de Saint-Laurent,* no. 17 (May 1970) (Saint-Laurent papers).

143. See e.g. *Journal de Saint-Laurent* no. 12–13 (December 1969-January 1970), p. 2; no. 16 (April 1970), p. 1 (Saint-Laurent papers).

144. Interview with Joël Sorin.

145. Interview.

146. Interview with Joël Sorin.

147. EDF CA: 281 (23 January 1970), 286 (26 June 1970), 287 (25 September 1970).

148. For a variety of perspectives on the choice between pressurized-water and boiling-water reactors, see Picard et al. 1985, Jasper 1990, and Lamiral 1988.

149. For more on these changes see Vallet 1986.

150. On French anti-nuclear protest see Touraine et al. 1980, Nelkin and Pollak 1981, Fagnani 1977, and Fagnani and Nicolon 1979.

151. Cited on p. 65 of Simmonot 1978.

152. Cited on p. 111 of Simmonot 1978.

153. Quoted on p. 123 of Simmonot 1978.

Conclusion

1. The conference facilities honored other men in Armand's circle too— there was, for example, a Gaston Berger Amphitheater.

2. The texts of these two speeches can be found in Badel 1996. Both speakers also discussed the new meanings of public service in a changing Europe and promoted specific development policies for the utility—indeed, these topics were clearly their main interest, and these speeches were clearly intended as policy statements made at the time of difficult transitions. These intentions make it all the more interesting that they chose to invoke EDF's history and mission as they did.

3. Alder 1997.

4. Edwards 1996; MacKenzie 1990.

5. See, notably, Smith 1983 and Russell 1996.

6. I demonstrate this point at greater length in Hecht 1996b.

7. Latour 1993.

Bibliography

Archival Sources

Archives and Catalogued Collections

Archives, Commissariat à l'Energie Atomique, Fontenay-aux-Roses: Papers from office of Haut Commissaire and Public Relations division; material on PEON, *guerre des filières*, Marcoule, nuclear development plans, relations with the press; photographs (boxes F3.24.25, M7.18.34, T5.01.65, F6.13.18, F6.13.19, F6.13.20, M4.01.13, DIV.215, DIV.216, DIV.218, DIV.219)

Archives confédérales, Confédération Française Démocratique du Travail: Personal and confederal papers (boxes 4P40, 5H80, 7H24, 7H268, 7H287)

Archives départementales du Gard, Nîmes: Procès-Verbaux des Séances et Délibérations du Conseil Général, 1955–1969; pamphlets; tourist brochures; reports

Archives, Electricité de France, Paris: In-house serials

Archives interfédérales, Confédération Française Démocratique du Travail: Procès-verbaux des réunions de la commission de coordination des comités d'hygiène et de sécurité (EDF)

Archives, Ministère de l'Industrie: Dossiers d'enquête pour les demandes de Déclaration d'Utilité Publique pour les sites de Chinon et Saint-Laurent-des-Eaux, série 9000957—870342/003; série 9000960—870342/27; série 9001464—870342/27; série 9000959—870342/026

Archives municipales, Avoine: Procès-verbaux, Conseil Municipal d'Avoine

Archives municipales, Bagnols-sur-Cèze: Procès-verbaux, Conseil Municipal de Bagnols-sur-Cèze; newsletters, flyers, pamphlets, photographs

Archives municipales, Bourgueil: Procès-verbaux, Conseil Municipal de Bourgueil

Archives municipales, Chinon: Procès-verbaux, Conseil Municipal de Chinon; newsletters, flyers, pamphlets

Archives municipales, Chusclan: Procès-verbaux, Conseil Municipal de Chusclan

Archives municipales, Codolet: Procès-verbaux, Conseil Municipal de Codolet

Archives municipales, Orsan: Procès-verbaux, Conseil Municipal d'Orsan

Archivi Storici delle Comunità Europee, Florence-Fiesole: Papers of Jules Guéron (box/folder: 27/2, 28/4, 28/6, 29/5, 40, 53/1, 53/2, 53/5, 87, 182, 185, 256, 362)

Audio archives, Maison de la Radio, Paris: Radio news programs, 1950–1970

Centre des Archives Contemporaine d'Indre-et-Loire, Chambray-lès-Tours: Conseil Général, Procès-Verbaux des Séances et Délibérations du Conseil Général, 1955–1969; Rapport du Préfet, 1955–1969

Inathèque, Institut National de l'Audiovisuel, Paris: Descriptive inventory of television and radio programs, 1950–1970; "Panorama" television news series (available for viewing)

Vidéothèque de Paris: Documentaries, newsreels, fiction films

Vidéothèque, Ministère de l'Equipement, La Défense: Promotional and documentary films

Personal Papers and Uncatalogued Collections

Centre Nucléaire de Chinon, Electricité de France: Operational guidelines for EDF2 and EDF3; safety report for EDF2; decommissioning report for EDF2

Centre Nucléaire de Saint-Laurent-des-Eaux, Electricité de France: Operational guidelines for EDF4; accident and cleanup reports on 1969 accident at EDF4; newsletters; minutes of committee meetings

Confédération Française Démocratique du Travail, Chinon: Minutes of Comité d'Hygiène et de Sécurité Chinon; minutes of Sous-CMP, Chinon; newletters; flyers

Confédération Française Démocratique du Travail, Saclay (CEA): Newsletters, reports, documentation on CEA strikes of 1968 and 1969

Confédération Française Démocratique du Travail, Saint-Laurent-des-Eaux: Newletters, flyers

Conseil d'Adminstration, Electricité de France, Paris: Minutes of meetings, 1955–1970

Fédération de l'Energie, Force Ouvrière: Minutes of EDF's Commission de Coordination des Comités d'Hygiène et de Sécurité

Office of Henri Loriers and Philippe Filhol, Cogéma: Minutes of EDF-CEA meetings; reports and memoranda prepared by CEA scientists and engineers on design and performance of natural uranium fuel rods

Personal papers, René Bernard: Documentation on strikes at Marcoule in 1968 and 1969

Personal papers, Claude Bienvenu: Proposals and blueprints for all EDF gas-graphite reactors; internal EDF memoranda; minutes of meetings of various EDF committees and joint EDF-CEA committees; private and professional correspondence

Personal papers, Paul Delpeyroux: Technical reports on uses of natural uranium fuel rods

Personal papers, Bernard Laponche: Newletters, flyers, pamphlets and other documentation on CEA strikes of 1968 and 1969.

Serials

Commissariat à l'Energie Atomique. *Bulletin d'Informations Scientifiques et Techniques.* 1950–1970.

Commissariat à l'Energie Atomique. *Rapport Annuel.* 1946–1970.

Confédération Française Démocratique du Travail. *Cahiers des groupes reconstruction.* 1953–1970.

Confédération Française Démocratique du Travail. *Formation.* 1950–1970.

Confédération Française Démocratique du Travail. *Rayonnement. Bulletin d'information du Syndicat National du Personnel du Commissariat à l'Energie Atomique.* 1959–1969.

Confédération Française Démocratique du Travail. *Syndicalisme.* 1950–1970.

Confédération Générale du Travail–Force Ouvrière. *Force Ouvrière Hebdomadaire.* 1957–1970.

Confédération Générale du Travail–Force Ouvrière. *Force Ouvrière Informations.* 1960–1965.

Confédération Générale du Travail–Force Ouvrière. *Force Ouvrière.* 1950–1970.

Confédération Générale du Travail. *La Vie Ouvrière.* 1950–1970.

Confédération Générale du Travail. *Le Peuple.* 1950–1970.

Electricité de France. *Contacts électriques.* 1950–1970.

Electricité de France. *Rapport Annuel.* 1954–1970.

énergie nucléaire. 1957–1965.

France-soir. 1953–1970.

L'Aurore. 1945–1955.

L'Echo du Midi. 1953–1970.

L'Express. 1953–1970.

l'Humanité. 1945–1970.

La Nouvelle République du Centre-Ouest. 1953–1970.

Le Canard Enchaîné. 1953–1970.

Le Midi-Libre. 1953–1970.

Le Monde. 1945–1970.

Le Parisien Libéré. 1945–1955.

Le Provençal. 1953–1970.

Government Documents

Assemblée Nationale. *Documents Parlementaires,* 1950–1958.

Assemblée Nationale. *Séances.* 1950–1969.

Assemblée Nationale. *Travaux Parlementaires.* 1959–1969.

Conseil de la République. *Documents Parlementaires.* 1950–1958.

Conseil de la République. *Séances.* 1950–1958.

Sénat. *Séances.* 1959–1969.

Interviews

Conducted by Gabrielle Hecht

Engineers, technicians, workers, and administrators, 1989–1990
Azam; P. Bacher; C. Bienvenu; Bertron; P. Boulin; B. Boudouresques; J. Bourgeois;
R. Brandt; M. Campani; R. Carle; M. Chabrillac; J. Chassany; J. Chérot; F. Cogné;
A. Combe; A. Crégut; J.-P. Cretté; Delarue; P. Delpeyroux; Delpla; M. Dürr; A.
Finkelstein; D. Gaussot; M.Gauthron; A. Gauvenet; B. Giraudel; Gloaguen; J.
Guéron; Guéry; J. Hébert; C. Heurteau; G. Jeanpierre; Joly; B. Laponche;
Leblond; C. Leduc; R. Le Maréchal; H. Loriers; R. Martin; A. Mergui; Mièvre; F.
Minnard; Nau; J. Pelcé; L. Patarin; J. Pottier; J.-P. Roux; B. Saïtcevsky; M. Surdin;
P. Tanguy; A. Teste du Bailler; J. Trélin; J. Weill; P. Zaleski; J.-C. Zerbib; R. Bernard,
M. Gallois, and A. Roger; S. Roullier, J.-C. Godineau, and P. Aubrier; S. Roullier F.
Mazier and J.-C. Contois; Jouquet and M. Brié; Guéry and Y. Pradel; J. Rastoin and
D. Bastien; Mureau, Marlet and Occhipenti

Residents of Touraine (1990)
M. & Mme. G. Arrault; J. Chamboissier; L. Chauvelin; G. Joubert; Laporte; Mme.
J. Lébert & Mme. M.-L. Feuillet-Bolies; M. & Mme. M. Marquet; M. & Mme.
Martin; M. Raffault; M. & Mme. Rathieuville; M. & Mme. P. Saint-Léger; J.-J. Van
Acker & M. Chauvelin; M. Vignaud; A. Voisin

Residents of Gard (1994)
L. Anglezan; J. Bonnaud; J. Canoby; M. J. Fabre; G. Jeanjean; R. Jéolas; M. Justamond; F. Lemesle; L. Menjaud; P. Menjaud; R. Menjaud; R. Plantevin; J. P. Ribière; M.-A. Sabatier

Engineers, technicians, workers, and administrators (1996)
M. Boiteux; P. Gaudin; G. Gellé; M.-L. Grémy; C. Heurteau; J.-P. Lecroc; J. Magnadas; G. Manceau; J. Sorin; J. Thomas; B. Tourillon

Conducted by Jean-François Picard, Alain Beltran, and Martine Bungener (transcripts courtesy of Jean-François Picard)

Engineers and administrators (1980–1982)
P. Ailleret; J. Andriot; E. Anzalone; P. Bacher; M. Banal; C. Bienvenu; M. Boiteux; J. Cabanius; J. Carteron; P. Caseau; A. Combe; J. Couture; A. Decelle; A. Dejou; Favez; R. Gaspard; J. Gaussens; R. Ginocchio; A. Giraud; P. Guillaumat; R. Hirsch; J. Horowitz; M. Hug; J.-M. Jeanneney; J. Lacoste; R. Le Guen; M. Luneau; P. Maerten; Malegarie; P. Massé; G. Morlat; M. Paul; R. Pauwels; L. Puiseux; A. Roux; J.-P. Roux; B. Saïtcevsky; M. Sagot; L. Saulgeot; M. Schuman; C. Tourgeron; J.-A. Vaujour

Conducted by André Finkelstein (transcripts courtesy of André Finkelstein)

CEA technologists (1985)
P. Guillaumat; M. Pascal

Published Primary Sources

N.B.: Most of the technical literature cited is available at the library of the Ecole Nationale Supérieure des Mines de Paris.

Ailleret, Pierre. 1955. Design for a dual purpose reactor G2. *Proceedings of the 1st Geneva Conference* 3.

Ailleret, Pierre. 1958. Atomic energy and the French problem of energy supplies. *Proceedings of the 2nd Geneva Conference* 1.

Ailleret, Pierre. 1972. *25 ans de vie technique et économique d'EDF.* Electricité de France.

Anonymous. 1957. Défense de la langue française: Le Langage Technique. *La Jaune et La Rouge* 107: 29.

Anonymous. 1958. Les vrais immortels. *Carrefour,* December 31: 5.

Aragon, M. 1957. La construction à Marcoule des ensembles nucléaires G1, G2, G3; Les problèmes d'entreprise générale du point de vue industriel. *Mémoires de la Société des Ingénieurs Civils de France,* tome 110, fasc. IV, July-August: 277–285.

Armand, Louis. 1958. Vues prospectives sur les transports. *Prospective* 1: 37–44.

Armand, Louis. 1967. Technocrates et Techniciens. *La Jaune et La Rouge* 216: 4–8.

Armand, Louis. 1986. *Quarante ans au service des hommes.* Charles-Lavauzelle.

Armand, Louis, and Michel Drancourt. 1961. *Plaidoyer pour l'avenir.* Calmann-Lévy.

Bacher, Pierre et al. 1958. Natural uranium-graphite lattices. *Proceedings of the 2nd Geneva Conference* 12.

Bacher, P., François Cogné, and Bernard Noc. 1964. Physique des piles à graphite. *Rapport CEA* 2699.

Barets, Jean. 1962. *La fin des politiques.* Calmann-Lévy.

Barets, Jean. 1963. Précisions sur l'objectivisme. *Revue de Défense Nationale*, May: 885–890.

Bauchard, Philippe. 1966. *Les Technocrates au pouvoir.* Arthaud.

Bauchet, Pierre. 1966. *L'Expérience française de planification.* Seuil.

Bauchet, Pierre. 1966. *La Planification Française: du premier au sixième plan.* Seuil.

Bellier, J., and M. Tourasse. 1959. Centre nucléaire de Marcoule. Caisson en béton précontraint des réacteurs G2 & G3. *Annales de l'Institut Technique du Batiment & des Travaux Publics, série béton précontriant* 29, July-August: 706–735.

Berger, Gaston. 1958. L'attitude prospective. *Prospective* 1: 1.

Berger, Gaston. 1959. Culture, qualité, liberté. *Prospective* 4: 5.

Berger, Gaston. 1961. L'idée d'avenir et la pensée de Teilhard de Chardin. *Prospective* 7: 131–153.

Berger, Gaston et al. 1958. *Politique et technique.* Presses Universitaires de France.

Bienvenu, C., and D. Gaussot. 1965. EDF 4, premier réacteur intégré. *Entropie* 5, September-October: 21–34

Bienvenu, C., et al. 1964. EDF 2, 3, 4. *Proceedings of the 3rd Geneva Conference* 5.

Biquard, Pierre. 1966. *Frédéric Joliot-Curie: The Man and His Theories.*

Bourgeois, J., and B. Saitcevsky. 1964. Développement des réacteurs à graphite et uranium naturel. *Proceedings of the 3rd Geneva Conference* 5.

Boussard, R. 1964. La fabrication en France de éléments combustibles. *Proceedings of the 3rd Geneva Conference* 11.

Boussard, R., et al. 1964. Expérience de fontionnement des réacteurs G2-G3 de Marcoule et . . . démarrage EDF 1. *Proceedings of the 3rd Geneva Conference* 5.

Bouzigues, H. 1963. Recherches de dépots de sel de Pu dans les batteries d'extraction du Pu de l'usine de Marcoule. *Rapport CEA* 2322.

Cabanius, J., and J. Horowitz. 1964. Le programme nucléaire français. *Proceedings of the 3rd Geneva Conference* 1.

Cambournac, Louis. 1959. Dix Années de Réalisations Techniques Françaises. *Mémoires de la Société des Ingénieurs Civils de France* 112, no. 1: 192–207.

Caquot, Albert. 1958. Le Rayonnement de la France au point de vue scientifique et economique du constructeur. *La Jaune et La Rouge* 112: 23–28.

Chenot, Bernard. 1956. *Les enterprises nationalisées. Que sais-je?* no. 695. Presses Universitaires de France.

Club de Grenelle. 1964. *Siècle de Damoclès: la force nucléaire stratégique.* Editions Pierre Couderc.

Club Jean Moulin. 1963. *La force de frappe et le citoyen.* Seuil.

Colson, Jean-Philippe. 1977. *Le nucléaire sans les Français: qui décide, qui profite?* François Maspero.

Combet, Georges. 1956. Défense de la langue française: Le Langage Technique. *La Jaune et La Rouge* 100: 24.

Commissariat à l'Energie Atomique. 1971. *The French nuclear fuel cycle industry.* Brun.

Commissariat à l'Energie Atomique. 1980. *L'industrie nucléaire française.*

Commissariat Général au Plan. 1961. *Quatrième Plan de Développement Economique et Social (1962–1965).*

Confédération Française Démocratique du Travail. 1975. *L'électronucléaire en France.* Seuil.

Confédération Française Démocratique du Travail. 1977. *Les dégâts du progrès.* Seuil.

Confédération Française Démocratique du Travail. 1980. *Le dossier électronucléaire.* Seuil.

Confédération Française Démocratique du Travail. 1984. *Le dossier de l'énergie.* Seuil.

Conférences du Centre d'Etudes Nucléaire de Saclay. 1956. *Initiation à l'énergie nucléaire.* Hachette.

Cot, Pierre. 1962. La Nouvelle Aérogare de l'Aéroport d'Orly. *La Jaune et La Rouge* 158: 10.

Cottier, Jean-Louis. 1959. *La Technocratie, Nouveau Pouvoir.* Edition du Cerf.

Couture, Pierre. 1960. Explosion de la première bombe A Française. *La Jaune et La Rouge* 136: 32–35.

Couture, Pierre. 1961. L'Energie Nucléaire: Problèmes scientifiques et techniques, perspectives économiques. *Perspectives Energétiques dans le Monde*, numéro spécial.

Darras, R. 1960. La corrosion par les gaz caloporteurs dans les réactuers nucléaires. *Rapport CEA* 1481.

Darras, R. 1963. La corrosion du zirconium et de ses alliages par le gaz aux températures élevées. *Industries Atomiques* 1–2.

Darras, R., and R. Caillat. 1958. Corrosion du magnésium et de certains de ses alliages dans les piles refroidies par gaz. *Rapport CEA* 983.

Darras, R., and H. Loriers. 1964. Problème de compatibilité des matériaux de gainage avec le gaz carbonique aux températures élevées. *Proceedings of the 3rd Geneva Conference* 9.

Debiesse, J. 1958. The training of scientists and technicians at the Saclay Nuclear Research Center. *Rapport CEA* 954.

Debiesse, J. 1961. L'énergie nucléaire dans le monde de demain. *Mémoires de la Société des Ingénieurs Civils de France* 11, November: 15–22.

Debiesse, J. 1963. La corrosion du zirconium et de ses alliages par le gaz aux températures élevées. *Industries Atomiques* 1–2.

Delouvrier, P., and R. Nathan. 1958. *Politique économique de la France.* Les Cours de droit.

de Rouville, Marcel. 1958. Le Centre de production de plutonium de Marcoule: sa place dans la chaine industrielle de l'énergie nucléaire. *Revue de l'Industrie Nucléaire* 40, June: 483–489.

de Rouville, Marcel. 1958. Experience obtained during 2 years operation of G1. *Proceedings of the 2nd Geneva Conference* 8.

de Rouville, Marcel. 1960. La voie du Pu passe par Marcoule. *Nucleus* 4, July-August: 225–330.

Devoret, Raymond. 1961. *Applications des sciences nucléaires: dangers des radiations atomiques.* Gauthier-Villars.

Duhamel, F. 1960. L'élimination des déchets radioactifs. *Rapport CEA* 1738.

Duhamel, F. 1961. Les déchets radioactifs. *Annales des Mines*, October: 765–786.

Dumanois, Ingénieur général de l'Air. 1957. L'Aéronautique Française, sujet de fierté et d'espoir. *La Jaune et La Rouge* 106: 27–28.

Duprat, Henri. 1964. Etude économique du site de Marcoule. *Rapport CEA* 254.

Durand, Claude et al. 1966. *Le travail ouvrier et progrès technique.* Armand Colin.

Ertaud, A., and G. Derome. 1958. Chargement et Déchargement. *Bulletin d'Informations Scientifiques et Techniques du CEA* 20: 69–88.

Faÿs, R., et al. 1964. Sûreté des piles à filière UNGG. *Proceedings of the 3rd Geneva Conference* 13.

Fondation Nationale des Sciences Politiques et Institut d'Etudes Politiques de l'Université de Grenoble. 1965. *La Planification comme processus de décision.* Armand Colin.

Fourastié, Jean. 1962. *Machinisme et Bien-Etre. Niveau de vie et genre de vie de 1700 à nos jours.* Third edition. Minuit.

Friedmann, Georges. 1956. *Le Travail en miettes. Spécialisation et Loisirs.* Gallimard.

Furet, Jacques. 1968. *Controle et electronique des réacteurs nucléaires.*

Gaussens, Jacques. 1965. Optimisation des cycles de combustibles. Valeurs marginales des pertes. *Rapport CEA* 2866.

Gaussens, Jacques. 1968. *Données et calculs économique de l'énergie nucléaire. Conférences faites à l'Institut National des Sciences et Techniques nucléaires.* Presses Universitaires de France.

Gaussens, Jacques, and J. Andriot. 1958. Cycles de combustible. Production d'uranium, programme de centrales électriques et effort financière correspondant. *Rapport CEA* 1010.

Gaussens, Jacques, and Henri Paillot. 1965. Etude des valeurs et des prix du plutonium à long terme. *Rapport CEA* 2795.

Gaussens, Jacques, et al. 1964. Quelques aspect économiques de la filière UNGG . . . coûts en France. *Proceedings of the 3rd Geneva Conference* 5.

Gaussens, Jacques, et al. 1964. Aspect économique des réacteurs produisant de l'électricité et de la chaleur industrielle. *Proceedings of the 3rd Geneva Conference* 6.

Gauzit, Maurice, and Théo Kahan. 1957. *Contrôle et protection des réacteurs nucléaires.* Dunod.

Giraud, André. 1983. Energy in France. *Annual Review of Energy* 8: 165–191.

Gors, André. 1959. Avant Propos. *Prospective* 2: 2.

Gournay, Bernard. 1960. Technocratie et Administration. *Revue française de science politique* 10: 881–890.

Grand, M. J., and M. J. Hurtiger. 1971. Aspect de radioprotection pendant les interventions de Saint-Laurent-des-Eaux. *Bulletin de l'Association Technique pour la production et l'utilisation de l'Energie Nucléaire* 91: 38–53.

Gurvitch, Georges. 1948. Industrialisation et Technocratie. In *Première Semaine Sociologique in Paris,* ed. G. Gurvitch. Jean Touzot.

Groupe 1985. 1964. *Refléxions pour 1985.* La Documentation Française.

Hirsch, Etienne. 1959. Les méthodes françaises de planification. *Mémoires de la Société des Ingénieurs Civils de France* 112, no. II: 81–94.

INSEE. 1966. Méthodes de Programmation dans le Ve Plan. *Etudes et Conjonctures* 21, no. 12.

Jeanneney, Jean-Marcel. 1959. *Forces et faiblesses de l'économie française, 1945–1959.* Armand Colin.

Jeanneney, Jean-Marcel, and Claude-Albert Colliard. 1950. *Economie et droit de l'électricité.* Domat-Montchrétien.

Joliot-Curie, Frédéric. 1954. *Cinq Années de lutte pour la paix: Articles, Discours, et Documents, 1949–1954.* Conférence Mondiale de la Paix..

Joliot-Curie, Frédéric. 1959. *La Paix, le Désarmement et la Coopération Internationale.* Editions Défense de la Paix.

Joliot-Curie, Frédéric. 1959. *Textes Choisis.* Editions Sociales.

Kieffer, J. 1963. La centrale de Marcoule: expérience, résultats et enseignements dans le domaine de la production d'électricité, *Energie nucléaire* 5, June.

Lamiral, Georges, et al. 1963. Une étape pour la construction soudée: la réalisation à Chinon des caissons de réacteurs de EDF 1 & EDF 2. *Supplément aux Annales de l'Institut du Bâtiment et des Travaux Publics* 183–184 (March-April). *Serie: construction métallique* 140: 396–381.

Lamiral, Georges, and M. Lancel. 1958. The pressure vessel and heat exchanger of Chinon nuclear power plant EDF-1. *Proceedings of the 2nd Geneva Conference* 7.

Léauté, André. 1958. Les Vertus Cardinales de l'Ingénieur de Grande Classe. *La Jaune et la Rouge* 120.

Le Brun, Pierre. 1958. Le point du vue d'un syndicaliste. In *Politique et technique,* ed. G. Berger et al. Presses Universitaires de France.

Lecarme, Jacques. 1958. Un triomphe nationale: 'la Caravelle'. *La Jaune et La Rouge* 118: 25–42.

Leo, Kaplan, and Segard. 1958. Problems of fuel loading and unloading in reactor EDF1. *Geneva Conference, 1958,* pp. 582–590.

Lescuyer, Georges. 1959. *Le contrôle de l'Etat sur les entreprises nationalisées.*

Mabile, J. 1956. L'Economie mondiale de l'uranium. *Revue Française de l'Energie* 80, November: 73.

Mallet, Serge. 1969. *La nouvelle classe ouvrière* (Seuil).

Marcoule et sa vocation dans le Languedoc-Rhôdanien. 1969. Document établi en commun entre la Direction et les Organisations Syndicales du Centre de Marcoule: CFDT, CGC, CGT, CGT-FO, SPAEN (October).

Martin, Charles-Noël. 1960. *Promesses et menaces de l'énergie nucléaire.* Presses Universitaires de France.

Massé, Pierre. 1959. Prévision et Prospective. *Prospective* 4: 91–120.

Massé, Pierre. 1959. Propos incertains. *Revue française de la recherche opérationelle*, 2e trim., no. 11: 60.

Massé, Pierre. 1965. *Le plan ou l'anti-hasard.* Gallimard.

Massé, Pierre. 1984. *Aléas et Progrès: Entre Candide et Cassandre.* Economica.

Mathias, C. 1981. Nuclear power development in France: Report to the Committee on Governmental Affairs. US Senate.

Meiffren, J. 1964. Le traitement des informations dans les réacteurs G2/G3. *Rapport CEA* 2433.

Merle, Marcel. 1958. L'Influence de la technique sur les institutions politiques. In Berger et al. 1958.

Meynaud, Jean. 1959. Les Mathématiciens et le Pouvoir. *Revue Française de Science Politique* 9, no. 2: 340–367.

Meynaud, Jean. 1960. *Technocratie et politique.* Etudes de Science Politique.

Meynaud, Jean. 1961. A propos de la Technocratie. *Revue Française de Science Politique* 11, no. 4: 671–683.

Meynaud, Jean. 1963. A Propos des Spéculations sur l'Avenir. Esquisse bibliographique. *Revue française de la science politique* 13: 666–688.

Meynaud, Jean. 1964, 1968. *Technocracy (Technocratie, Mythe ou Réalité?)* Free Press.

Migeon, Henri. 1958. *Le Monde après 150 ans de technique.* New edition.

Mitrani, Nora. 1955. Reflexions sur l'opération technique, les techniciens et les technocrates. *Cahiers internationaux de sociologie* 19.

Mitrani, Nora. 1957. Les mythes de l'énergie nucléaire et la bureaucratie internationale. *Cahiers internationaux de sociologie* 21: 138–148.

Mitrani, Nora. 1960. Attitudes et symboles techno-bureaucratiques: refléxions sur une enquête. *Cahiers internationaux de sociologie* 24: 148–166.

Naville, Pierre. 1956. *Essai sur la qualification du travail.* Marcel Rivière.

Naville, Pierre. 1961. *L'automatisation et le travail ouvrier.* CNRS.

Naville, Pierre. 1963. *Vers l'automatisme sociale.* Gallimard.

Naville, Pierre. 1964. *La Classe ouvrière et le régime gaulliste.* Questions du Socialisme. Etudes et Documentation Internationales.

Organisation Européenne de Coopération Economique. 1958. *L'industrie devant l'énegie nucléaire. Exposés présentés au cours de la deuxième conférence d'information sur l'énergie nucléaire pour les dirigeants d'entreprises.*

Papault, R. 1957. Le Centre de Production de Plutonium et d'Energie Electrique d'Origine Nucléaire de Marcoule (Gard). *Le Génie Civil* 134, October: 389–398.

Pasquet, M. 1957. La construction à Marcoule des ensembles nucléaires G1, G2, G3. Les problèmes d'entreprise générale du point du vue industriel, Réalisation de G2 et G3. *Mémoires de la Société des Ingénieurs Civils de France,* tome 110, fasc. IV, July-August: 287–293.

Perrin, Francis, ed. 1968. *Génie Atomique: Cours fondamental. Tôme II.* Presses Universitaires de France (Saclay, ISTN).

Perroux, François. 1963. *Le IVe Plan Français (1962–1965)* (2e édition; 1e éd. 1962). *Que Sais-Je?* 1021. Presses Universitaires de France.

Reine, Philippe. 1969. *Le problème atomique. VII: Applications civiles de l'énergie nucléaire.* Berger-Levrault.

Reis, Thomas. 1958. *Aspects économiques des applications industrielles de l'énergie nucléaire.* Dunod.

Roux, Jean-Pierre. 1957. La Centrale Nucléaire EDF1 de Chinon. *Mémoires de la Société des Ingénieurs Civils de France,* tome 110, fascicule IV, July-August: 294–309.

Roy, Maurice. 1960. Progrès et Tradition. *La Jaune et la Rouge* 140: 34–45.

Rustant, Maurice. 1959. *L'automation: ses conséquences humaines et sociales.* Editions ouvrières.

Saïtcevsky, B., and D. Gaussot. 1964. Les appareils de chargement dans l'UNGG. *Proceedings of the 3rd Geneva Conference* 8.

Sanguinetti, Alexandre. 1964. *La France et l'arme atomique.* Julliard.

Sauvy, Alfred. 1958. Lobbys et Groupes de Pression. In *Politique et technique,* ed. G. Berger et al. Presses Universitaires de France.

Schmitt et al. 1958. Système uranium naturel-graphite-gaz. *Rapport CEA* 968.

Siegfried, André. 1958. Le Problème de l'Etat au XXe siècle en fonction des transformations de la production. In *Politique et technique,* ed. G. Berger et al. Presses Universitaires de France.

Société Française de Radioprotection. 1968. *Congrès International sur la Radioprotection du Milieu devant le Développement des Utilisations Pacifiques de l'Energie Nucléaire.*

Teste, Yvan. 1957. Les Installations de Production d'Energie de Marcoule et la Centrale Nucléaire de Chinon. *Mémoires de la Société des Ingénieurs Civils de France,* tome 110, fascicule II, March-April.

Touchard, Jean, and Jacques Solé. 1965. Planification et Technocratie: Esquisse d'une analyse idéologique. In *La Planification comme processus de décision.* Armand Colin.

Touraine, Alain. 1955. *L'Evolution du travail ouvrier aux Usines Renault.* CNRS Centre d'Etudes Sociologiques.

Touraine, Alain. 1965. *Les travailleurs et les changements techniques.* OCDE.

Veraldi, Gabriel. 1958. *L'humanisme technique.*

Villiers, Georges. 1959. Industrie, Technique et Culture. *Prospective* 4: 21–32.

Waline, Marcel. 1958. Les Résistances techniques de l'administration au pouvoir politique. In *Politique et technique,* ed. G. Berger et al. Presses Universitaires de France.

Weill, Jacky. 1955. Complete automation of the operation of nuclear reactors. *Proceedings of the 1st Geneva Conference* 3.

Yvon, J. 1957. Les piles atomiques en France. *Rapport CEA* 747.

Secondary Sources

Adam, Gérard, Frédéric Bon, Jacques Capdevielle, and René Mouriaux. 1970. *L'Ouvrier Français en 1970: enquête nationale auprès de 1116 ouvriers d'industrie.* Armand Colin.

Adas, Michael. 1989. *Machines as the Measure of Men: Science, Technology, and Ideologies of Western Dominance.* Cornell University Press.

Agar, Jon. forthcoming. *Science and Spectacle: The Work of Jodrell Bank in Postwar Britain.* Harwood.

Akrich, Madeleine, and Vololona Rabeharisoa. n.d. *Les conseils en économies d'énergie.* Working paper, Centre de Sociologie de l'Innovation.

Albonetti, Achille. 1972. *Europe and Nuclear Energy.* Atlantic Institute for International Affairs.

Alder, Ken. 1995. A Revolution to Measure: The Political Economy of the Metric System in France. In *The Values of Precision,* ed. N. Wise. Princeton University Press.

Alder, Ken. 1997. *Engineering the Revolution: Arms and Englightenment in France, 1763–1815.* Princeton University Press.

Anderson, Benedict. 1991. *Imagined Communities: Reflections on the Origins and Spread of Nationalism,* revised edition. Verso.

Andrieu, Claire, Lucette LeVan, and Antoine Prost. 1987. *Les Nationalisations de la Libération: de l'utopie au compromis.* Presses de la Fondation Nationale des Sciences Politiques.

Asselain, Jean-Claude. *Histoire économique de la France du XVIIIe siècle à nos jours, vol. 2: De 1919 à la fin des années 1970.* Seuil, 1984.

Auriol, Vincent. 1970. *Mon septennat (1947–1954)*. Gallimard.

Babelon, Jean-Pierre. 1992. Les châteaux de la Loire. In *Les lieux de mémoire, III: Les France, 3. de l'archive à l'ensemble*, ed. P. Nora. Gallimard.

Badel, Laurence, ed. 1996. *La nationalisation de l'électricité en France: Nécessité technique ou logique politique?* Presses Universitaires de France.

Balogh, Brian. 1991. *Chain Reaction: Expert debate and public participation in American commerical nuclear power, 1945–1975.* Cambridge University Press.

Barjonet, André. 1968. *La CGT: Histoire. Structure. Doctrine.* Seuil.

Baudoui, Rémi. 1987. Raoul Dautry: La Conscience Du Social. *Vingtième Siècle* 15: 45–58.

Baumgartner, Frank R. 1989. *Conflict and Rhetoric in French Policymaking.* University of Pittsburgh Press.

Baumgartner, Frank R., and David Wilsford. 1994. France: Science within the State. In *Scientists and the State: Domestic Structures and the International Context*, ed. E. Solingen. University of Michigan Press.

Bédarida, F., and J.-P. Rioux, eds. 1985. *Pierre Mendès France et le mendèsisme.* Fayard.

Belhoste, Bruno, Amy Dahan Dalmedico, and Antoine Picon, eds. 1994. *La formation polytechnicienne, 1794–1994.* Dunod.

Beltran, Alain, and Patrice A. Carré. 1991. *La fée et la servante: la société française face à l'électricité, XIXe–XXe siècle.* Belin.

Bergounioux, Alain. 1975. *Force ouvrière.* Seuil.

Bergounioux, Alain. 1984. The Trade Union Strategy of the CGT-FO. In *The French Workers' Movement*, ed. M. Kesselman. Allen & Unwin.

Berlanstein, Lenard R., ed. 1993. *Rethinking Labor History: Essays on Discourse and Class Analysis.* University of Illinois Press.

Bernstein, Barton. 1993. Seizing the Contested Terrain of Nuclear History. *Diplomatic History* 17, no. 1: 35–72.

Berstein, Serge. 1989. *La France de l'expansion. I. La République gaullienne, 1958–1969.* Seuil.

Berstein, Serge, and Pierre Milza. 1991. *Histoire de la France au XXe siècle. Tome III: 1945–1958.* Editions Complexe.

Biernacki, Richard. 1995. *The Fabrication of Labor: Germany and Britain 1640–1914.* University of California Press.

Bijker, Wiebe. 1993. Do Not Despair: There Is Life after Constructivism. *Science, Technology, and Human Values* 18, no. 1: 113–138.

Bijker, Wiebe, Thomas P. Hughes, and Trevor Pinch, eds. 1987. *The Social Construction of Technological Systems*. MIT Press.

Birnbaum, Pierre. 1982. *The Heights of Power: An Essay on the Power Elite in France*. University of Chicago Press.

Bloch-Lainé, François, and Jean Bouvier. 1986. *La France restaurée, 1944–1954: Dialogue sur les choix d'une modernisation*. Fayard.

Boltanski, Luc. 1982. *Les cadres: la formation d'un groupe social*. Minuit.

Bonin, Hubert. 1987. *Histoire économique de la IVe république*. Economica.

Botelho, Antonio José J. 1994. The Industrial Policy That Never Was: French Semiconductor Policy, 1945–1966. *History and Technology* 11: 165–180.

Bourdieu, Pierre. 1975. The specificity of the scientific field and the social conditions of the progress of reason. *Social Science Information* 14, no. 6: 19–47.

Bouvier, Jean, et al. 1982. *Histoire économique et sociale de la France. Tome IV: L'ère industrielle et la société d'aujourd'hui*. Presses Universitaires de France.

Boyer, Paul. 1985. *By the Bomb's Early Light: American Thought and Culture at the Dawn of the Atomic Age*. Pantheon.

Branciard, Michel. 1990. *Histoire de la Confédération Française Démocratique du Travail: soixante-dix ans d'action syndicale*. La Découverte.

Braverman, Harry. 1974. *Labor and Monopoly Capital: The Degradation of Work in the Twentieth Century*. Monthly Review Press.

Briais, Bernard. 1984. *La vallée de la vienne*. CLD.

Bron, Jean. 1973. *Histoire du Mouvement Ouvrier Français. Tome III: La lutte des classes aujoud'hui, 1950–1972*. Editions Ouvrières.

Brun, Gérard. 1985. *Technocrates et Technocratie en France*. Albatros.

Buffotot, Patrice. 1987. Guy Mollet et la Défense: du socialisme patriotique au socialisme atlantique. In *Guy Mollet, un camarade en république*, ed. B. Ménager et al. Presses Universitaires de Lille.

Bupp, Irvin C., and Jean-Claude Derian. 1978. *The Failed Promise of Nuclear Power: The Story of Light Water*. Basic Books.

Burawoy, Michael. 1979. *Manufacturing Consent: Changes in the Labor Process under Monopoly Capitalism*. University of Chicago Press.

Burawoy, Michael. 1985. *The Politics of Production: Factory Regimes under Capitalism and Socialism*. Verso.

Caldwell III, Bill S. 1980. The French Socialists' Attitudes Toward the Use of Nuclear Weapons, 1945–1978. PhD. dissertation, University of Georgia.

Callon, Michel, and John Law. 1982. On Interests and their Transformation: Enrolment and Counter-Enrolment. *Social Studies of Science* 12: 615–625.

Callon, Michel, John Law, and Arie Rip, eds. 1986. *Mapping the Dynamics of Science and Technology.*

Callon, Michel. 1980. The state and technical innovation: A case study of the electric vehicle in France. *Research Policy* 9: 358–376.

Callot, Jean Pierre. 1993. *Histoire et Perspective de l'Ecole Polytechnique.* C. Lavauzelle.

Campbell, John L. 1986. The state, capital formation, and industrial planning: Financing nuclear energy in the United States and France. *Social Science Quarterly* 67: 707–721.

Carlson, W. Bernard. 1988. Academic Entrepreneurship and Engineering Education: Dugald C. Jackson and the MIT-GE Cooperative Engineering Course, 1907–1932. *Technology and Culture* 29: 536–567.

Caron, François. 1981. *Histoire économique de la France, XIXe–XXe siècle.* A. Colin.

Caute, David. 1964. *Communism and the French Intellectuals.* Macmillan.

Cazals, Thierry. 1982–83. La résistance au nucléaire dans la région du Gard Rhôdanien. Mémoire sous la direction de M. Bentz. Université d'Aix-en-Province.

Cazes, Bernard. 1991. Un Demi-Siècle de Planification Indicative. In *Entre l'Etat et le marché: L'économie française des années 1880 à nos jours,* ed. M. Lévy-Leboyer and J.-C. Casanova. Gallimard.

Chadeau, Emmanuel. 1985. Etat, Industrie, Nation: la formation des technologies aéronautiques en France (1900–1950). *Histoire, Economie et Société* 4, no. 2: 275–299.

Chapman, Herrick. 1991. *State Capitalism and Working Class Radicalism in the French Aircraft Industry.* University of California Press.

Chapsal, Jacques. 1981. *La vie politique sous la Ve République. Tome 1: 1958–1974.* Presses Universitaires de France.

Chardot, Jean. 1983. *Le gaullisme d'opposition, 1946–1958.* Fayard.

Charvolin, Florian. *The greening of public policy analysis—The construction of the content of two public debates: Canadian Biosecurity and French Environment.* ECPR. 1991.

Chilton, Paul, ed. 1985. *Language and the Nuclear Arms Debate: Nukespeak Today.* Francis Pinter.

Clark, T. J. 1984. *The Painting of Modern Life: Paris in the Art of Manet and His Followers.* Princeton University Press.

Cohen, Elie. 1992. *Le Colbertisme High Tech: Economie des Telecom et du Grand Projet.* Hachette.

Cohen, Lizabeth. 1990. *Making a New Deal: Industrial Workers in Chicago, 1919–1939.* Cambridge University Press.

Cohendet, Patrick, and André Lebeau. 1987. *Choix stratégiques et grands programmes civils.* Economica.

Colson, Jean-Philippe. 1977. *Le nucléaire sans les Français: qui décide, qui profite?* François Maspero.

Comaroff, Jean. 1985. *Body of Power, Spirit of Resistance.* University of Chicago Press.

Corbel, P. 1969. *Le Parlement français et la Planification.* Cujas.

Corbin, Alain. 1996. Paris-Province. In *Realms of Memory, I: Conflicts and Divisions,* ed. P. Nora. Columbia University Press.

Corn, Joseph. 1983. *The Winged Gospel: America's Romance with Aviation, 1900–1950.* Oxford University Press.

Corn, Joseph, ed. 1986. *Imagining Tomorrow: History, Technology, and the American Future.* MIT Press.

Coutrot, Aline. 1983. La Création du Commissariat à l'Energie Atomique. In *De Gaulle et la Nation face aux problèmes de défense, 1945–46.* Plon.

Cowan, Ruth Schwartz. 1983. *More Work For Mother.* Basic Books.

Cowans, Jon. 1991. French Public Opinion and the Founding of the Fourth Republic. *French Historical Studies* 17, no. 1: 62–95.

Cowans, Jon. 1993. Wielding the People: Opinion Polls and the Problem of Legitimacy in France since 1944. Ph.D. dissertation, Stanford University.

Crawford, Stephen. 1989. *Technical Workers in an Advanced Society: The Work, Careers, and Politics of French Engineers.* Cambridge University Press.

Crozier, Michel. 1964. *The Bureaucratic Phenomenon.* University of Chicago Press.

Dalmedico, Amy Dahan. 1994. Rénover sans se renier: L'Ecole polytechnique de 1945 à nos jours. In *La formation polytechnicienne, 1794–1994,* ed. B. Belhoste et al. Dunod.

Danielsson, Bengt, and Marie-Thérèse Danielsson. 1986. *Poisoned Reign: French Nuclear Colonialism in the Pacific.* Penguin.

Daviet, Jean-Pierre. 1995. Pierre Guillaumat et l'enrichissement de l'uranium, 1952–1962. In *Pierre Guillaumat: La passion des grands projets industriels,* ed. G.-H. Soutou and A. Beltran. Editions Rive Droite.

Davis, Mary D. 1988. *The Military-Civilian Nuclear Link: A Guide to the French Nuclear Industry.* Westview.

Debeir, Jean-Claude, Jean-Paul Deliage, and Daniel Hémery. 1986. *Les Servitudes de la puissance: une histoire de l'énergie.* Flammarion.

Debord, Guy. 1977. *Society of the Spectacle*. Black and Red.

de Carmoy, Guy. 1971. *Le dossier européen de l'énergie*. Editions d'Organisation.

de Gaulle, Charles. 1954. *Mémoires de Guerre I. L'Appel, 1940–1942*. Plon.

de Gravelaine, Frédérique, and Sylvie O'Dy. 1978. *L'Etat -EDF*. Alain Moreau.

Descostes, M., and J.-L. Robert. 19484. *Clefs pour une histoire du syndicalisme cadre*. Editions Ouvrièrs.

Dorget, François. 1984. *Le choix nucléaire français*. Economica.

Douglas, Mary. 1966. *Purity and Danger*. Routledge.

Douglas, Mary. 1982. Environments at Risk. In *Science in Context: Readings in the Sociology of Science*, ed. B. Barnes and D. Edge. Open University Press.

Douglas, Mary. 1992. *Risk and Blame: Essays in Cultural Theory*. Routledge.

Douglas, Mary, and Aaron Wildavsky. 1982. *Risk and Culture: An Essay on the Selection of Technological and Environmental Dangers*. University of California Press.

Downey, Gary. 1986. Risk in Culture: The American Conflict Over Nuclear Power. *Cultural Anthropology* 1: 388–412.

Downs, Laura Lee. 1993. If "Woman" Is Just an Empty Category, Then Why Am I Afraid to Walk Alone at Night? Identity Politics Meets the Postmodern Subject. *Comparative Studies in Society and History* 35, no. 2: 414–437.

Downs, Laura Lee. 1995. *Manufacturing Inequality: Gender Division in the French and British Metalworking Industries, 1914–1939*. Cornell University Press.

Dreyfus, Michel. 1995. *Histoire de la CGT: Cent ans de syndicalisme en France*. Editions Complexe.

Dunlavy, Colleen. 1994. *Politics and Industrialization*. Princeton University Press.

du Tertre, Christian, and Giancarlo Santilli. 1992. *Automatisation du travail: utopies, réalités, débats des années 50 aux années 90*. Presses Universitaires de France.

Duval, Marcel. 1995. Pierre Guillaumat et l'arme atomique. In *Pierre Guillaumat: La passion des grands projets industriels*, ed. G.-H. Soutou and A. Beltran. Editions Rive Droite.

Duval, Marcel, and Dominique Mongin. 1993. *Histoire des Forces Nucléaires Françaises depuis 1945*. Presses Universitaires de France.

Earle, E. M., ed. 1964. *Modern France: Problems of the Third and Fourth Republics*. Russell and Russell.

Edmonson, James M. 1987. *From Mécanicien to Ingénieur*. Garland.

Edwards, Paul N. 1996. *The Closed World: Computers and the Politics of Discourse in Cold War America*. MIT Press.

Electricité de France. 1979. *Recherches sur la formation de l'opinion publique: à propos de l'énergie nucléaire considéré comme thème d'expression.*

Electricité de France. 1983. *Images d'une centrale nucléaire.*

Electricité de France. 1984. *Histoire de la Direction de l'Equipement: Evolution des effectifs de la direction de l'équipement depuis 1948.*

Electricité de France. 1985. *Le Programme électro-nucléaire français.*

Elgey, Georgette. 1968. *La République des contradictions, 1951–1954.* Fayard.

Etner, François. 1987. *Histoire du calcul économique en France.* Economica.

Fagnani, Francis, ed. 1977. *Le Débat nucléaire en France: Acteurs sociaux et communication de masse.* Université des Sciences Sociales, Institut de Recherche Economique, Grenoble.

Fagnani, Francis, and Alexandre Nicolon, eds. 1979. *Nucléopolis: matériaux pour l'analyse d'une société nucléaire.* Presses universitaires de Grenoble.

Fagnani, Jeanne, and Jean-Paul Moatti. 1984. The Politics of French Nuclear Development. *Journal of Policy Analysis and Management* 3, no. 2: 264–275.

Fassin, Eric. 1995. Fearful Symmetry: Culturalism and Cultural Comparison after Toqueville. *French Historical Studies* 19, no. 2: 451–460.

Floquet, Pierre-Henri. 1995. *Histoire de la centrale nucléaire des ardennes.* Association pour l'histoire de l'électricité en France.

Forbes, Jill, and Michael Kelly, eds. 1995. *French Cultural Studies: An Introduction.* Oxford University Press.

Ford, Caroline. 1993. *Creating the Nation in Provincial France: Religion and Political Identity in Brittany.* Princeton University Press.

Ford, Daniel. 1982. *The Cult of the Atom: The Secret Papers of the Atomic Energy Commission.* Touchstone.

Foucault, Michel. 1977. *Discipline and Punish: The Birth of the Prison.* Pantheon Books.

Fourgous, J. M., J. F. Picard, and C. Raguenel. 1980. *Les français et l'énergie. Recueil d'enquêtes et de sondages d'opinion effectués sur des thèmes se rapportant à l'énergie en France de 1945 à nos jours.* CNRS and EDF.

Fourquet, François. 1980. *Les Comptes de la Puissance: histoire de la comptabilité nationale et du plan.* Recherches.

Frank, Robert. 1994. *La hantise du déclin. La France, 1920–1960: finances, défense, et identité nationale.* Belin.

Frost, Robert. 1985a. La Technocratie au pouvoir . . . avec le consentement des syndicats: la technologie, les syndicats et la direction à l'Electricité de France (1946–1968). *Le Mouvement Social* 130, January–March: 81–96.

Frost, Robert. 1985b. The Flood of Progress: Technocrats and Peasants at Tignes (Savoy), 1946–1952. *French Historical Studies* 14, no. 1: 117–140.

Frost, Robert. 1988. Labor and Technological Innovation in French Electrical Power. *Technology and Culture* 29: 865–887.

Frost, Robert. 1991. *Alternating Currents: Nationalized Power in France, 1946–1970.* Cornell University Press.

Fox, Robert, and George Weisz, eds. 1980. *The Organization of Science and Technology in France, 1808–1914.* Cambridge University Press.

Fumaroli, Marc. 1992. Le génie de la langue française. In *Les lieux de mémoire, III: Les France, 3. de l'archive à l'ensemble,* ed. P. Nora. Gallimard.

Gagnon, Paul A. 1976. La Vie Future: Some French Responses to the Technological Society. *Journal of European Studies* 6: 172–189.

Gaudy, René. 1978. *Et la lumière fut nationalisée.* Editions Sociales.

Gaudy, René. 1982. *Les Porteurs d'énergie.* Temps Actuels.

Geiger, Reed G. 1984. Planning the French Canals: The "Becquey Plan" of 1820–1822. *Journal of Economic History* 44, June: 329–339.

Georgi, Frank. 1995. *L'Invention de la CFDT, 1957–1970.* Editions de l'Atelier/Editions Ouvrières.

Gildea, Robert. 1994. *The Past in French History.* Yale University Press.

Gildea, Robert. 1996. *France since 1945.* Oxford University Press.

Gilpin, Robert. 1968. *France in the Age of the Scientific State.* Princeton University Press.

Goldschmidt, Bertrand. 1962. *L'Aventure Atomique.* Fayard.

Goldschmidt, Bertrand. 1967. *Les Rivalités Atomiques 1939–1966.* Fayard.

Goldschmidt, Bertrand. 1980. *Le complexe atomique.* Fayard.

Goldschmidt, Bertrand. 1987. *Les pionniers de l'atome.* Stock.

Gordon, Robert B. 1988. Who Turned the Mechanical Ideal into Mechanical Reality? *Technology and Culture* 29, no. 4: 744–778.

Graham, Loren R. 1993. *The Ghost of the Executed Engineer: Technology and the Fall of the Soviet Union.* Harvard University Press.

Grelon, André, ed. 1986. *Les Ingénieurs de la Crise: titre et profession entre les deux guerres.* Editions de L'Ecole des Huates Etudes en Sciences Sociales.

Guedeney, Colette. 1973. *L'Angoisse Atomique et les centrales nucléaires.* Payot.

Guéry, Alain. 1992. L'Etat, l'outil du bien commun. In *Les lieux de mémoire, III: Les France, 3. de l'archive à l'ensemble,* ed. P. Nora. Gallimard.

Guigeno, Vincent. 1994. Une figure contestée: l'officier-ingénieur (1920–1943). In *La formation polytechnicienne, 1794–1994*, ed. B. Belhoste et al. Dunod.

Guillaume, Sylvie. 1987. Leon Gingembre défenseur des PME. *Vingtième Siècle* 15: 69–80.

Gusterson, Hugh. 1991. Testing Times: A Nuclear Weapons Laboratory at the End of the Cold War. PhD. thesis, Stanford University.

Häckel, Erwin, Karl Kaiser, and Pierre Lellouche. 1980. *Nuclear Policy in Europe: France, Germany and the International Debate*. Forschungsinstitut der Deutschen Gesellschaft für Auswärtige Politik.

Halbwachs, Maurice. 1992. *On Collective Memory*. University of Chicago Press.

Hall, Peter. 1986. *Governing the Economy: The Politics of State Intervention in Britain and France*. Oxford University Press.

Hall, Stuart. 1990. Cultural Identity and Diaspora. In *Identity, Community, Culture, Difference*, ed. J. Rutherford. Lawrence and Wishart.

Hall, Stuart, and Paul du Gay, eds. 1996. *Questions of Cultural Identity*. Sage.

Hatch, Michael T. 1986. *Politics and Nuclear Power: Energy Policy in Western Europe*. University Press of Kentucky.

Hecht, Gabrielle. 1992. Living with large-scale technology. *Techniques et culture* 19: 73–101.

Hecht, Gabrielle. 1993. Constructing Competitiveness: The Politics of Engineering Work in the French Nuclear Program, 1955–1969. In *Technological Competitiveness: Contemporary and Historical Perspectives on the Electrical, Electronics, and Computer Industries*, ed. W. Aspray. IEEE Press.

Hecht, Gabrielle. 1996a. Rebels and Pioneers: Technocratic Ideologies and Social Identities in the French Nuclear Workplace, 1955–1969. *Social Studies of Science* 26, no. 3: 483–530.

Hecht, Gabrielle. 1996b. Le regard américain sur la politique et la technique dans la nationalisation française. In *La nationalisation de l'électricité en France: Nécessité technique ou logique politique?* ed. L. Badel (Presses Universitaires de France).

Hecht, Gabrielle. 1997a. Enacting Cultural Identity: Risk and Ritual in the French Nuclear Workplace. *Journal of Contemporary History* 32, October: 483–507.

Hecht, Gabrielle. 1997b. Peasants, Engineers, and Atomic Cathedrals: Narrating Modernization in Postwar France. *French Historical Studies* 20 3: 381–418.

Hecht, Gabrielle. forthcoming. Planning a Technological Nation: Systems Thinking and the Politics of National Identity in Postwar France. In *The Spread of the Systems Approach*, ed. T. Hughes. University of Chicago Press.

Herblay, Michel. 1977. *Les hommes du fleuve et de l'atome*. La pensée universelle.

Hilgartner, Stephen, Richard Bell, and Rory O'Connor. 1982. *Nukespeak: The Selling of Nuclear Technology in America.* Penguin.

Hoffman, Stanley. 1956. *Le Mouvement Poujade.* Armand Colin.

Holter, Darryl. 1992. *The Battle for Coal: Miners and the Politics of Nationalization in France, 1940–1950.* Northern Illinois University Press.

Hounshell, David A. 1995. Hughesian History of Technology and Chandlerian Business History: Parallels, Departures, and Critics. *History and Technology* 12: 205–224.

Howorth, Jolyon, and Patricia Chilton, eds. 1984. *Defence and Dissent in Contemporary France.* St. Martin's Press.

Hughes, Thomas P. 1969. Technological Momentum in History: Hydrogenation in Germany, 1898–1933. *Past and Present* 44, August: 106–132.

Hughes, Thomas P. 1979a. Emerging Themes in the History of Technology. *Technology and Culture* 20: 697–711.

Hughes, Thomas P. 1979b. The Electrification of America: The System Builders. *Technology and Culture* 20: 124-161.

Hughes, Thomas P. 1983. *Networks of Power: Electrification in Western Society, 1880–1930.* Johns Hopkins University Press.

Hughes, Thomas P. 1989. *American Genesis: A Century of Invention and Technological Enthusiasm.* Viking.

Hughes, Thomas P. 1994. Technological Momentum. In *Does Technology Drive History?* ed. M. Smith and L. Marx. MIT Press.

Humphreys, George G. 1986. *Taylorism in France, 1904–1920: The Impact of Scientific Management on Factory Relations and Society.* Garland.

Hunt, Lynn, ed. 1989. *The New Cultural History.* University of California Press.

Infométrie. 1974. *L'Energie nucléaire, 1945–1974.* EDF/Infométrie.

Jasanoff, Sheila, et al., eds. 1995. *Handbook of Science and Technology Studies.* Sage.

Jasper, James. 1990. *Nuclear Politics: Energy and the State in the United States, Sweden, and France.* Princeton University Press.

Jones, Joseph. 1984. *The Politics of Transport in Twentieth-Century France.* McGill-Queen's University Press.

Kaijser, Arne. 1992. Redirecting Power: Swedish Nuclear Power Policies in Historical Perspective. *Annual Review of Energy Environments* 17: 437–462.

Kaplan, Steven Laurence, and Cynthia Koepp, eds. 1986. *Work in France: Representations, Meaning, Organization, and Practice.* Cornell University Press.

Kolodziej, Edward A. 1987. *Making and Marketing Arms: The French Experience and its Implications for the International System.* Princeton University Press.

Kosciusko-Morizet, Jacques-A. 1973. *La Mafia Polytechnicienne.* Seuil.

Krakovitch, Raymond. 1994. *Le Pouvoir et La Rigueur.* Publisud.

Kranakis, Eda. 1989. Social Determinants of Engineering Practice. *Social Studies of Science* 19, no. 1: 5–70.

Kranakis, Eda. 1997. *Constructing a Bridge: An Exploration of Engineering Culture, Design, and Research in Nineteenth-Century France and America.* MIT Press.

Kuisel, Richard F. 1967. *Ernest Mercier; French Technocrat.* University of California Press.

Kuisel, Richard F. 1973. Technocrats and Public Economic Policy: from the Third to the Fourth Republic. *Journal of European Economic History* 2: 53–99.

Kuisel, Richard F. 1981. *Capitalism and the State in Modern France: Renovation and Economic Management in the Twentieth Century.* Cambridge University Press.

Kuisel, Richard F. 1991. Coca-Cola and the Cold War: The French Face Americanization, 1948–1953. *French Historical Studies* 17, no. 1: 96–116.

Kuisel, Richard. 1993. *Seducing the French: The Dilemma of Americanization.* University of California Press.

Kuisel, Richard F. 1995. American Historians in Search of France: Perceptions and Misperceptions. *French Historical Studies* 19, no. 2: 307–319.

Lacouture, Jean. 1985. *De Gaulle.* Volume 2: *Le politique, 1944–1959.* Seuil.

Lacouture, Jean. 1986. *De Gaulle.* Volume 3: *Le souverain, 1959–1970.* Seuil.

Lacouture, Jean. 1981. *Pierre Mendès France.* Seuil.

Lagadec, Patrick. 1981. *La civilisation du risque: Catastrophe technologiques et responsabilité sociale.* Seuil.

Laird, Robin. 1985. *France, the Soviet Union, and the Nuclear Weapons Issue.* Westview.

Lamiral, Georges. 1988. *Chronique de Trente Années d'Equipement Nucléaire à Electricité de France.* Association pour l'Histoire de l'Electricité en France.

Lamont, Michèle. 1995. National Idenity and National Boundary Patterns in France and in the United States. *French Historical Studies* 19, no. 2: 349–365.

Lanthier, Pierre. 1979. Les dirigeants des grandes entreprises électriques en France, 1911–1973. In *Le Patronat de la seconde industrialisation,* ed. M. Lévy-Leboyer. Editions Ouvrières.

Larkin, Maurice. 1988. *France Since the Popular Front: Government and People, 1936–1986.* Clarendon.

Latour, Bruno. 1983. Give Me a Laboratory and I will Raise the World. In *Science Observed*, ed. K. Knorr-Cetina and M. Mulkay. Sage.

Latour, Bruno. 1987. *Science in Action*. Harvard University Press.

Latour, Bruno. 1993. *We Have Never Been Modern*. Harvard University Press.

Latour, Bruno. 1996. *Aramis or the Love of Technology*. Harvard University Press.

Laumier, Pierre. 1979. *L'Electronucléaire: faits et chiffres*.

Laurent, Philippe. 1978. *L'aventure nucléaire*. Aubier.

Law, John, ed. 1991. *A Sociology of Monsters: Essays on Power, Technology, and Domination*. Routledge.

Lebovics, Herman. 1992. *True France: The Wars over Cultural Identity, 1900–1945*. Cornell University Press.

Leclerq, Jacques. 1986. *L'Ere nucléaire: le monde des centrales nucléaires*. Hachette.

Le Goaziou, Véronique. 1988. Les réactions de la presse française au moment de l'explosion nucléaire d'Hiroshima et Nagasaki. Paper prepared for course Technologie et Société, CNAM.

Lenoir, Timothy. 1994. Was the Last Turn the Right Turn? The Semiotic Turn and A. J. Greimas. *Configurations* 1: 119–136.

Lerman, Nina E. 1997. 'Preparing for the Duties and Practical Business of Life': Technological Knowledge and Social Structure in Mid-19th-Century Phildelphia. *Technology and Culture* 38: 31–59.

Lerman, Nina E., Arwen Palmer Mohun, and Ruth Oldenziel. 1997a. Versatile Tools: Gender Analysis in the History of Technology. *Technology and Culture* 38: 1–8.

Lerman, Nina E., Arwen Palmer Mohun, and Ruth Oldenziel. 1997b. The Shoulders We Stand On and the View From Here: Historiography and Directions for Research. *Technology and Culture* 38: 9–30.

Levin, Miriam. 1989. The Eiffel Tower Revisited. *French Review* 62, May: 1052–1064.

Lévy-Leboyer, Maurice, and Jean-Claude Casanova, eds. 1991. *Entre l'Etat et le marché: L'économie française des années 1880 à nos jours*. Gallimard.

Lipartito, Kenneth. 1994. When Women Were Switches: Technology, Work, and Gender in the Telephone Industry, 1890–1920. *American Historical Review* 99, no. 4: 1075–1111.

Loeb, Paul. 1982. *Nuclear Culture: Living and Working in the World's Largest Atomic Complex*. Coward, McCann and Geoghegan.

Loyrette, Henri. 1992. La Tour Eiffel. In *Les lieux de mémoire, III: Les France, 3. de l'archive à l'ensemble*, ed. P. Nora. Gallimard.

Lucas, N. J. D. 1979. *Energy in France: Planning, Politics and Policy.* Europa Publications Limited.

Lucas, Nigel. 1985. *Western European Energy Policies: A Comparative Study of the Influence of Institutional Structure on Technical Change.* Clarendon.

MacKenzie, Donald. 1984. Marx and the Machine. *Technology and Culture* 25: 473–502.

MacKenzie, Donald. 1990. *Inventing Accuracy: A Historical Sociology of Nuclear Missile Guidance.* MIT Press.

Maier, Charles. 1970. Between Taylorism and Technocracy: European ideologies and the vision of industrial productivity in the 1920s. *Journal of Contemporary History* 5, no. 2: 27–61.

Massey, Andrew. 1988. *Technocrats and Nuclear Politics: The Influence of Professional Experts in Policy-Making.* Avebury/Gower.

Maza, Sarah. 1996. Stories in History: Cultural Narratives in Recent Works in European History. *American Historical Review* 101, no. 5: 1493–1515.

Mazuzan, George T. 1986. 'Very Risky Business': A Power Reactor for New York City. *Technology and Culture* 27: 262–284.

Mazuzan, George T., and J. Samuel Walker. 1984. *Controlling the Atom: the Beginnings of Nuclear Regulation, 1946–1962.* University of California Press.

McArthur, John H., and Bruce R. Scott. 1969. *Industrial Planning in France.* Division of Research, Graduate School of Business Administration, Harvard University.

McDougall, Walter. 1985a. *The Heavens and the Earth: A Political History of the Space Age.* Basic Books.

McDougall, Walter. 1985b. Space-Age Europe: Gaullism, Euro-Gaullism, and the American Dilemma. *Technology and Culture* 26: 179–203.

McEvoy, Arthur F. 1995. Working Environments: An Ecological Approach to Industrial Health and Safety. *Technology and Culture,* supplement to April issue: S145–S173.

McGaw, Judith A. 1987. *Most Wonderful Machine.* Princeton University Press.

Menahem, Georges. 1976. *La science et le militaire.* Seuil.

Ministère de la Culture et de la Communication. 1989. *Cultures du Travail: Identité et savoirs industriels dans la France contemporaine.* Editions de la Maison des sciences de l'homme.

Miquel, Pierre. 1994. *Les polytechniciens.* Plon.

Misa, Thomas J. 1994. Retrieving Sociotechnical Change from Technological Determinism. In *Does Technology Drive History?* ed. M. Smith and L. Marx. MIT Press.

Mongin, Dominique. 1997. *La Bombe atomique française, 1945–1958.* Bruylant.

Mounier-Kuhn, Pierre-E. 1994. French Computer Manufacturers and the Component Industry, 1952–1972. *History and Technology* 11: 195–216.

Mouriaux, René. 1984. The CFDT: From the Union of Popular Forces to the Success of Social Change. In *The French Workers' Movement: Economic Crisis and Political Change,* ed. M. Kesselman. Allen & Unwin.

Nathanson, Charles. 1988. The Social Construction of the Soviet Threat. *Alternatives* 13: 443–483.

Nau, Henry. 1974. *National Politics and International Technology: Nuclear Reactor Development in Western Europe.* Johns Hopkins University Press.

Nelkin, Dorothy, and Michael Pollak. 1981. *The Atom Besieged: Antinuclear Movements in France and Germany.* MIT Press.

Nevers, Jean-Yves. 1983. Du Clientelisme à la technocratie: cent ans de démocratie communale dans une grande ville, Toulouse. *Revue Française de Science Politique* 33, no. 3: 428–454.

Noble, David F. 1984. *Forces of Production: A Social History of Industrial Automation.* Knopf.

Noiriel, Gérard. 1990. *Workers in French Society in the 19th and 20th Centuries.* Berg.

Noiriel, Gérard. 1996a. French and Foreigners. In *Realms of Memory, I: Conflicts and Divisions,* ed. P. Nora. Columbia University Press.

Noiriel, Gérard. 1996b. *The French Melting Pot: Immigration, Citizenship, and National Identity.* University of Minnesota Press.

Nora, Pierre, ed. 1992. *Les lieux de mémoire, III: Les France, 3. de l'archive à l'ensemble.* Gallimard.

Nora, Pierre, ed. 1996a. *Realms of Memory, I: Conflicts and Divisions.* Columbia University Press.

Nora, Pierre. 1996b. Gaullists and Communists. In *Realms of Memory, I: Conflicts and Divisions,* ed. P. Nora. Columbia University Press.

Nye, David. 1990. *Electrifying America: The Social Meanings of a New Technology.* MIT Press.

Nye, David. 1994. *American Technological Sublime.* MIT Press.

Nye, Mary Jo. 1980. N-rays: An episode in the history and psychology of science. *History and Philosophy of Science* 11, no. 1: 125–156.

Oldenziel, Ruth. 1997. Boys and their Toys: The Fisher Body Craftsman's Guild, 1930–1968, and the Making of a Male Technical Domain. *Technology and Culture* 38: 60–96.

Pace, David. 1991. Old Wine—New Bottles: Atomic Energy and the Ideology of Science in Postwar France. *French Historical Studies* 17, no. 1: 38–61.

Papin, Jean-Philippe. 1996. *Les syndicats d'EDF, 1946–1996.* Association pour l'histoire de l'électricité en France.

Papon, Pierre. 1978. *Le pouvoir et la science en France.* Centurion.

Péan, Pierre. 1982. *Les Deux Bombes: Comment la France a donné la bombe à Israel et à l'Irak.* Fayard.

Péan, Pierre. 1995. Pierre Guillaumat, l'homme action. In *Pierre Guillaumat: La passion des grands projets industriels,* ed. G.-H. Soutou and A. Beltran. Editions Rive Droite.

Perroux, Michel, ed. 1983. *Energie et Société.* Pergamon.

Pfaffenberger, Bryan. 1990. The Harsh Facts of Hydraulics: Technology and Society in Sri Lanka's Colonization Schemes. *Technology and Culture* 31, July: 361–397.

Pfaffenberger, Bryan. 1992. Technological Dramas. *Science, Technology and Human Values* 17, no. 3: 282–312.

Picard, J. F. 1980. *Les français et l'énergie: recueil d'enquêtes et de sondages d'opinion effectués sur des thèmes se rapportant à l'énergie en France de 1945 à nos jours, note de synthèse.* CNRS/CDHS, EDF/DER.

Picard, Jean-François. 1987. *Recherche et Industrie: Témoignages sur quarante ans d'études et de recherches à Electricité de France.* Eyrolles.

Picard, Jean-François. 1990. *La République des Savants: La recherche française et le CNRS.* Flammarion.

Picard, Jean-François, and Jean-Michel Fourgous. 1977. *La Grande Presse dans le débat nucléaire, 1967–1976.* Association pour le développement de l'informatique dans les sciences de l'homme et Institute de recherche économique et de planification.

Picard, Jean-François, Alain Beltran, and Martine Bungener. 1985. *Histoire(s) de l'EDF. Comment se sont prises les décisions de 1946 à nos jours.* Dunod.

Picon, Antoine. 1992. *L'Invention de l'Ingénieur Moderne: L'Ecole des Ponts et Chaussées, 1747–1851.* Presses de L'Ecole Nationale des Ponts et Chaussées.

Picon, Antoine. 1994. L'Ecole polytechnique, une école d'ingénieurs? In *La formation polytechnicienne, 1794–1994,* ed. B. Belhoste et al. Dunod.

Pignon, Dominique. 1981. *Enquête au cœur des centrales nucléaires.* F. Nathan.

Pignon, Dominique, et al., eds. 1975. *Questions sur le nucléaire: Des risques d'accidents dans les centrales nucléaires.* Christian Bourgeois.

Pinch, Trevor, and John Law, eds. 1992. *Shaping Technology/Building Society*. MIT Press.

Pipon, Pascal-Michel. La Gestion des Hommes au S.P.T (1946–1992). Mémoire de Maîtrise, sous la direction de M. le Professeur François Caron, Université Paris IV-Sorbonne, 1993–1994.

Plantey, Alain, ed. 1994. *De Gaulle et Les Médias*. Plon.

Pointreau, Abel. 1989. *La Loire: les peuples du fleuve*. Horvath.

Porter, Theodore M. 1991. Objectivity and Authority: How French Engineers Reduced Public Utility to Numbers. *Poetics Today* 12, no. 2: 245–266.

Porter, Theodore M. 1992. Quantification and the Accounting Ideal in Science. *Social Studies of Science* 22: 633–652.

Porter, Theodore M. 1986. *The Rise of Statistical Thinking, 1820–1900*. Princeton University Press.

Porter, Theodore M. 1995. *Trust in Numbers: The Pursuit of Objectivity in Science and Public Life*. Princeton University Press.

Pringle, Peter, and James Spigelman. 1981. *The Nuclear Barons*. Holt, Rhinehart and Winston.

Puiseux, Louis. 1986. *Crépuscule des atomes: Les vrais risques du nucléaire*. Hachette.

Puiseux, Louis. 1981. *Le Babel Nucléaire*. Galilée.

Puiseux, Louis, and Dominique Saumon. 1977. Actors and Decisions in French Energy Policy. In *The Energy Syndrome*, ed. L. Lindberg. Heath.

Pursell, Carroll. 1993. The Construction of Masculinity and Technology. *Polhem* 11: 206–219.

Rabeharisoa, Vololona. 1990. Mesures 'Techniques', Mesures 'Morales': De l'institution d'un habitant raisonnable face aux economies d'energie. *Techniques et culture* 16: 63–82.

Rabinbach, Anson. 1990. *The Human Motor: Energy, Fatigue, and the Origins of Modernity*. Basic Books.

Rabinow, Paul. 1989. *French Modern: Norms and Forms of the Social Environment*. MIT Press.

Radcliffe, Barrie. 1989. Bureaucracy and Early French Railroads: the Myth and the Reality. *Journal of European Economic History* 18: 331–370.

Rayner, Steve. 1986. Management of Radiation Hazards in Hospitals: Plural Rationalities in a Single Institution. *Social Studies of Science* 16: 573–591.

Reclus, Philippe. 1987. *La République impatiente: le Club des Jacobins, 1951–1958*. Sorbonne.

Reid, Donald. 1993. Reflections on Labor History and Language. In *Rethinking Labor History*, ed. L. Berlanstein. University of Illinois Press.

Renou, Jean. n.d. Le Commissariat à l'Energie Atomique ou La Persévérance (1945–1980). Unpublished manuscript, courtesy of Maurice Guéron.

Richards, Thomas. 1990. *The Commodity Culture of Victorian England: Advertising and the Spectacle, 1851–1914*. Stanford University Press.

Rioux, Jean-Pierre. 1980. *La France de la Quatrième République, vol. 1: l'ardeur et la nécessité*. Seuil.

Rioux, Jean-Pierre. 1983. *La France de la Quatrième République, volume 2: L'expansion et l'impuissance, 1952–1958*. Seuil.

Roberts, Mary Louise. 1994. *Civilization without Sexes: Reconstructing Gender in Postwar France, 1917–1927*. University of Chicago Press.

Rogers, Susan Carol. 1991. *Shaping Modern Times in Rural France: The Transformation and Reproduction of an Aveyronnais Community*. Princeton University Press.

Roqueplo, Philippe. 1983. *Penser la technique: Pour une démocratie concrète*. Seuil.

Ross, Kristin. 1995. *Fast Cars, Clean Bodies: Decolonization and the Reordering of French Culture*. MIT Press.

Rousso, Henry, ed. 1986a. *De Monnet à Massé: Enjeux politiques et objectifs économiques dans le cadre des quatre premiers Plans*. CNRS.

Rousso, Henry. 1986b. Le Ministère de l'Industrie dans le Processus de Planification: Une Adaptation Difficile (1940–1969). In *De Monnet à Massé: Enjeux politiques et objectifs économiques dans le cadre des quatre premiers Plans*, ed. H. Rousso. Editions CNRS.

Russell, Diana, ed. 1989. *Exposing Nuclear Phallacies*. Pergamon.

Russell, Edmund P. III. 1996. "Speaking of Annihilation": Mobilizing for War Against Human and Insect Enemies, 1914–1915. *Journal of American History* 82, March: 1501–1530.

Sahlins, Peter. 1989. *Boundaries: The Making of France and Spain in the Pyrenees*. University of California Press.

Sanguinetti, Alexandre. 1964. *La France et l'arme atomique*. Julliard.

Shapin, Steve, and Simon Schaffer. 1984. *Leviathan and the Air Pump: Hobbes, Boyle, and the Experimental Way of Life*. Princeton University Press.

Scheinman, Lawrence. 1965. *Atomic Energy Policy in France Under the Fourth Republic*. Princeton University Press.

Scott, Joan W. 1988. *Gender and the Politics of History*. Columbia University Press.

Scott, Joan W. 1991. The Evidence of Experience. *Critical Inquiry* 17: 773–791.

Scott, Joan W. 1993. "The Tip of the Volcano." *Comparative Studies in Society and History* 35, no. 2: 438–443.

Scranton, Philip. 1988. None-Too-Porous Boundaries: Labor History and the History of Technology. *Technology and Culture* 29: 722–743.

Scranton, Philip. 1994. Determinism and Indeterminacy in the History of Technology. In *Does Technology Drive History?* ed. M. Smith and L. Marx. MIT Press.

Sewell, William. 1980. *Work and Revolution in France: The Language of Labor from the Old Regime to 1848.* Cambridge University Press.

Sewell, William. 1990. Review of *Gender and the Politics of History. History and Theory* 29, no. 1: 71–82.

Sewell, William. 1993. Toward a Post-materialist Rhetoric for Labor History. In *Rethinking Labor History: Essays on Discourse and Class Analysis,* ed. L. Berlanstein. University of Illinois Press.

Shinn, Terry. 1980a. From corps to profession: The emergence and definition of industrial engineering in modern France. In *The Organization of Science and Technology in France, 1808–1914,* ed. R. Fox and G. Weisz. Cambridge University Press.

Shinn, Terry. 1980b. *Savoir Scientifique et Pouvoir Sociale: L'Ecole Polytechnique, 1794–1914.* Presses de la Fondation Nationale des Sciences Politiques.

Sigal, Leon V. 1984. *Nuclear Forces in Europe: enduring dilemmas, present prospects.* Brookings Institution.

Simmonot, Philippe. 1978. *Les nucléocrates.* Presses Universitaires de Grenoble.

Smith, Cecil O. 1990. The Longest Run: Public Engineers and Planning in France. *American Historical Review* 95, June: 657–692.

Smith, Merritt Roe, ed. 1985. *Military Enterprise and Technological Change: Perspectives on the American Experience.* MIT Press.

Smith, Merritt Roe. 1994. Technological Determinism in American Culture. In *Does Technology Drive History?* ed. M. Smith and L. Marx. MIT Press.

Smith, Merritt Roe, and Leo Marx, eds. 1994. *Does Technology Drive History? The Dilemma of Technological Determinism.* MIT Press.

Smith, Michael L. 1994. Recourse of Empire: Landscapes of Progress in Technological America. In *Does Technology Drive History?* ed. M. Smith and L. Marx. MIT Press.

Smith, Michael L. 1983. Selling the Moon: The US Manned Space Program and the Triumph of Commodity Scientism. In *The Culture of Consumption,* ed. R. Fox and T. Lears. Pantheon.

Snyder, Lynne Page. 1994. The Death-Dealing Smog over Donora, Pennsylvania: Industrial Air Pollution, Public Health Policy, and the Politics of Expertise, 1948–1949. *Environmental History Review*, spring: 117–139.

Souchon, Marie-Françoise. 1968. *Le Maire: Elu local dans une société en changement.* Cujas.

Soutou, Georges-Henri. 1991. La logique d'un choix: le CEA et le problème des filières électronucléaires. *Relations Internationales* 68: 351–377.

Soutou, Georges-Henri. 1995. Pierre Guillaumat, le CEA et le nucléaire civil. In *Pierre Guillaumat: La passion des grands projets industriels*, ed. G.-H. Soutou and A. Beltran. Editions Rive Droite.

Soutou, Georges-Henri, and Alain Beltran, eds. 1995. *Pierre Guillaumat: La passion des grands projets industriels.* Editions Rive Droite.

Staudenmaier, John. 1985. *Technology's Storytellers.* MIT Press.

Staudenmaier, John. 1990. Recent Trends in the History of Technology. *American Historical Review* 95, June: 715–726.

Stoffaës, Christian. 1991. La Restructuration Industrielle, 1945–1990. In *Entre l'Etat et le marché: L'économie française des années 1880 à nos jours*, ed. M. Lévy-Leboyer and J.-C. Casanova. Gallimard.

Stoffaës, Christian. 1995. Présentation. In *Pierre Guillaumat: La passion des grands projets industriels*, ed. G.-H. Soutou and A. Beltran. Editions Rive Droite.

Suleiman, Ezra. 1974. *Politics, Power, and Bureaucracy in France: The Administrative Elite.* Princeton University Press.

Suleiman, Ezra. 1978. *Elites in French Society.* Princeton University Press.

Suleiman, Ezra. 1995. Politique et technostructure. In *Pierre Guillaumat: La passion des grands projets industriels*, ed. G.-H. Soutou and A. Beltran. Editions Rive Droite.

Tambiah, Stanely Jeyaraja. 1985. *Culture, Thought, and Social Action: An Anthropological Perspective.* Harvard University Press.

Thépot, André. 1979. Le Corps des Mines. In *Le Patronat de la seconde industrialisation, Cahiers du Mouvement social*, ed. M. Lévy-Leboyer. Editions Ouvrières.

Thépot, André, ed. 1985. *L'Ingénieur dans la société française.* Editions Ouvrières.

Thépot, André. 1986. Images et réalité de l'ingénieur entre les deux guerres. In *Les Ingénieurs de la Crise: titre et profession entre les deux guerres*, ed. A. Grelon. Editions de L'Ecole des Hautes Etudes en Sciences Sociales.

Thévenot, Laurent. 1990. La Politique des statistiques: Les Origines des enquêtes de mobilité sociale. *Annales: Economies, sociétés, civilisations* 6: 1275–1300.

Thoenig, Jean-Claude. 1987. *L'Ere des Technocrates: le cas des Ponts et Chaussées.* L'Harmattan.

Thompson, E. P. 1993. *Customs in Common: Studies in Traditional and Popular Culture.* New Press.

Touchard, Jean. 1980. *La gauche en France depuis 1900.* Seuil.

Touraine, Alain, Zsuzsa Hegedus, Francois Dubet, and Michel Wieviorka. 1980. *La prophétie anti-nucléaire.* Seuil.

Touraine, Alain, Michel Wieviorka, and François Dubet. 1984. *Le mouvement ouvrier.* Fayard.

Traweek, Sharon. 1988. *Beamtimes and Lifetimes: The World of High Energy Physics.* Harvard University Press.

Tudesq, André-Jean. 1982. Systeme d'information et contenu politique: l'évolution de la presse quotidienne en France au XXe Siècle. *Revue d'Histoire Moderne et Contemporaine* 29, July–September: 500–507.

Turner, Paul, and David Pitt. 1989. *The Anthropology of War and Peace: Perspectives on the Nuclear Age.* Begin and Garvey.

Turner, Victor. 1966. *The Ritual Process.* Routledge.

Vaïsse, Maurice. 1995. L'indépendance nationale, d'une république à l'autre. In *Pierre Guillaumat: La passion des grands projets industriels,* ed. G.-H. Soutou and A. Beltran. Editions Rive Droite.

Vallet, Bénédicte M. 1986. The Nuclear Safety Institution in France: Emergence and Development. Ph.D. dissertation, New York University.

Van Gennep, Arnold. 1909. *The Rites of Passage.* Routledge and Kegan Paul.

Vansina, Jan. 1985. *Oral Tradition as History.* Madison: University of Wisconsin Press.

Vincenti, Walter G. 1986. The Davis Wing and the Problem of Airfoil Design: Uncertainty and Growth in Engineering Knowledge. *Technology and Culture:* 717–758.

Wakeman, Rosemary. 1992. La Ville en Vol: Toulouse and the Cultural Legacy of the Airplane. *French Historical Studies,* spring: 769–790.

Wakeman, Rosemary. 1997. *The Modernization of the Provincial City: Toulouse, 1945–1975.* Harvard University Press.

Walker, J. Samuel. 1990. Reactor at the Fault: The Bodega Bay Nuclear Plant Conroversy, 1958–1964: A Case Study in the Politics of Technology. *Pacific Historical Review* 59: 323–348.

Walker, Samuel. 1992. *Containing the Atom.* University of California Press.

Weart, Spencer. 1979. *Scientists in Power.* Harvard University Press.

Weart, Spencer. 1988. *Nuclear Fear: A History of Images.* Harvard University Press.

Weber, Eugen. 1976. *Peasants into Frenchmen: The Modernization of Rural France, 1870–1914.* Stanford University Press.

Weiss, John. 1982a. *The Making of Technological Man: The Social Origins of French Engineering Education.* MIT Press.

Weiss, John. 1982b. The Lost Baton: The Politics of Intra-professional Conflict in Nineteenth-Century French Engineering. *Journal of Social History* 16 (1): 3–20.

Weiss, John. 1983. Changing Contours of the Social History of Science and Technology in Industrializing France. *History of Education Quarterly* 23, no. 2: 237–259.

White, Richard. 1995. *The Organic Machine: The Remaking of the Columbia River.* Hill and Wang.

Wieviorka, Olivier. 1990. Charles de Gaulle, la technique et les masses. In *De Gaulle en son siècle. Journées internationales organisées par l'Institut Charles de Gaulle* in Paris (G0004: 1–13).

Willard, Claude, ed. 1995. *La France ouvrière: Histoire de la classe ouvrière et du mouvement ouvrier français.* Editions de l'Atelier/Editions ouvrières.

Williams, Philip, and Martin Harrison. 1971. *Politics and Society in De Gaulle's Republic.* Longman.

Williams, Philip. 1972. *Crisis and Compromise: Politics in the Fourth Republic,* fourth edition. Longmans, Green.

Winner, Langdon. 1977. *Autonomous Technology: Technics-out-of-Control as a Theme in Political Thought.* MIT Press.

Winner, Langdon. 1985. Do artifacts have politics? In *The Social Shaping of Technology,* ed. D. MacKenzie and J. Wajcman. Open University Press.

Winner, Langdon. 1986. *The Whale and the Reactor: A Search for Limits in an Age of High Technology.* University of Chicago Press.

Winner, Langdon. 1993. Upon Opening the Black Box and Finding it Empty: Social Constructivism and the Philosophy of Technology. *Science, Technology and Human Values,* summer: 362–378.

Woolgar, Steve. 1983. Irony in the Social Study of Science. In *Science Observed,* ed. K. Knorr-Cetina and M. Mulkay. Sage.

Wright, Gwendolyn. 1991. *The Politics of Design in French Colonial Urbanism.* University of Chicago Press.

Wylie, Lawrence. 1974. *Village in the Vaucluse.* Harvard University Press.

Zonabend, Françoise. 1989. *La presqu'île au nucléaire.* Odile Jacob.

Index